T0197998

IRIDESCENT
SKIN

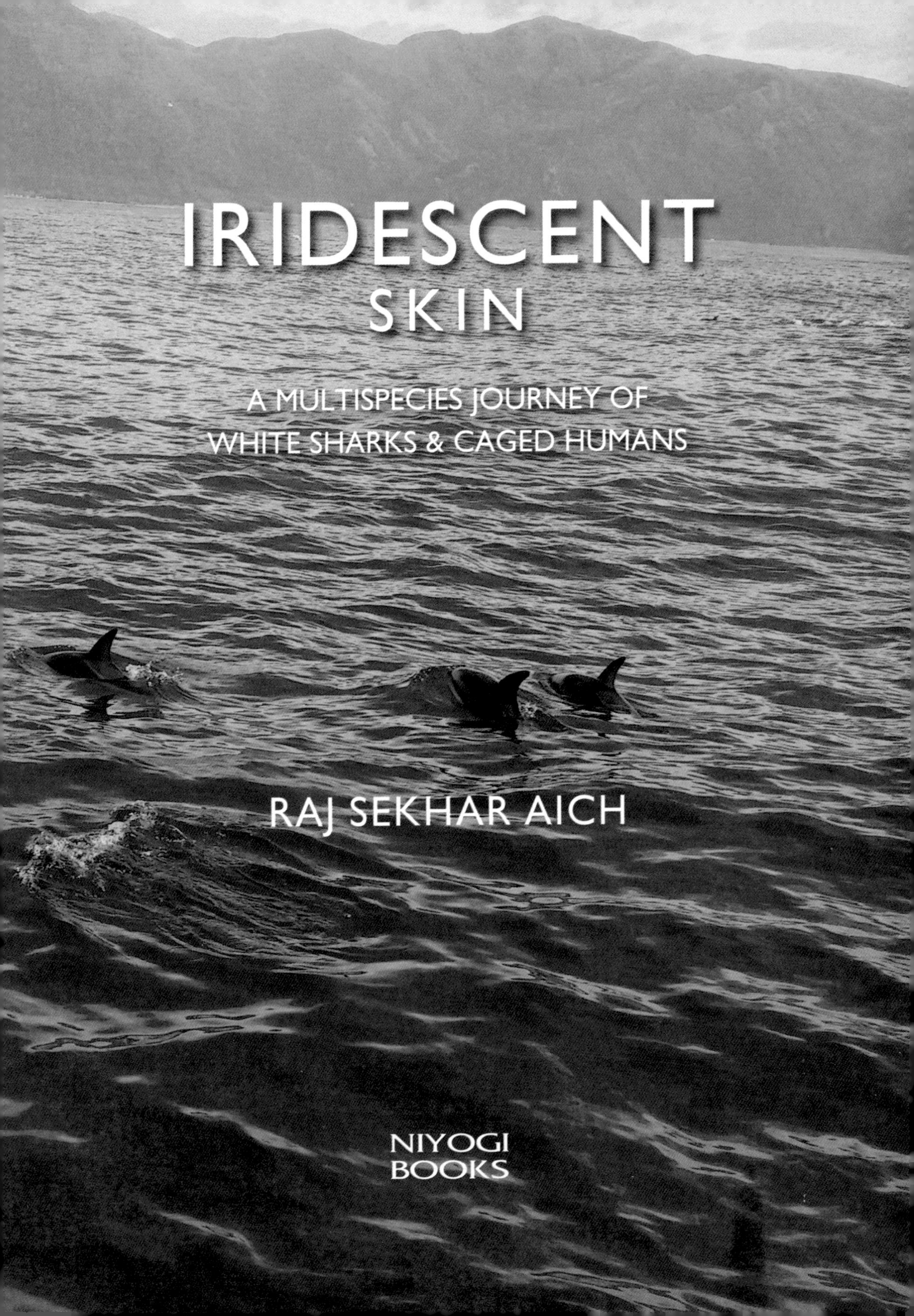

IRIDESCENT SKIN

A MULTISPECIES JOURNEY OF WHITE SHARKS & CAGED HUMANS

RAJ SEKHAR AICH

NIYOGI
BOOKS

Published by
NIYOGI BOOKS
Block D, Building No. 77
Okhla Industrial Area, Phase-I
New Delhi-110 020, INDIA
Tel: 91-11-26816301, 26818960
Email: niyogibooks@gmail.com
Website: www.niyogibooksindia.com

Editor: Arunima Ghosh
Design: Shashi Bhushan Prasad

ISBN: 978-93-91125-53-0
Publication: 2022

Printed at: Niyogi Offset Pvt. Ltd., New Delhi, India

...to my ma

Image source: https://unsplash.com/

Contents

MILAN: THE MEET

Foreword

When human beings jump in the marine environment, they know they are not at the top of the food chain. This knowledge, along with the darkness and murkiness of the environment, create the scope for conjuring up monsters who are waiting to pounce on us just beyond the realm of our vision.

White Sharks are not monsters, nor are humans, but both offer much to fear and admire. Two very different beings who overlap rarely and when they do, it often is in the form of a conflictual, mutually preferentially avoided context. However, there is one space wherein humans and these great fish do come face-to-face in a different setting. It is in the Shark's home space, mediated through a very particular human-controlled environment: cage diving. The opening line frames this experience in a manner I am sure every reader can identify with, even if only in one's imagination. But in the pages of this book, we come close to sensing that the Sharks might also be suspecting something similar about the whole endeavour. This context alone, as revealed via the sections of text focused on before, during, and after the cage diving encounters and activities, makes *Iridescent Skin* worth reading. But the cage diving interface is actually only a small part of the adventure.

Dr Raj Sekhar Aich, a marine social scientist and interdisciplinary multispecies researcher, invites the reader along for extended fieldwork at the intersection of two sentient species, Humans and White Sharks. This frame lets us know that we are in for a voyage. And what an expedition it is. The book you hold in your hands is an enticement into the experiential journey of a scholar/artist/narrator who starts out with the goal of appreciating/assessing a multispecies entanglement, but ends up traversing a tangle of personal, social, historical, ecological, and perceptual water/land/species-scapes. On the surface, this is a narrative about a brief slice of time, place, and context with a focus on one town, one ship, one cage diving enterprise, and two species. But in the end, this is not simply a tale of sharks and people, of cage-diving or human-other being conflict. Nor is it just another academic analytic. This is an expedition weaving (auto) ethnographic and multispecies chronicles with imaginative place-making and narrative journeying. Prepare for a deep dive into various modes of perception, sensation, and analyses, invoking a deep entanglement of the entire sensorium. This is an immersive multispecies anthropology.

You are about to take the plunge yourself, so I'll not keep you too long, but here are a few key thoughts offering a little signposting and wayfinding.

The shark diving narrative forms the most traditional core of the text. An anthropological frame facilitates an introduction to and experience in the town of Bluff, its history and the peoples in it. This frame carries through on the water and off of it. The cage-diving enterprise is situated in a particular history, context, economy, and politic. The setting, the background, and the quotidian vignettes are welcome elements enabling the text to be as effective as it is. Core to the ethnographic moment are the descriptions of the cage-diving enterprise, the boat, the characters who people the boat, and centrally, the White Sharks, who swim up to 'interact' with Raj, the boat, and the various humans involved. Here the narrative works on multiple fronts, enabling us to envision and feel the experience. There is excitement in meeting the sharks, even if all too briefly, revealing for a moment what being a few metres away (or closer!) to a massive creature like the white shark is like. Each shark is a stranger, but each is also an individual, like the people on board the boat and in the town. We know so little about the sharks, yet we get to share space with them and in doing so, push against a certain aspect of human hubris. Here is where the amazing imagery comes into play. The photos of the sharks, even when we only get glimpses of parts of their bodies, are powerful. There is always the sense of movement, immersion in a foreign life-medium (ocean), and the sense that maybe this interface is not such a good idea all around. But it is a truly fascinating reality; spend time with it.

Surrounding the ethnographic narrative and analyses, and moulded to it in often surprising fashion, are the artistic, personal, reflective, often vulnerable, thoughts and experiences of Raj as he undertakes the journey of PhD research; friendships, isolation, camaraderie, loneliness, and all the other chaos and confusion of a deeply moving and often non-linear life experience that he found in Bluff. Histories of childhood in India, artistic imaginings, and personal revelations move and spread across the pages offering one a sense of the experience, the mental landscapes, and the emotional processes this research experience fostered. These are not the typical contents of the products of doctoral theses, and yet they act to make the intellectual contribution more real, visceral, and convincing. Let them wash over you as you read and don't worry about the details or force specific logics in the connections; follow Raj's lead.

On many levels, this book feels like a multi-modal sensorium even though it literally consists only of text and images. There is an accompanying documentary that I recommend if you can access it, that only adds to this sensation. The power in this book rests in the respect Raj has for the sharks, people, and places he spends time with. There emerges a depth well beyond the descriptive and beyond the analytic in the reflections and unfolding narrative. The honesty of conflict, confusion, and challenge, of grace and humanity/animality, and of care enliven the overall impression one obtains thinking with Raj, the sharks, the sea, Bluff, and beyond.

Agustín Fuentes
Primatologist and Biological Anthropologist
Princeton University

Introduction

'Nothing is "proper" to the rogue animals called beasts. The beast escapes an ontology of presence, because it does not properly belong to any taxonomy, and therefore, cannot "be" in any simple sense... Excess marks its monstrous nature. A monster is necessarily cryptic: enigmatic, concealed, camouflaged' (Bubandt, 2019, p. 225). This is not a 'shark book', although it revolves around sharks in the classical sense of the manner. It is not merely an academic book, although the basis of it is six years of detailed academic investigation. It is not a detailed exploration of shark behaviour or ecology, although they are certainly part of it, and no, it is not even a book about shark attacks, although shark bites are featured here. On the surface, it is a sensory narrative of human and white shark encounters, but beneath the ripples, it is a story of love, among humans, rivers, oceans, bricks, stones, and sharks. It is a story rising from the desire to meet among eternal lovers, stories of people who have passed on but are vibrantly alive, and the shark is the symbol of this love.

When human beings jump into the marine environment, they know they are not at the top of the food chain. This knowledge, along with the darkness and murkiness of the environment, creates the scope for conjuring up monsters who are waiting to pounce on us just beyond the realm of our vision. The white shark is a cryptic, enigmatic, and camouflaged beast in the mass public imagination, even though sharks, and specifically white sharks, have been the villains of the sea for a long time in human history. A particular Hollywood film of 1975 gave our fear of the marine environment

The shadow at the corner of your eye, 2017

and the unknown a tangible form, and the contemporary myth of the great white shark as the monster was created. In the course of the last 40 years, the myth created by the film and ever-promoted by global media, has produced a hyperreal image of the white shark, the fish—an image, which has become a marketable commodity and has embodied the symbol of our deepest fears. This hyperreal image has generated multibillion-dollar industries, from television to toys, to movies, and even jewellery. One of the most dynamic industries it has given rise to is the act of diving inside a cage suspended in ocean waters and providing people the opportunity to come face to face with the monster of their imagination—white shark cage diving. White shark cage diving takes place in five places on Earth, the southernmost one being off the coast of Bluff, New Zealand (NZ).

The premise of this book is my expedition as a marine anthropologist/social scientist living and working among the great white sharks, 1,600 km from Antarctica, at the very end of New Zealand. My intention is a holistic exploration of the interconnected lives of humans and white sharks, created from the practice of cage diving. To my knowledge, this is the first extensive white shark cage diving ethnography in the world, let alone in New Zealand. This book is also a meditation on the methodological dimension of marine anthropological research; its aim is not only to elaborate on the equipment I used, but to provide material for future marine social scientists to consider in their expeditions. The book further meditates on the realities of fieldwork, because it provides future researchers exposure to varied psychological, physical, and logistical challenges in the field, as it can be encouraging to show the researcher's own insecurities and vulnerabilities, which are not merely burdens of imperfect fieldworkers, but a deep facet of the human condition.

To describe the focus of my research interest, I choose to use the term 'marine anthropology'. There is a subfield of anthropology, which has been termed 'maritime anthropology', a lot of which deals with maritime community, ocean-going business, and study of ship wreck (Poggie, 1980), and fishing and human engagement and other marine resources. The international journal of maritime anthropology, *Maritime Studies* (MAST), focuses on varied disciplines of research on maritime and coastal matter, realizing the connection of human culture with the marine space. Its purpose is studying nature and diversity of relations among humans and maritime areas. As in the case of biological sciences, there is the recognized subgroup of marine biologists, whose interest lies in studying marine life forms and the environment; in the same sense of the term, I use the term 'marine anthropology'. Coming from a multispecies background, I operationally define marine anthropology as—the holistic study of culturescapes and naturescapes shaped by physical and symbolic interaction of humans with marine environment and life forms.

There are four significant points to clarify regarding the text. First, I am neither for, nor against cage diving; I have done the best I can to be as much objective as possible while documenting it and its relation to the humans and the sharks. Second, all the contemporary photographs have been captured by me, unless specified otherwise. Most of the other ones have been taken by Soosan Lucas, and one image by my brother, Raj Kamal Aich. Third, to refer to these sharks in the text, I have used the terms 'white sharks', 'great white sharks', and 'white pointers' (the local term for them); they are also known as blue pointers in South Africa. And finally, I have used the terms 'shark attack' and 'shark bite' interchangeably throughout the text to indicate incidents where human beings were bitten by sharks. This I have done basing on the subjective perception of the terms in the particular context. 'Attack' refers to the perception that humans are specifically targeted for a malevolent action and is generally considered to be unprovoked. 'Shark bite' on the other hand does not hold any such emotionally charged tone. In an ideal world, the term 'shark attack' should perish from human lexicon, but in the reality, it is still the most prevalent term in public consciousness among the two; hence, I travel between them. I want to be clear: both terms objectively refer to the same thing, and the underlying meaning is always a 'bite', because a shark bite can occur from myriads of behavioural reasoning in a shark, be it for an enquiry about the object in front of them (sometimes it is a human), to fend off intruders from their territory, for aggression during procreation, or actually targeted bites when they want to consume the prey. But I use them interchangeably as deemed necessary for the emotions such incidents raise in the public psyche in each context.

When I started on the journey of writing this book, it was purely intended as a social scientific ethnographical exploration of white shark cage diving in New Zealand (WCN). However, as I progressed, I realized that the book itself wanted a different life, and it would be naïve of me not to allow it such. In a painting, god and human works together to materialize a vision of the artist (and by 'god' I mean beautiful accidents). The genesis of the vision may be from within or beyond her mind and body, but an accidental drop of pink paint might give it a different direction than the artist imagined going towards. The job of the artist is to acknowledge that sometimes, more than what she wants in the painting, it may be what the painting wants that has to be adhered to. Similarly, as the book progressed, I realized this book needed to be a sensory book more than anything else. At the end, this book is in search of the self as much as answers for the specific research questions and objectives in hand. It is as much about family, as broken-down dwelling and loves across generations and continents, and yes, the great white sharks. So, who is this book for? I leave it to the informed reader to decide.

Crossing the Lookout Point into the windward side of the Bluff hills, 2017

ABHISHAR

The Assignation

Raj getting ready to clean the Southern Isle, 2016

Underneath the Southern Isle

There was a storm in the evening; the electric transformer near my house has blown and the faults men announced, 'We won't see any light till the morning.' It is calm now; warm, fat, and polluted plump drops of rain start falling, infused with the aroma of wild *shiuli*, and trickle down the stems of the tree of sorrow. Small puddles are gathering below, slowly around the soft soil, and one can hear the sound of a Kolkata street dog running after a wet cycle rider. His bark, paws, and unmanicured jagged nails splash in the water, as the rusted cycle chain just about holds on in anticipation of a rabies-ridden snippet.

I can't sleep, my bed for the night is a mattress that I dragged out of my dad's godown. It smells of turpentine and saw dust, which does not bother me much; however, the moist heat and the canopy of mosquitoes above are flirting with the edge of unbearability. I get up, get a small cup of Horlicks from my flask and an electric racket with which I vanquish the mosquitoes, light a hurricane lantern and 'dhuno', and transform an arid mud-smothered sculpting plinth in the corridor into a writing desk for the transient night. I am leaving again tomorrow, so I decided to spend the night here—probably not the brightest idea I have ever had.

Sitting here today, almost six years since I started this journey, back in Bengal, India, and in my old house, which is about to be demolished next month, I have a moment to reflect on how I had gotten that far. I do the egregious indulgence of reminiscing that journey leading to the start of my research and narrate the incidents that followed, and the knowledge gained from my fieldwork among the sharks in the following unruly pages. I urge the reader to consider nothing in this chapter or beyond as a figment of my imagination, which might seem like the case at certain instances, even though I might have taken advantage of the perceived circularity of time.

I open my horse-leather bag, and in it lies an eclectic collection of fountain pens, a swiss army knife, a *monjira* (an Indian percussion instrument), a black leather hip flask with the engraving 'rock star', a bottle of blue beard colour, and a weathered field journal. Softly, I press my fingers among the sweaty cotton pages of the journal, and a story jumps out at me.

Journal entry
20 November 2015
7:30 am
Bluff, New Zealand

Today, Mike invited me to lend a hand in cleaning the Southern Isle, our boat and shelter from the Antarctic winds. We removed the cage from the diving platform and submerged it in the water a few feet behind. Maca, the dive instructor and shark wrangler of the boat, suited up and attached the new 65-feet air hose to a regulator.

'There are two regulators, why don't you come down with me?' Maca asked. I said, 'Maybe!'

'If you don't wanna come, fuck you man; I am going down.' Hesitantly, I replied, 'You know what—I will come! Let me just suit up!'

'But we only have one set of fins, and we need a surface support too; so, you suit up, but stay on the boat. Make sure that the hose is not tangled, and every so often give it a tug to check whether I am alright. If I am in distress, I will pull it four times, then you have to come in to assist me'—saying this, Maca went down, under the boat anchored in the Bluff wharf to clean her hull.

As I was diligently taking care of the hose that was feeding Maca, Mike asked me to give him a hand with welding some loose grills in the cage. Mike said, 'It is crucial to re-attach these grills; really, it had no effect on safety, but it is vital for the 'feeling' of safety for the cage divers in the water when there are white sharks around.' So, I divided my time between looking after the hose and the welding, where the erupting embers reminded me of fire crackers in Kolkata Kali Pujo.

I had an extra set of goggles; so, I was constantly putting my head under the water, trying to see what Maca was actually doing under the boat. After a while, Mike suddenly asked... 'You want to go in? It seems, you really want to go.' Before I could say anything, he said, 'It's okay, Raj. Maca will come up—you go in.' And just like that, I had volunteered to dive in these waters for the first time! These waters were particularly interesting, because this was Bluff, and off the coast of Bluff is Foveaux Strait, one of the known five white shark hotspots in the world, and 20 km out in this unpredictable strait is Edwards Island, where we will be heading off to every morning for our cage diving trips. As an anthropologist investigating the impact of human and white shark encounters through cage diving, this was my field site. But little did Mike know, besides being obsessed with sharks and researching them, I was intensely afraid of them! Although I was told by a Department of Conservation (DOC) official, 'Bluff has as many white sharks as my bath tub,' I was not convinced. As a matter of fact, I decided to lose a lot of weight before the fieldwork, because I did not want to look like a plump seal underwater, especially in a black neoprene suit, if I ever had to dive in these waters. What if today is one of those days when a shark decides to show up in the wharf and I am underwater, cleaning the boat?

I knew the trick was not to think about it, and just jump in. Instinctively, as an experienced scuba diver, I suited up—put on the gloves, booties, weight harness, Maca's fins, cleared and defogged my mask with spit and sea water, clipped on the hose to the clip of the harness, took a deep breath—and dove in. A sudden rush of bubbles engulfed my face. When the long second passed, the first thing I saw underwater was the cage, but from outside! My mind went—'This is not the right perspective! I am not supposed to be outside the cage!' I had to come up, but as I surfaced, I could imagine a huge towering fin gently coming towards me. Panicking, I dove down again into the murky water.

The hit of adrenaline in my body meant that I was not yet feeling cold, just disorientated and petrified. I kept on moving forward with a slow inverted

Mike Haines out in the ocean, 2017

crawl on the hull. Even though I was just a few feet underwater, I was isolated from anyone and anything that was known to me; it was a longer distance than the 8,000 km to a place I call home, India...Kolkata, to be precise. All I could hear now were my bubbles, and the sound of my heart thumping, almost bursting out of my chest. Understanding my own pathology did not help me control the hyper-activation of my sympathetic nervous system. My body was stiffening up and it was getting hard to flap my fins. I was taking fast and shallow breaths, and an irrational feeling of drowning—even though I had my regulator (much like when someone learns to scuba dive for the first time)—was engulfing me. My rational brain was saying, 'It's all in your mind and it is all controlled fear.' My body on the other hand adhering to the instinctual urges shouted: 'Get out of here, get out of here!'

I tried to distract myself with the work at hand, and for the first time, I was able to touch the propeller and the cold underbelly of the Southern Isle. The hull was like a living, breathing organism, it moved with the short waves in the wharf, and when a boat passed by, it moved even faster and higher. The entire belly was filled with algae and barnacles. On the port side of the boat, there were more algae, as it always faced the sun when docked, as opposed to the starboard side, which was facing the wharf for most of the time. These organisms not only corroded the metal, but also slowed her down by as much as up to three knots—hence, the reason for my current predicament.

Using the scraper tied to my wrist with a lanyard, I started releasing the algae, beginning with the propeller. The poor visibility was getting worse as I scraped with strong strokes. A sense of self-loathing ensued, and I kept on telling myself, 'Don't look back... You're just being stupid.' I started singing underwater, to calm myself, which always constituted songs of the Boss or Tagore, but like all the other times I used this strategy of catharsis underwater, I was just mumbling rhymeless bubbles. I kept on scraping, and fighting the fear—all I wanted was to get out of the water! But no matter what, I could not go up without finishing the job, and neither could I show my fear.

This was my week of recce in the boat, and an important rite of passage for me from being an outsider to becoming a trusted and hardworking insider of the cage diving industry, as in the next two months I would be finally starting my fieldwork, working on Mike's boat. From then, on across two years, I was supposed to be a participant observer, living, working, and learning not only from being amongst the white shark cage divers and the operators facilitating it, but also from the sharks and the Bluff community. Participant observation creates an avenue to learn about people and practices in their natural environment, gives the researcher an opportunity to be part of the day-to-day activities, get first-hand accounts of the community or practice in focus to gain novel insights, and quite literally, observe while participating. It was the only way to create an in-depth knowledge about the human and white shark interaction facilitated through cage diving.

Mike is the owner and operator of Shark Experience Bluff, currently the only surviving cage diving facility in New Zealand. He had been a charter fisherman, a dive instructor, and a technical diver, who had travelled the world, and past experience with Indians had made him believe that they all just eat curry and are not the most forthcoming lot; they were already having a jest about my lack of interest in eating fish; so, I had to prove him wrong. Furthermore, Mike had always complained to me about shark researchers' lack of a hands-on approach. I had to be the exception and prove my worth. Having never worked with anything related to the sea and having never been out in the ocean for more than a few hours at any time, I had to do everything I could to make sure that I earned my place among the hard-working, ocean-going sea-people.

With each hard stroke, I finally finished cleaning the boat, end to end, and swam back. I surfaced, ironically welcomed by the sound of Bruce Springsteen's 'Born to Run' blaring on the radio, and I thought, I had wanted to run...run away from home, and the obvious trajectory my life had been moving along (possibly following my dad's path in painting, or acting); hence, I was here in New Zealand, under a cage diving boat. But I was coming to realize that I was not the fearless maverick of my imagination, who would jump into any water and take any chance. As a matter of fact, everything I had done since I had been in New Zealand petrified me. I could not cope with the dissonance of my ideal self vs the real self! Now I was starting to feel the cold and my heart was racing fast, but I was glad to be out of the water, while at the same time, feeling foolish as to why I could not control my bodily reactions to the fear and maybe, even phobia. Most importantly, as I caught up with my breath, I could not understand why I was so petrified; I knew that one had more chances of dying by being hit by a coconut rather than being on the receiving end of a shark attack, and many other strange statistics like that. I was truly baffled by how emotion was overpowering my knowledge of these shark attack statistics, and I was considering whether this lifelong fear had transformed into a full-grown phobia. I came to the realization that this dissonance was created, because even now the idea of the shark and its mindless man-eater image was more 'true' to my intrinsic psychological schema than my overt and apparent understanding of them, an image that could not be dispelled by mere information. Now the question to myself really was: Would I be able to complete this research in such close proximity to these animals, knowing that I was burdened with an untrue, but very 'real' imagined image of them in my mind?

The Trampoline Cat

After my *dadu* (grandpa), Susil Bhushan Aich, retired, he started a business and pursued his passion for fish. I have been told that he was the first person to breed aquarium angelfish in Bengal. I remember him having huge fish ponds and aquariums all through the house. For whatever reason, I was always interested in the marine world and wanted to see underwater. One day, when I was really little, I took a motorcycle helmet that my dad's friend had brought. Standing in front of one of our fish ponds, I put the full-faced helmet on my head, closed the lid attempting to make a makeshift scuba diving helmet, and dumped my head in the water to see the fish. I cannot be certain if, even momentarily, I was able to see any fish with the helmet on, but upon hearing a big splash, my father had to come and rescue me, once he realized that it was not indeed a big fish in the pond.

Dadu had to emigrate from Bangladesh when he was 17 and come to India. He was in the army for a while and travelled all across the country, finally getting some land in Gangulibagan in Kolkata to settle down. But he never lost the longing for his homeland. When he got this piece of land, he and his newly-wedded wife made a house of *hogla pata* (elephant grass). The first concrete room too, he made with his own hand, laying brick after brick, with my father, uncle, and aunt supplying the mortar to him.

However, this home of my *dadu* followed me beyond the Indian Ocean, the Tasmanian Sea, and manifested in the form of the kind humans who became patrons of my foolish aspirations and the desire embedded in my quarks to meet the sharks. And this home, my only constant, was

The apple tree, University of Canterbury, Soosan Lucas, 2015

sublimated from a sudden stumble on my path onto a new-found identity... which gave my hands and my blood meaning.

Journal entry
4 February 2016
7:15 am
Timaru, New Zealand

Received an SMS: 'Mate, no point in you coming down to Bluff now, we have not seen any sharks over two weeks.'

I sat down on the couch, I could not believe my eyes. I rang Mike immediately.

'Mike, what's going on?'

Mike replied, 'Well, the Stewart Islanders are killing the sharks, and we have not seen them for two weeks now; so, I don't know if you would want to come or not at this point.'

I immediately replied, 'I am coming, Mike—no matter what.'

For some reason, tears started running down my eyes—for an animal I had never met in my life till that point, but I was all too acquainted with the idea of them. I started shouting out to Soosan: 'Soosan, they are killing them... They are killing the sharks!' I was dumbfounded. At that moment, I was not really thinking about my research; rather, I was distressed about people killing the sharks for a mere grudge. I felt helpless and tried contacting anyone I could to find out if there was something I could do (even at my university), but no one could help me. After a while, I calmed down and decided to go to Bluff first and see what was going on. So, I waited impatiently and got ready to make my move to Bluff.

My fieldwork was supposed to start around the second week of February 2016. However, before reaching there, I got the news that the cage diving operators had not seen any sharks for two weeks. The operators believed that certain sections of the neighbouring community were killing the sharks and disposing the bodies near the cage diving site. And the cage divers believe if there is a dead shark in the waters, other white sharks avoid the region—hence the absence of the sharks. This absence, though intensely frustrating at times, created a unique conundrum for the whole of my doctoral thesis.

For the last few months, I had been running and prepping for this project, and as if suddenly, everything had stopped. When I got the message, I had left Christchurch and was based in Timaru, a quaint little town in South Island, New Zealand—at Soosan's house, getting ready for the fieldwork down in Bluff. Soosan was kind enough to share her little house with me, which she was already sharing with her little daughter (Aurora Slater), whom I had named 'the Minion', and 'the Trampoline Cat'—named so because of his affinity to jump on the Minion's backyard trampoline. Often, Soosan would rescue me from near-homelessness, unknown to the university, when I used to stay and sleep under my desk. I had so many apple pits and seeds under there, Soosan was convinced that there must have been an apple tree growing under my desk. She would invite me down to Timaru, for hot meals and stories that we shared with each other. She helped me move, time and time again, with all my luggage from Christchurch, and stayed awake night after night, copy-checking my thesis. Yes, it is true, I should have had a job in a café or at any other place as others do while studying...but I was doing my research, and only my research; I didn't have time for other things. I would much rather be a naked pauper than work on anything else, and concentrate all day, every day, only on my actual work.

It all started with a chance encounter with a tall blonde Brit anthropologist and artist at a Christchurch art gallery, who was to become one of my closest friends, and later, my research assistant. At that time (2015), I had just come to New Zealand and was trying to figure out how I can do my research, considering I had no scholarship. I started doing odd and un-odd jobs. At one point, I received $1,200 from a research project, and I could have used the money to buy food for more than a month; however, I ended up buying a white digital piano. One day, Soosan got a message on her phone... 'Want to go shark cage diving?' Soosan took me up on the offer, and we set off on an 800-km-journey to the very last point of South Island (Christchurch to Bluff), New Zealand, in May 2015. But while on our way, Soosan got a message from Mike—'Trip's off—50 knots winds'. Sitting down—disappointed—with a cup of coffee near Balclutha, we decided to continue on to Bluff, because we had already booked a cottage down by the water, from the handful of accommodation options available. Lisa, the kind owner of the cottage was an interesting individual in her own right. She talked about how common it was for the cage diving trips to be cancelled because of the weather, about the sharks, and the new set of false teeth she had got recently. We settled in and stared out from the balcony at the slow dazzles of the vermilion waters in front, and for the entire night, I sat there and thought: 'Why could we not go out in the ocean when it looks so calm from here?' Little did I know (what I would learn eventually) what it was really like being out in Foveaux in 50 knot winds.

But no matter what I was doing, I was preoccupied! I had moved from India to New Zealand a few months back for a PhD on a multispecies research at the University of Canterbury, Department of Anthropology. But I had been struggling with the research due to lack of funding and other logistical considerations, a research which was supposed to take me to the shores of Africa and to African elephants. Although I must admit, at the heart of my hearts, I had the wish to use this knowledge gained from my multispecies research with elephants to work with sharks someday, but it was too far off a dream.

Everything was falling apart, and quitting seemed like the only option. But something happened on that unsuccessful shark cage diving trip! On the way back from Bluff, I told Soosan that I had decided on going to Bluff to work with the sharks for my ethnography—I didn't know how, but I knew distinctly, that was what I was going to do. Her answer was 'ok'! We spoke of this on a Tuesday night, in the middle of May 2015; I sat down and wrote a new 1000-word proposal for my PhD and submitted it the next day to the university. Initially, it was laughed at, but after months of cajoling, I was able to convince the university and my supervisors that such a research was indeed possible. In the end, it all came down to if I can convince any of the two cage diving operators in Bluff to let me be on their boat as a participant observer.

I was a nobody; no one knew me or wanted to know about my research interest, and the two cage operators in Bluff wanted nothing to do with me. At that point, Soosan came to the rescue and quite literally, salvaged the project. Besides her expertise in anthropology and art, Soosan is a PR queen, having worked in London for *Vogue*, *GQ*, and other big media houses. She contacted Mike Haines, drove us to Bluff again, and we sat down over cups of hot coffee at Anchorage, the very last restaurant at the southernmost end of South Island. Soosan worked her magic and we were able to convince Mike and his partner, Carwyn, and finally, it was settled that I will be on their boat for the upcoming season as a participant

Aurora (the Minion), Soosan, and Raj, 2018

observer, learning about the practice, and the human–shark interaction facilitated through it.

Without Soosan, none of the adventures or research that followed would have existed. As an academician, I officially acknowledge her contribution to this project. Literally, she is as deserving of the PhD degree achieved from this, as am I.

So now that everything was fixed, I still had to come in terms with my phobia of sharks, something I had hidden from everyone, possibly even from myself. But now as I sit here, I cannot help but ask myself, why was I so fascinated and yet afraid of them at the same time? Why indeed was this untrue but very 'real' imagined image of them in my mind more potent than my knowledge of them?

Allegory of the Cage

Prisoners shackled in a cave and forced to gaze at a blank wall.
There is a bright fire lit behind them, and everything that passes in front of the fire
leaves giant distorted shadows on the cave wall.
And these prisoners—you and I,
perceive these shadows of reality as reality herself

—Plato, 381 BC

The winds swim pass through the empty windows of the house, slapping the crumbling cement and the Marandi frames, echoing the deep rumbles of a bamboo forest. The living room had black walls, with my sword collection, photographs of Springsteen, Tagore, and Calvin and Hobbes at one side, and a baby grand in the other corner. The walls seem darker now, and the floor feels covered in wet mouldy grey ash. As you step on it, the ash parts show the marble underneath and on this, you can see unruly marks in some cryptic dialects. The swords are gone, so is Springsteen; what remain are only the remnants of deep geometric cuts in their place. The piano has disappeared, leaving only a stain on the marble, and there was a giant charcoal drawing of Batman on one of these walls—he seems to have flown away too. A distant lightning gently emblazes the maroon sky. A lizard passes over some creeper vines—his shadow on the wall materializes into a giant megalania crossing the Australian open forest. The light disappears, and I stand again in front of ambiguous, procreating shadows scattered on the naked walls whose modesty is only hidden by their stooped-over elongated penumbra. These shadows and their offsprings not only mimic the real, but sometimes create images, which transcend the real. Much like in our dreamscape, our perceived reality is often a metamorphosis of such

The white shark, a contemporary global symbol of fear, 2016

images, where the real and the fictional renditions of the real blend together seamlessly, murking up the distinction between reality and simulation of reality. One such image and global marketable commodity in the last three or four decades has been the image of the poster boy of all sharks, the great white shark (*Carcharodon carcharias*) or the *makō* (or *mangō*) *taniwha* in Māori, and I had been eternally caught between the real fish, the global simulacrum of them, and the one in my own imagination and sensations.

There is a giant dusty mirror where I had made a 7-feet etching of a shark jaw. As I look into its gape—it is about to swallow me alive while reminding me all the way how sharks have come back for me again and again as my obsession and my dread. They have even been my muse explored through words, monochromatic scars on my skin, and an entire exhibition

of paintings. Sharks in blue, orange, and gold; some dematerializing in the blankness of handmade paper, while others playing hide and seek with oil and phosphorescent pigments on canvas. But how they found me, the answer alludes me to this moment.

I can remember the first book I bought (besides comics) was a book on sharks. I am dyslexic, and suffer from chronic ADHD (although I did not know about this until I was 35); so, the picture of the blue ocean and the bluer sharks enticed me, and all I could think of was their vibrant skin. During this research, I was also able to retrieve one of my first memories of being in a movie hall, which had a deep impact on me. I watched the movie, *The Shark Hunter* (1979), with my *dadu*, when I was eight or nine years old. I remember being glued to the screen when a man was para-sailing

Mural painted for Alpine Energy, Timaru, Soosan lucas, 2019

above the ocean, and suddenly, upon seeing a shark, jumped in with a knife, battled with it, and came out victorious. Bizarre as the premise may be now to consider, I can't help but wonder if this experience had anything to do with my love for knives and sharks. But I do know, in my formative years, sharks, particularly the white shark, resembled what I considered to be the best traits of a human 'male'—strong, sharp, silent, beautiful, and the master of his realm. As I grew up, sharks became my personal totem, engraved on my mind and on my skin, symbolized by multiple shark tattoos, and by the paintings and sculptures I made of them.

But I have been deeply fearful of them too. My brother Kuntal and I used to go for swimming in the Ramgarh *pukur* (pond) in our neighbourhood, after hours of hapkido practice, late in the evening. The orange halogen light partially lit the green water and everything that went on the nearby ghats. Someone would be busy in late-night dishwashing, with the suds floating on the surface and the inquisitive fish coming up to play with the bubbles and nibbling on the permeating wet rice. A red saree-ed mother would have come to invite Ganga to her son's wedding by praying to the holy waters and the faces of the wedding entourage would light up with dashes of sindoor and trembling lucent flames from the mustard oil-filled terracotta diya, which they would float on the calm waters. And more often than none, groups of young boys would be sitting and enjoying a shared smoke acquired from pooling a few coins, excess from their autorickshaw fare, while bunking their maths tuition. But no matter what, there would always be an empty ghat where we could get in to wash our frequently bloodied noses and cleanse our aching and sweaty shoulders. We would swim to the middle of the large dark pond to cool ourselves, and compete to declare, once and for all, who could do the longest underwater stretch. Even then, every once in a while, I would suddenly start panicking in the water with a sensation that something was watching me and would swim frantically to the ghat. I had nightmares of sharks often—of being lost in the ocean, and a white shark, with its giant gape, coming for me. I would wake up sweaty and flustered, with the unmistakable knowledge that something had devoured me alive.

There is something about them that makes them such a potent element of fear—certain behavioural and aesthetic elements that have helped create this global image of dread. Maybe, it is related to the fact that they are the only ones among the large predators who still prey upon humans, even though these are statistically negligible incidents in comparison to the number of people who use the water space globally, and you have more chance of dying after being hit by a coconut, or in any other such non-spectacular way (Kasprak 2017; *The Daily California*, 2017). According to the international shark attack files, from 2013–2017, annually, there have been 84 cases of shark bites. In 2018, there were a total of 130 shark–human 'interactions', with 66 confirmed unprovoked attacks on humans, with four fatalities (Florida Museum, 2020). The number of white shark attacks within that group would be understandably another smaller cross section of the same. But this fear of sharks goes beyond the knowledge of some statistics, like Michel Maccentire said: when he was in Australia, some friends invited him to get in the water, and they said 'you have more chance of being hit by a car.' ... 'Not when I 'm fucking swimming, I don't. I'll be staying here on the beach' (Micher Mcintyre; Showman, 2020).

Sernert (2011) elaborated on fear relation with predators in Sweden—bear, wolf, lynx, wolverine, moose, and wild boar—in relation to attacks on humans. They explained that fear is an evolutionary response to avoid and anticipate pain and death, creating a fight or flight response in our body. 'Fear can be reduced with knowledge and experience, but this takes time and does not necessarily imply a positive attitude... The change of attitude and fear is also presided with a peak of negative feelings which can be so strong that the predator populations are not given a chance to re-establish themselves in the environment...' (p. 14). They further argued that fear is often related to myth and stories and negative symbolic values like that with wolves in the area. They dealt with varied factors, like proximity to the animals and 'how long the predators have occupied the area, if they are seen as something natural that has always been there or something new that does not really belong' (p. 17). Media also plays an important part in creating the fear of animals. 'The objective view disappears with editing, twisting of information and pressure from advertisers which is dependent on where in the country you live, different countries get different information' (p. 17). ÖHman (1986) applied a functional-evolutionary perspective to fear in the context of encounters with animals and threatening humans. They noted,

> It is argued that animal fear originates in a predatory defence system whose function is to allow animals to avoid and escape predators. Animal stimuli are postulated to be differentially prepared to become learned elicitors of fear within this system. Social fears are viewed as originating in a dominance/submissiveness system. The function of submissiveness is to

> avert attacks from dominating conspecifics. Signs of dominance paired with aversive outcomes provide for learning fear to specific individuals. It is argued that responses to evolutionary fear-relevant stimuli can elicit the physiological concomitants of fear after only a very quick 'unconsciousness' or preattentive stimulus analysis. (p. 123)

In Cousteau's *Great White Shark* (Cousteau and Richards, 1992), Jean-Michel Cousteau commented:

> Why sharks? There are other fierce animals on earth. One answer might be that the other potential predators of humanity have almost uniformly been extinguished, tamed, or banished beyond easy contact by industrialized society. The sabretoothed tiger is gone. The African lion is corralled by gamereserve boundaries. The grizzly bear survives only in the faraway north country. The Timber wolf paces nervously along zoo fences. In many parts of the tropics, the crocodile is routinely raised in farm ponds, as sequestered as chicken in poultry shed. We have grown comfortable, even protective, of the other beasts capable of devouring us whole, we are the masters of virtually every biological tract of earth, but for one. We remain potential food for creatures patrolling the liquid wilds encircling our terrestrial kingdoms. (p.16)

Cousteau pointed to some more specific factors of the fear, such as the randomness of their appearance, the lack of warning, the violence associated with their attack, and their silence as opposed to other big predators. To which I might add, most predators who generally feed on humans have a distinct smell, which is helpful for the prey to detect and evade them. The white shark has no specific smell, at least not one that is perceivable by humans. There is also the environment of their predation, the seas and oceans, which humans at best try to negotiate with clumsy strokes and sophisticated mass-produced equipment. But is there more to this fear?

There are dreadful stories about sharks in general, and in particular, about white shark bites from many corners of the world. The first documented case of shark bite in the Americas, according to archaeological evidence, was sometime between AD 789 and 1033 (Florida Museum, 2018). One of the first recorded witnessed bites was in 1749, when a British merchant sailor was bitten in Havana harbour, and lost his right foot (Cama, 2012). The Jersey Shore bites in the US are infamous; in 1916, four people were killed within a week; and then, the final fatality happened in the Matawan Creek, where a shark swam up a narrow creek, killing a 12-year-old boy (Blake, 2010). Probably the most famous is the USS Indianapolis incident of 1945, which delivered important components for the atom bomb. It was hit by a Japanese torpedo, 900 men went in the water, and after four days in water, it is estimated that as many as 150 sailors were devoured by sharks, possibly by Oceanic White Tips (Geiling, 2013). These stories—stories of monsters in the ocean recalled by European sailors, and stories in ocean-going cultures that warned of the toothed monster in the water—have become part of the man-eater narrative and global myths of human and shark encounters. The Matawan Creek incident is supposed to have inspired Peter Benchly to write *Jaws* (*HistoryvsHollywood.Com,* n.d.).

Then, there are personal stories...

It was 5:30 am. The year was 1970. Young Samir was out with his friends from Kolkata, for a couple of days to Canning, 50 odd kilometres away from the city, near the Sundarbans delta. It was a beautiful morning; the sun gently blazed over the maroon Matla River and the boys jumped into the cool water for a fun dip. All proficient swimmers, the boys were splashing around in the murky water and slowly going more and more into the heart of the river. Suddenly, a *majhi* (boatsman) spotted this and started shouting at them to immediately get out of the water. Not knowing what was going on, they frantically swam to get up onto the boat feeling they were in grave peril. The young boys got on to the boats unhurt, but when they asked the majhi what had happened, the majhi did not reply and dropped them to the shore. While returning on his boat, he uttered quietly, '*Kamot...*'

Kamot is the colloquially used Bengali term for shark fish in the river, Ganga, particularly in the Sundarban delta region of West Bengal. The limited knowledge we have predicates that *kamot* is either the Gangetic Shark, or the Bull Shark, or an idea created from the combination of both, or maybe, some other unidentified shark. Be as it may, that story always stayed with me, even if I did not know what *kamot* was when I was young (as a matter of fact, most Bengalis are unaware of sharks in the Ganga), but maybe that story added to my fear of 'something' in our water.

Then, there are the overtly dramatized tellings of the contemporary white shark 'attacks', and the familiar spaces they take place in—spaces both the species share.

> Humans and animals do not just interact with each other, they also interact with the landscape [in this case, oceanscape]. This might, for example, be with regard to the availability of food sources or to the creation or removal of physical barriers that attract or repel humans or wild animals. These entanglements of human, wild animal...are drawn together through their prolonged shared experiences. (Boonman-Berson, Turnhout, and Carolan, 2016, p. 194)

'Predator-prey interactions can be represented as a "space race", where preys try to minimize and predators try to maximize spatial overlap... Prey must identify space where they can obtain sufficient resources to live (e.g., food, water, cover, etc.) and avoid predators' (Muhly, Semeniuk, Massolo, Hickman, and Musiani, 2011, p. 1). In the open water space, the position of prey and predator between white sharks and the human is an ever-evolving intersection; unless we are specifically hunting a white shark, with controls of the premeditated logistics and technology involved, we are below them in the food chain. The beach and the ocean immediately near it are geographically and culturally constructed spaces claimed by humans primarily for their leisure activities, away from the uncertainty of the open ocean; hence, this creates a feeling of safety from the unpredictable dangers of the open ocean too. However, beyond the odd incidents of shark attack out in the open sea (Geiling, 2013), in the past few decades, these attacks/bites have been primarily associated with beaches and beach-going activities (as a huge number of people partaking in them increases the chance of human–shark interactions, and possibly, shark bite incidents). Beaches are products of the hyper-capitalist market, selling the idea of leisure and relaxation. They feign the image of true nature, and create social and physical boundaries of Edenic places, where human beings are safe. Not only are they symbols in relaxation, but they may also symbolize achievement and even happiness in a human life (Baldacchino, 2010). And human beings have always tried to create boundaries between themselves and the wild in such spaces of interactions (Sage, Justesen, Dainty, Tryggestad, and Mouritsen, 2016).

Hence, an attack here is a shock to the social expectations and activities, for example, the Mick Fanning episode in July 2015 at the Jay bay open surfing competition in South Africa (Howard, 2015). This attack—as it is called—can be more realistically described as, at best, an encounter with a juvenile shark. However, panic ensued, because it was also televised live. 'If we take otherness to be the privileged vantage from which we defamiliarize our "nature", we risk making our forays into the nonhuman a search for ever-stranger positions from which to carry out this project. Nature begins to function like an "exotic" culture' (Eduardo Kohn in: Kirksey and Helmreich, 2010, p. 562).

Consequently, in response to such attacks, though limited in number, American, Australian, and even New Zealand governments sanctioned nets on the beach to keep the dreadful 'other' and the exotic culture out of our sacred place of pleasure and the known and accepted decorum in our culture. To a shark, it may be a normal day, travelling, feeding, and living their life, as we do in our daily lives. But to us humans, it is a dramatic interjection to the civilized society, it is a homicidal activity, which should result in separating the civilized and uncivilized by boundaries, prison walls, or shark

The shark going for the bait (screen capture from video footage), 2017

nets. In extreme cases though, depending on the intensity of the emotional charge, the action rises in the local public psyche—capital punishment, poison injection, or harpoon. The only difference, in the case of a human criminal, is that there is a trial and the alleged murderer gets to plead their case, but the fish does not, unless their murdered, still carcass is cut open.

In New Zealand, from the beginning of human history, there were no predators hunting humans; the only plausible predator I can think of doing such was the shark—in particular, the great white sharks (besides other humans of course). Even though there have been merely five confirmed white shark attack fatalities in New Zealand in recorded history (Muriwai in North Island, 2013; Clark Island in South Island, 2009; Otago Harbour in South Island, 1968; St. Kilda Beach in South Island, 1967; St. Clair Beach, Dunedin in South Island, 1964), the fear of white shark attack in the region is perceived as quite real, especially if there are operations trying to bring them towards themselves for people in cages to see them—situated less than 10 km away from where children swim in the water. And then, there is effect of the global image of the white shark, 'In addition to specific local relationships between animals and people, another type of relationship is becoming increasingly important—the globalized, urbanized, Western view of wildlife' (Pooley et al., 2017, p. 518). Can this fear also be related to shark aesthetics presented through the mass media? Most of the time when we see images of white sharks, they are in their 'attacking' posture with their mouth wide open. One cannot help but wonder, why do these creatures always want to attack the camera? The simple answer is, these shots are almost always taken when the sharks are lunging towards a bait in cage diving expeditions, and this particular 'produced moment' creates some interesting aesthetics.

Most of the first-person perspective shots in the public forum is of sharks, with their jaws open, going for the bait and, since their predatory method includes independent movement of their jaws, the redness of the extended jaws is very striking (possibly due to biting the bait provided during the shooting, or the general redness in any jaw), giving the perception that they have just arrived from a bloody feast, and are

The dreaded triangular fin, 2017

possibly looking for more, or worst yet—in the process of eating 'you'. Maybe, it is the hard angles of their face/head, which potentially is related to fear or aggression perception, as indeed, it seems, acute angular cues of facial features have more of an aggressive effect on the observers than rounded ones (Aronoff, Woike, and Hyman, 1992; Barrett, Mesquita, andand Gendron, 2011; Fink, Neave, and Seydel, 2007; Larson, Aronoff, Larson, Aronoff, Sarinopoulos, and Zhu, 2009).

The other symbol that we associate sharks with is the raised dorsal fin—a perfect triangle, if ever there was in the animal kingdom. So much so that, the scuba diving hand symbol of shark is two palms making a triangle over the head of the diver, signalling to their buddy. Even there has been comparison made between the triangular shape of the fin of the shark in *Jaws* and images from Hitchcock's movies which were influential upon it:

> In both films, the swath of violence is symbolized by a triangular shape. It is the sign of Hitchcock's Avenger killer, and it also signifies death in *Jaws*, in the shape of the shark fin.... Needless to say, connotations of sexual violence are inscribed in the symbol, the triangular 'venus delta' invoking both the sexual subtext of Hitchcock's take on the 'Jack the Ripper' myth and the ambiguous role of the shark as both phallic aggressor and 'vagina dentata' (cf. Pollock 42). (Griselda, 1976; Schwanebeck, 2017, p. 252)

Furthermore, many times, I have had people tell me (in Bluff, during cage diving, and in different professional and personal interactions) that they thought that white sharks were disproportionate, whereby they were thinking primarily of the open jaw of an animal. Yet again, the very representation of this angle of photography can explain this perception. But what about the claustrophobic void that can be seen in their open abyssal gape in almost all of these photographs? Can that have any effect on the observer?

While spelunking out in a sea cave of the West Coast of New Zealand, I came face to face with 'void' as an organic form for the first time. Every dark turn, every squeeze in the unknown depth was more intimidating than the one before. It was the first time I had reached my psychological limits.

Journal entry
26 October 2016
3 pm
Punakaiki, New Zealand

I remember being on my chest and looking into the darkness tunnelling afront, and a void looking back at me. There was one cave in particular, called the Wriggly Snake, not necessarily a very deep one, but as you start exploring it, the namesake is realized. The cave started to get smaller and smaller, tighter and lower with every winding squeeze, and half submerged in water, which you had to crawl through. And just like a serpent, as you breathed out to squeeze in deeper into the cave, it was as if the cave squeezed back on you.

Eventually, that claustrophobia was unbearable, and I reached the limits of my mental strength and had to take refuge in a much larger air pocket, accompanied by three glow worms, and then crawl back. I started the slow aided climb to the cave entrance 65 feet above, which was literally a hole in a ground among giant flax bushes, with a muddy face and a few bruises to show on my ego. Hanging from the rock face, I looked down into the void and I realized that the claustrophobia I was escaping from has a stark resemblance with the void of a shark's gape about to engulf me.

Caving in West coast New Zealand, 2015

My inquisitiveness concerning this experience led me to an interesting book called *Deadly Powers: Animal Predators and the Mythic Imagination* (Trout and Ehrenreich, 2011). In the book, Paul Trout (a retired associate professor of English at Montana State University) discussed how many myths in different cultures possibly evolved from our instinctual fear of predators and the environments they hunted in. He elaborated on some physical factors of predators and sensory elements of the environment that create dread in us, even if we are potentially not in danger of being devoured by them in this day and age; some of those factors can be related to the fear of white sharks. Trout talked about the predator's stare (Trout and Ehrenreich, 2001, Chapter 3, part 1, para 6), and who can forget that from the iconic resemblance of a 'dolls eyes' from the movie *Jaws* (1975). Truly, when a 17-feet white shark puts its eyes out of the water while spyhopping, the sense of insecurity and even inferiority that this creates in any watching human is vividly apparent.

He also mentioned the fear of darkness (Trout and Ehrenreich, 2011, Chapter 3, part 1, para 67) where predators often hunt us, and of course the fear of teeth (Trout and Ehrenreich, 2011, Chapter 3, part 1, para 17). However, what really caught my attention was the discussion exploring the fear of 'predators' gape', tapping into our fear of being devoured:

> In Polynesia, for example, death is equated with being 'devoured'. 'For when a person died it was customary to say he was eaten alive by the gods...which is apt to the extremely literal [myth], that the soul of the dead were devoured by Miru [the goddess who is in Polynesian and New Zealand myths, often referred to as the deity of death (*Encyclopedia Mythica*, 1997)].' In Greek mythology, time, the universal devourer, as depicted as a giant python who...closes its 'envenomed jaws and little by little consume[s] all things in our lingering death.' (Smith, 2003, p. 334; and Ovid, 2004, p. 605; from Trout and Ehrenreich, 2011, Chapter 3, part 1, para 15)

There is an overpowering void and darkness in the gape of the white shark, which may be related to our fear of darkness as Trout points out, along with the fear of the gape itself. White pointers have a unique gape in the animal kingdom; every other predator who predates on humans has a thinner feeding tract than the white shark's gape, and none have that observable 'darkness' and void proceeding into the 'belly of the beast' that can devour us in entirety. Crocodiles have a palatal valve and whales have the epiglottis to stop from ingesting water while hunting underwater; hence, we cannot see directly into their stomach, but we can in sharks. Along with this open abyss of a mouth, we are also confronted with the 300 3-inch serrated teeth that decorate its entrance. This is an especially powerful image in case of a white shark, because of their huge girth, as compared to most other predatory sharks; for example, the largest known living white shark, Deep Blue, is estimated to have a girth of 8 feet across (Lampen, 2019). Even if other sharks are quite long, no predatory shark has such big a girth, which gives them such an imposing presence. It is even a different visual experience compared to the largest two fish on earth—the Whale Shark and the Basking Shark, because they do not possess large predatory teeth, even if they have a large gape (adding to the fact that they do not move and attack as fast as a white shark). So, in this fear of being eaten alive by a giant fish in an alien environment, our demise is symbolized by the giant claustrophobic hollow that one sees when looking into their open mouth.

Can deeper explorations of human cognition explain more about our fear of the white sharks? Our fear of sharks may be related to the availability of heuristics, which is where the brain creates shortcuts for us to recognize fear to assess any potential risk of a situation, no matter however improbable (Kahneman et al., 1982; Ropeik, 2010). It seems emotions have prevalence over cognition while making decisions. Our decisions and actions in the world are affected more by our emotions than mere objective analysis of the data around the visuals that are created after a shark bite, and the media presentation of them become more potent in affecting our decisions while engaging with the ocean and the topic of

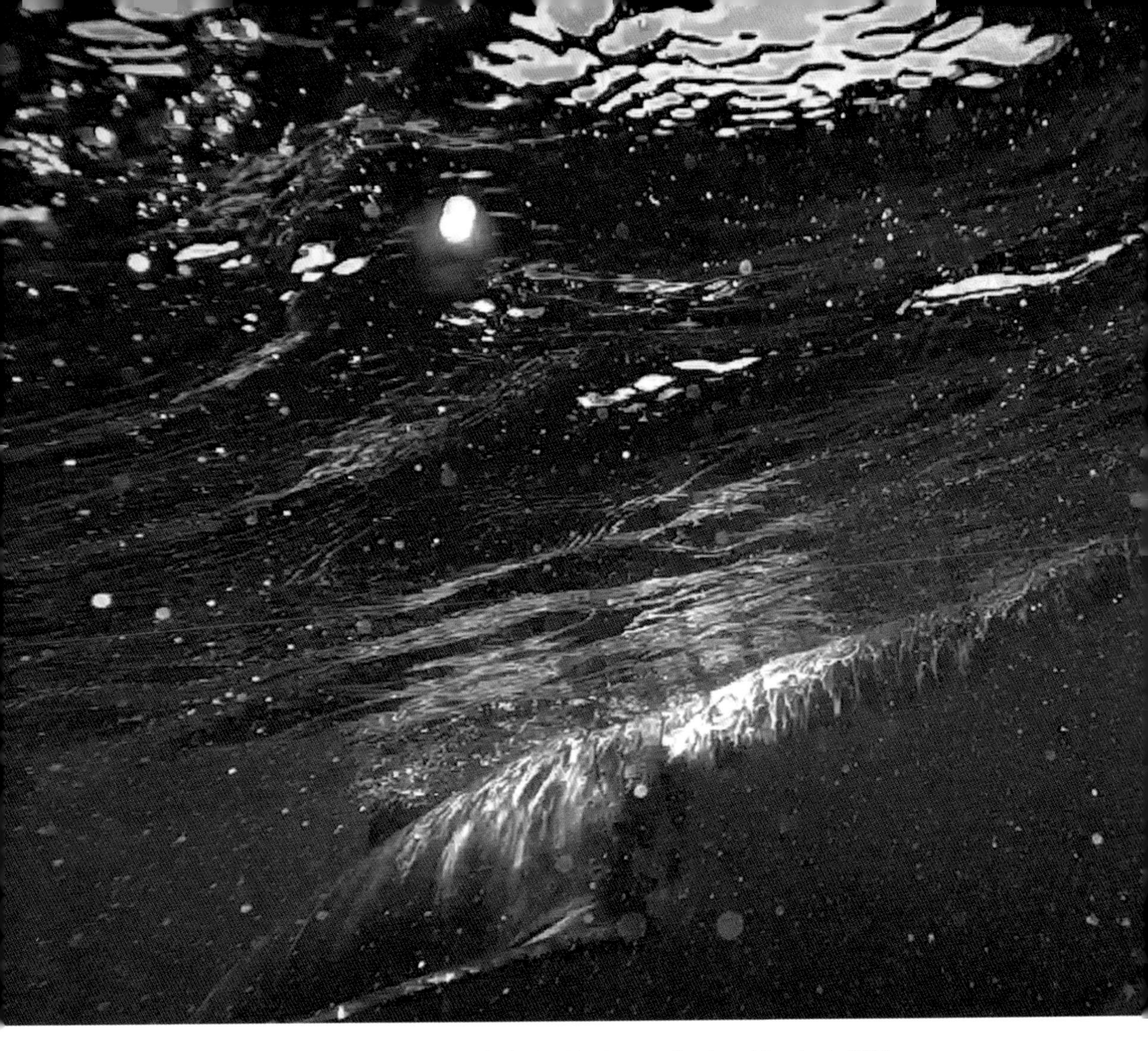

The darkness and the claustrophobia inside the mouth of a white shark, 2016

sharks, than the shark bite statistics present. The fear we have of sharks is more potent in our reaction to them and decision regarding actual and potential encounters with them. Indeed, the emotional and physical states of a human being have effect on their decision making (Schwarz, 2000; Dolan, 2002). Even in the choice of consumerism, the decision of purchase is often related to affective reaction to a product, rather than one's entire reliance on cognition (Shiv and Fedorikhin, 1999) and political scientists note the effect of emotional issues on policy formation even in case of human–shark relationships (Pepin-Neff, 2019). While discussing a person's fear of shark in a swimming pool, psychologist Dr Simon, argued that

primarily because in that particular case, the person was afraid of sharks only when he was alone in the darkness of a pool, maybe it was more related to the unknown danger outside our awareness, the fear of uncertainty, and the lack of control than the fear of the shark itself (Simon, 2014).

Maybe, this claustrophobia and fear of darkness is also associated with the murky waters these sharks are often associated with. As Trout (Trout and Ehrenreich, 2011) indicates, the fear of darkness may be also linked with our fear of the predators of the dark environment that can predate on us. 'Sensory deprivation—e.g., limited visibility, murkiness, loss of spatial orientation—can also cause illusions, particularly when visibility is impaired. Anxiety associated with this environment can cause heightened suggestibility and result in mistaking fish, other divers and objects for sharks or other threatening entities' (Cempbell, n.d.).

Perhaps, we can further relate this effect of sensory deprivation with the concept of sensed presence. Sensed presence is a cognitive theory, which is related to evolutionary agency detection mechanisms in times of distress and uncertainty (Granqvista, et al., 2005; Cheyne, 2001; Shermer, 2010; McAndrew, 2015). 'The sensed presence is the experiential component of a threat detection mechanism that gives rise to interpretive efforts to find, identify and elaborate sources of threat' (Cheyne, 2001, p. 2). Such presence may be felt in different kinds of situations of distress and extreme unease, such as in darkness, in isolation, in a state of fatigue, hunger, dehydration, and fear. These situations lower the threshold of alertness of our flight or fight mode, a mode related to hyper-activation of our sympathetic nervous system, and glandular secretion such as adrenaline (Seligman, Walker, and Rosenhan, 2001). In such situations, the brain can create a narrative from cues in the surrounding environment when there is lack of sufficient stimuli to analyze any potential threat. Cheyne notes that such a presence is often related to a reflection on a monitoring presence of the 'other' and relates it to the sensation of helplessness and vulnerability that a prey may feel while being stalked by a predator—'A threatening, malignant or evil intent is frequently ascribed to the presence' (Cheyne, 2001, p. 144). The white shark is the other, waiting for us in the darkness of the abyss in our perceived isolation.

While there are all these approaches to dissect our fear of these animals, I must yet admit that the reasoning for this fear is far from clear; it is as murky as the environment they often hunt in. Furthermore, it is probably different for different individuals, but it is certainly more complicated than it is generally considered to be. Having said that, the reality is that an encounter with a white shark is a rare thing. There are none in any aquarium in the world at the current time (2020) and the chance of actually ever seeing one of them in the flesh, underwater, is quite minuscule anywhere in the world; hence, it is safe to say that the contemporary image of the white shark has become a far more significant part of public (and for few, personal) mythology than the creatures themselves.

Maybe, the world was ready for a monster, a monster which was close enough to reality to exist, but resembled all the worst fears we could conjure up—cue 1975—the movie *Jaws* (Spielberg, 1975). The story has certain elements of Hitchcock-like horror in the treatment of the film (Schwanebeck, 2017) and the music-initiated creation of this monster (Biancorosso, 2010; Nosal et al., 2016). Finally, mass media found an entity in the form of an animal, who was the perfect candidate for creating an undirected moral panic in global consciousness. The myth of the great white shark was manufactured as a mindless man-eater; in fact, it was even termed as the 'teeth of the sea' (Kermode, 2015). Whether this was as influential as Hollywood, which has a far-reaching effect on the global popular culture, is speculative, but a hyperreal monster was born.

In the early 80s, Baudrillard coined the term 'hyperreality' to describe places and things that feel more real than the real world by blending an existing environment with simulated sensations (Baudrillard, 1994; Baudrillard, 1988; Baudrillard, Smith, and Clarke, 2015; Bonanni, 2006, p. 135). In hyperreality, the original version of the objects, ideas, or signs have no significance to the general consumers, because 'it belongs to a different realm and therefore loses its referential value... Reality, in this sense, dies out' (Wolny, 2017, p. 76). Reality is meticulously reduplicated, and the context and environment and settings are built around it to create the illusion of reality, much like in creation of fronts while creating personal social images (Goffman, 1959). In this reality, meaning is created through systems of signs, and images interacting and working against and with each other. These meanings are understood through their differences, and in relation to other objects and ideas and what they are not (Wolny, 2017, p. 76), much like the accepted reality that the beach is not a space for the wild and the unpredictability associated, but a place for controlled human existence.

Hyperreality is composed of two significant components, simulation and simulacrum. Simulacrum is the image that is simulated to resemble something else without constituting the inherent property of it, and is merely an imitation (OED from Sandoz, 2003). Simulation is the process or practice of simulating or imitating; through the process of simulation a simulacrum takes on their function (Wolny, 2017). But in hyperreality, 'the simulacrum has no more association with reality whatsoever and while offering images of the real, allow reality to evaporate into hyperreality as copies proliferate' (Bougen and Young, 2012. p. 391; Macintosh, Shearer, Thornton, and Welker, 2000). The image or sign bears no relation to any reality whatsoever; in effect, it is its own simulacrum (Baudrillard, 1994, p. 6), which is hyperreality.

I think that the present image of the great white shark that the public is most acquainted with, may be best expressed as a hyperreal (an image which is more than real) or para-real (an image which exists parallel to the reality) avatar of the white shark, the real fish. In other words, the hyperreal or para-real are not 'non-real', but are either more than (hyper) or exist besides/alongside (para) the real, and the images produced by mass media (circulated through multitudes of printed and virtual outlets) of the shark are grander than the reality, or they exist parallel to the reality.

Another way to consider this draws upon Hindu thoughts. The word 'avatar' has its genesis in Hindu mythology; one of the definitions Merriam-Webster uses to describe it is: 'A variant phase or version of a continuing basic entity' (Merriam-Webster, n.d.). For example, Lord Vishnu has 10 avatars or images; each of the avatars has a different form or purpose, and according to the needs of the humans and the situations, the different avatars materialize. The white shark also has different avatars as recognized by humans, which perhaps alters from time to time and place to place. To fishers, who go to the sea and encounter them each day, they are often just another fish or fin in the water, and sometimes, another entity who competes with their catch. On the other hand, for fishers who aim at catching them for their products, they are just prey and an economic resource. By the same token, when shark hunters caught them after *Jaws*, they were treated as trophy and it was more a matter of pride than anything else. For biologists, they are often a symbol of a healthy ecosystem. For people who associate with them, they are a symbol of beauty, strength, and vitality. But for a large section of the global society, since *Jaws*, they may have been a symbol of terror. And often, these borders between these varied avatars of the same fish are permeable, and the identities morph and transmit among different groups.

'Consumer society in the Western urban "core" passes into hyperreality when being itself—our most human impulses and desires—is experienced and expressed through the commodity-form of consumer spectacle' (Miller and Del Casino, 2018, p. 664). Social media, the fake news, photoshopped images of celebrities, manufactured wars—hyperreality predates this all. The constant epidemic barrage of 'think positive and achieve that nice car' mantra, blaring by our ears, points its fingers to our eyes, reminding us the epoch of the hyperreality we are all a part of, where images of lifestyle, sexuality, and prosperity are feigned to entice us to purchase more stuff. Almost everything we perceive around us in this industrialized 'connected' world is no longer 'real', but an exaggerated representation of the real, often created by a concoction of different elements of different ideas, and more than real. And as the real is no longer what it has been or used to be '...nostalgia assumes its full meaning. There is a proliferation of myths of origin and signs of reality; of second-hand truth, objectivity and authenticity' (Baudrillard, 2001, p. 174).

> Simulation collapses the real and the imaginary. Representation seeks to absorb simulations as false representations. Even knowing an image is fake has little importance. For example, the striking image of a great white shark breaking the ocean's surface went viral on social media in 2016. The chief National Geographic photographer, Bob Burton, was awarded the society's 'photo of the year'. The only trouble was that there is no such award and no such photographer.[6] The image is in fact digitally manipulated by a Russian 3D graphic artist called Alexyz3d, readily found on their Shutterstock profile. Simulation absorbs representation as a simulacrum, to the point where, ultimately, identities, not just of people but also fields such as outdoor learning, images, models, and the code, determine the understanding and experience of these things. (Leather and Gibson, 2019, p. 81)

In a lecture from 1993, philosopher Rick Roderick discussed hyperreality and the various images created and enacted in our day-to-day life. Roderick mentioned that different animals, rather images of animals like dinosaurs and even the shark in *Jaws* were hyperreal concepts.

> When I was a kid, I used to dream about dinosaurs. Why would I need to dream about dinosaurs now? Steven Spielberg has made them, filmed them, they are more real than real dinosaurs, they are hyperreal. You would be disappointed if you saw a Tyrannosaurus Rex after the movie; you would be disappointed, it won't be as noisy, as scary, as frightening. Same is true with

> *Jaws*. I have caught some rather large sharks, I like to fish, but those real experiences are so boring...compared to *Jaws*... *Jaws* is a hyperreal experience. (*The Partially Examined Life*, 2012)

I propose, if *Jaws* was a simulacrum of the real shark, almost everything that has been presented by the media in the last 30–40 years after that is a simulacrum of that simulacrum of *Jaws*, hence the contemporary image (rather than the shark in *Jaws*) is a hyperreal concept. That is, can we think back before *Jaws*, which created the widespread irrational negative image of the sharks, that the global connected society recognizes them with since then? This image has very serious 'real life' implications on the lives of the sharks, and human–shark interaction. As Beryl Francis from Murdoch University puts it:

> The media has played a dominant role in shaping [the] attitude towards shark conservation, shark attack representations and government policies around the world. When Steven Spielberg's film *Jaws* about a 'rogue shark' was released in 1975, very little was known about sharks and the media exploited the public fear that it evoked for commercial gain. The media hype of sharks and shark incidents gave rise to an explosion of shark hunters, shark game fishing, shark tournaments, shark artefact sales and commercial fishing. The result was a plunge in shark populations worldwide. (Francis, 2012, p. 59)

Christopher Neff made similar observations and proposed the term *'the Jaws effect'* (Neff, 2015; Neff and Hueter, 2013). '*Jaws* effect' is when significant actors use fictional narratives like that from *Jaws* to explain real life events, and even effect policy creations. They further raised issues with the rogue shark hypothesis, which was proposed by Australian surgeon Victor Coppleson, and attributed the human–shark conflicts in Australian waters in the 1950s to a single shark. This theory got its footing furthermore from *Jaws* and created the enemy of the 'other' of intentionally targeting humans. In Australia's case, these false narratives were used as political tools to create legislation and policy. These policy creations in western Australia were based on 'movie mythology than evidence-based science. Indeed, fiction was used to overwhelm competing scientific evidence' (Neff, 2015, p. 125). So, the fictional image was more potent than the 'real' fish. Consequently, I present the hyperreal image of the white shark, an image which keeps on feeding the mass media, even now, in the form of exposure and revenue.

This hyperreal 'monster' that is the white shark comes in many forms. Like Rob Van Ginkel (2015) argued, the image of whales in general public imagination is of a hyperreal creature that resembles all the good and positive aspects of the human condition (the key events and characteristics that compose the essentials of human existence) and is a combination of multiple species of whales. In comparison, the white shark may personify all that is to fear about the unknown associated with the human condition, and is created by the amalgamation of multiple real sharks, and all those imagined ones—such as sharks which can swim in fresh water, which are bigger than ships, ones which are entirely black, and one which has an especially defined taste for human flesh. There is the white shark shown in the National Geographic documentaries, which relates to a lot of concepts and presentation of the sharks we can consider reality, to the 'legendary' killer 35-feet shark in a Discovery Channel docufiction, which is in between reality and fiction, and finally the vanishing of any reality in the form of the *Sharknado*, and *Sand Shark*, and in a series of books where a 'gay shape-shifting shark' takes the form of a man, to have homosexual sex with human males (White, 2014). An analysis of human-shark interaction in major publications from Australia and America during 2000–2010 depicted that the majority were emphasizing on shark attacks even if the accounts of shark bites were quite minuscule compared to the number of humans entering the water (Muter et al., 2013).

However, *Jaws* was such a success because of its *'injection of reality'* (The Partially Examined Life, 2012). The extreme unreal (for example, *Sharknado*, where sharks are falling from the sky and eating people in a freak

tornado), which most people realize is fiction, can make the unreal look more real as in the docufiction, '*Shark of Darkness*' (Glover, 2014)—a reference to Joseph Conrad's novel, *Heart of Darkness* (1899). The beginning of the film gave this narration: 'Submarine is a legendary shark first sighted off the coast of South Africa in 1970. Eyewitness accounts say it is 35 feet long. Its existence is highly controversial. Events have been dramatized, but many believe Submarine exists to this day.' Even there was a disclaimer with it, 'WARNING—this program features disturbing images of shark attacks, viewer discretion is strongly advised' (Glover, 2014). This disclaimer had nothing to do with any shark attack that was 'real', and all the attacks were graphical and practical effects, in a fictional rescue. So much effort was put in making it all seem real—that they introduced a shark biologist in the film called Mel Thurmond, belonging to a shark research institute called the South African Marine Research Centre. This biologist was also presented as having a few YouTube videos expressing his frustration that no one believed his claims about a monster shark out there, no matter how much 'proof' he presented. However, a quick internet search will reveal that not only was there no academic of this name researching such a topic, nor was there any research organization of that name. But it was a successful endeavour, most people did not recognize the significance of this small part of the disclaimer '...events have been dramatized...' or in other words—completely made up! As a matter of fact, I had university students come up to me even in 2017, asking my opinion on the big shark that is out there, which they have seen in the 'documentary'.

Hyperreality and simulacra not only help us to conceptualize reality, but our perception of what is also imagined (Bougen and Young, 2012). With the advent of new technology, that which is real today becomes unreal tomorrow, because unreal ideas can be made more real tomorrow through technological advancements. The advancement of technology shapes the

The shark on a sunny day, 2017

contemporary ideas of the unreal and the real, for it is argued '...idea of the real and the real world is a cultural construction, certainly linked to the birth of the science and technology' (Smith, 2010, p. 95). One evening, Soosan's 14-year-old daughter and her friends ended up seeing *Jaws*, because they wanted to see a horror classic. However, when they actually saw Bruce (the shark in *Jaws*), they started laughing, because he looked so 'fake', and were not petrified by the film like the generations before them. Understandably, they belong to a generation that has grown up seeing sophisticated computer-generated (CG) characters, and the animatronic puppet, which had such a pivotal role in shaping the contemporary image of the white shark, and it made no impact on them because of its 'unreal-ness', and the question may also be raised if viewing it on TV creates a difference in affective impact on the perception of the shark on the viewer as opposed to in a cinema hall, which would be a more engrossing experience.

So, a hyperreal commodity transcending cultural paradigms was born. This hyperreal image proliferated via media outlets, utilizing a creature that was primed to be the 'real' monster of our times, gave rise to many industries including billion-dollar TV properties, and movies:

'The public is aware of shark bites and their interest has led to a cottage industry of entertainment and media to reinforce attention to these events. In motion pictures and on television, the portrayal of sharks is big business. For instance, a leading film website lists *Jaws* (1975) as the seventh highest grossing film of all time (adjusted for inflation), at more than $1 billion' (Box Office Mojo, 2014). Meanwhile, the Discovery Channel's *Shark Week* has generated 'hundreds of millions of dollars in ad revenue' (Tapper, 2013). In fact, no other animal, on land or in the water, generates the entertainment income that shark species do. From the book and motion picture *Jaws*, which manufactured a public panic, to the more than 25 television seasons of *Shark Week*, which keep the fears and fascination alive, the human–shark relationship presents a well-known story predicated on a primal battle for survival between human and shark (Neff, 2015, p. 115).

However, one of the most unique industries this image gave birth to was a ticket to see this monster in flesh—the great white shark cage diving industry. Consequently, people from all over the globe congregate on small boats in four or five corners of the world, pay a lot of money, and get inside an aluminium cage, for one chance to behold this alleged monster.

Social Science of a Big, Biting Fish

Beyond their hyperreal image, sharks are enigmatic ancient creatures, often apex predators of the sea. Their history can be traced back to 450 million years ago (MYA) (Skomal and Caloyianis, 2008), and as of now, we have identified more than 500 species of sharks (Star, 2018). There are three super class of fish: Agnatha (jawless fish), the Osteichthyes (true bony fish), and the Chondrichthyes (cartilaginous fish). Under the class Chondrichthyes (1,200 known living species) are the subclass Holocephali containing the Chimaeriformes or chimaeras (about 50 known and living species), and the Elasmobranchii, containing the Selachii—the sharks (about 500 species) and the Batoidea—rays and skates (about 650 species) (Ebert, Fowler, and Dando, 2015). In the subdivision Selachii, there are two superorders, the Squalomorphi, and the Galeomorphi, containing a total of eight orders of sharks. White sharks are classified here under the order Lamiformes or mackerel sharks (Klimley, 2013).

First fossil records go back to the Silurian period, but fossilized recorded tooth of true sharks are only found from the early Devonian period, about 400 MYA. The Carboniferous period (360—286 MYA) is called the golden age of sharks, with large proliferation in shark species diversity; however, about 250 MYA ago, there was a mass extinction causing the loss of large number of marine species, including sharks. The Jurassic period ranging 200 to 145 MYA was when the first modern sharks, like the Palaeospinax (Klimley, 2013; Skomal and Caloyianis, 2008), appeared.

The flexible cartilaginous skeleton helped them to change direction and accelerate rapidly, providing them a high rate of prey capture. Their multiple rows of teeth were also really helpful, because unlike other fish, when one row was lost, another one took its place. Furthermore, early sharks were fertilized internally, which meant when the sharks were out of the womb, they were already active predators (Kimley, 2013). Their torpedo shape also helped them to propel fast in the water. Among the first extant sharks to evolve were the Horn sharks and the Cow sharks; by 100 MYA, most modern sharks had appeared. The great white shark might have evolved in the middle Eocene about 65 MYA, and the modern saw-toothed white sharks have evolved from smooth-toothed sharks (Martin, n.d.). The modern serrated-toothed white sharks probably appeared 11 MYA (Klimley, 2013).

Globally, humans and sharks have had a long history of encounter in ocean-going communities. Sharks have not only been seen as predators who at times prey on us, but also as a competitor in food gathering, as food themselves (fish and chips are often colloquially called 'shark and taties' in parts of New Zealand), and even as entities of beauty and power to be admired and worshipped (Skomal, 2008; Cunningham Day, 2001). There is a significant body of ethnographic and archaeological literature supporting the historical use of body parts of sharks and other similar cartilaginous fish as material resources in various corners around the globe, from the Americas to Asia and Oceania (Akerman, 1995; Hunt,1981; Kosuch, 1993; Lima, 1999–2000; Lo´ pez Mazz, 1994–1995; MacQuitty, 1993; Tricas et al., 1997). Shark products such as teeth, fins, vertebra, meat, oil, and skin have been employed for weapons, tools, ornaments, and food (Boessneck and von den Driesch, 1992; Barbosa and Franco, 1991; Lima, 1999–2000; Kosuch, 1993) for a long time.

Polynesians, the ancestors of Māori, thought of sharks as guardian spirits. Many Hawaiian families had an Aumakua, or shark protector. Sharks and rays, along with other animals living in the sea, were considered to be the children of the ugly god, Punga. Punga was the child of Tangaroa, the god of the sea, and hence, sharks were linked genealogically to the god of the ocean. In the far north of Aotearoa, the ocean taniwha Ruamano took the form of a mako shark. If a waka (canoe) overturned, the crew called upon Ruamano to deliver them safely to land. In another legend, when the canoe, captained by Tamatekapua was voyaging towards New Zealand, it met Te Parata, an ocean creature who almost swallowed the canoe and its crew. They were saved by a shark, and in its honour, the crew renamed the canoe and their tribe as Te Arawa (shark) (Jon and Aich, 2015; Doak, 1975). In Murihiku tradition, during storms at sea, wise men could call great fish to provide protection to the canoe and even had visions where sharks warned them of upcoming danger. Even the fierce shark, the *mako-ururoa* could be made to obey the will of the tohunga (Māori priest).

Moremore was a kaitiaki (guardian) of the Ngai Tamanuhiri of the east coast of the North Island. He was the son of Pania, a sea maiden, and was considered to be a shark with no tail who warned the locals of any danger. Then there was the story of Tūtaeporoporo who began his life as a shark and was initially the pet of a chief named Tuariki. But as he grew, he became a scaly monster with shark's teeth. When the chief was killed, he set out to avenge his death, and got the taste of human flesh. He continued his reign of terror in the Whanganui River, until finally slayed by Ao-Kehu, a renowned taniwha slayer (Keane, 2007). One more myth can be found where the shark and the tuatara (an endemic New Zealand reptile) descended from heaven. 'These two then argued about where to go next. The shark favouring the sea, the tuatara the land. In the end the tuatara said, "Go to the ocean if you wish, and be caught with a hook and served up with cooked vegetables as food for man." To this, Mango [the shark] retorted that if the tuatara remained on land, he would be loathed by all and destroyed by fire when the dry bracken was burned off. Both these prophecies subsequently came true' (Cox and Francis, 1997, p. 39).

Sharks were an important part of New Zealand Māori culture (John and Aich, 2015). They were an important part of the diet for certain Māori communities, and each year, the Te Rarawa tribe of the far north of New Zealand set aside two days for shark fishing. The first day was close to the full moon in January; the second was two weeks later, and people catching sharks outside of these days were stripped of their property. As the article notes, even if the sharks may not have been as significant to the southern tribes as the northern ones, they were still good for eating, and were caught in line with line and hooks. We found that the body of the shark was significant in Māori material culture too. Shark teeth were used to make jewellery, such as pendants and earrings, and shark liver oil was mixed with red ochre to make the distinctive paint used on carvings. Materials made from sharks were used for varied purposes, for example, a weapon called Wahaika, which was used to emasculate sexual offenders. There is also evidence of an ancient practice among the southern Māori, where the corpse were preserved by a process called *whakataumiro*, prepared with the mixture of shark oils, wooden oils, and red clay known as *maukoroa*, and left in a dark and cool cave.

The image of the sharks was a significant part of aesthetic culture, shark images can be seen on certain stone artefacts, and sharks also became an important part of traditional Māori tattooing. 'From their tales of creation, we learn that the godhead Māui set the mythic shark Te Māngōroa across the sky—as the Milky Way. A symbolic māngōroa pattern in the Māori visual aesthetic is used in *tukutuku* panels throughout New Zealand, representing the Milky Way—and hence the place of this sacred shark' (John and Aich, 2015, p. 178).

Great whites are known in Māori as *mangō taniwha*, *mangō ururoa* and, to people of the Ngāi Tahu tribe, as *tupu*. Taniwha, being linked with mythical monsters often residing in water. Under the Treaty of Waitangi settlements, the *mangō tuatinifa* or white shark is considered as a taonga (or treasured) species to the Ngāi Tahu iwi (John and Aich, 2015). Māori likened their warriors to sharks, invoking them in battle cries such as: '*Kia mate uruora tātou, kei mate-ā-tarakihi'a* (let us die like white sharks, not tarakihi fish).' They are also invoked in place names, for example, Keane records that 'The northern bay of Mangōnui (meaning great shark) [I have been told by some of my Māori brothers that this was a white shark] is named after a guardian taniwha, in the form of a giant shark, that accompanied the canoe Riukaramea into the harbour' (Keane, 2007, p. 5). There may be even indications to myths about white sharks in Foveaux strait: 'There is the folkloric shark Kaitiaki-o-Tukete (the Guardian of Tukete) which was thought to live within Foveaux Strait, and particularly around the passage between Codfish Island and Stewart Island—in the far south of New Zealand' (John and Aich, p. 179). Kaitiaki-o-Tukete was either a shark or a monster fish, and was spiritually connected to the Kati-Mamoe Chief, Tukete. And after the death of Tukete in a battle on Stewart Island, the beast haunted the coast.

Even if they have been significant culturally and as material resource to varied ocean going communities, our knowledge of them have been quite limited till recently. Sharks were extremely hard to study, even in biological sciences (Castro, 2016),

> Several factors made sharks difficult to study. First, they are fast moving and far-ranging fishes in a vast marine environment, and the technology for studying such free-roaming aquatic animals did not exist. Second, it was difficult to obtain study specimens, owing to the historical lack of shark fisheries caused by their low commercial value. Third, the often-large size of sharks made it difficult to preserve museum or reference specimens. Fourth, ichthyologists had a general lack of scientific interest in cartilaginous fishes, and few people studied sharks. Consequently, sharks long remained a poorly known group. (Castro, 2016, p.14)

But the same film which made them a monster in public imagination actually created interest in studying them, and this went hand in hand with the interest in military research on them for use in weaponization, interest in shark fisheries and modern tagging system (Castro, 2016).

Sharks and other animals have been historically considered to be out of the sphere of anthropological investigation. They were generally understood to be as resources, or symbolic significance or being merely part of the national oceanscapes, and there was no acknowledgement of their affective agency in shaping human culture. Recently, there has been interest in social scientific research with sharks (Neff, 2015; Neff and

Hueter, 2013; Pepin-Neff and Wynter, 2018; L. Gibbs and Warren, 2015; Muter, Gore, Gledhill, Lamont, and Huveneers, 2013; Neff and Hueter, 2013; Pepin-Neff and Wynter, 2018; Eovaldi, Thompson, Eovaldi, and Eovaldi, 2016; O'Bryhim and Parsons, 2015; Apps, Dimmock, and Huveneers, 2018; Apps, Dimmock, Lloyd, and Huveneers, 2016; Richards et al., 2015; Techera and Klein, 2013; Topelko and Dearden, 2005; Torres, Bolhão, Tristão da Cunha, Vieira, and Rodrigues, 2017; Adorni, 2017; Hammerton and Ford, 2018). However, even now, in most of these research endeavours, the social scientists actually focused on the humans, while those engaged in the biological sciences focused specifically on the sharks even in relation to such activities as cage diving. In effect, maintaining the tradition of separating these two species by different epistemological paradigms, and not studied as an assemblage of affective entities. That is where research such as this attempts to create a holistic outlook at these two species engaged in a relationship where they live: how the oceanscapes of the sharks are connected to that of the humans sharing it with them. Both impacting each other's physical and social interaction exemplified by this particular human practice of cage diving.

My research adopts the 'more than human' approach in contemporary social scientific research, recognizing the limits of classical anthropological investigations of cultures and the animals around them. It questions the humanist epistemology of classical ethnography and aims to bring sharks into the overview of the ethnographic study, using an overarching methodology of multispecies ethnography (ME), were the humans and non-human animals are considered to be part of an entangled existence (Dashper, 2020; Ellis et al., 2018; Gannon and Gannon, 2017; Gillespie and Narayanan, 2020; Hohti and Tammi, 2019; Kohn, 2018; Press, 2019; Mason, 2016; Satsuka, 2018; Sepie, 2017; Vanutelli and Balconi, 2015; Wilkie, 2015). ME argues life to be an ever-evolving process, which is shaped and morphed by intersecting the needs and demands of varied species. 'The ethnographic of multispecies ethnography writes the human as a kind of corporeality that comes into being relative to multispecies assemblages, rather than a biocultural gives' (Ogden et al., 2013, p. 6).

ME is a 'more than human approach to ethnographic research and writing' (Locke, and Muenster, 2015, para 1), and '...draws attention to humanness as an impure, synthetic cultural category. It considers how beings, species, and categories of nature/culture get made through multispecies engagements' (Faier and Rofel, 2014, p. 371). In effect, ME challenges the scholarly tradition of human exceptionalism that presents humans as separate from other life forms and, therefore, possessing exclusive agency for understanding and having significant impact in shaping the sociocultural, political, and economic environment of the lived worlds we are all part of (Haraway, 2008). Classically, in cultural and social anthropology, non-human animals were primarily regarded as being part of human practices like rituals, part of cosmologies as symbols, and material resource. ME explores the 'mutual dependencies, influences, and hybrid ontologies involving human and nonhuman actors' (Faier and Rofel, 2014, p. 371), and in the process, questions the 'privileged ontological status', where humans were held at the top of the chain and other animals were only considered significant 'to the realm of fables, folklore and tales; to things symbolic, allegorical and mythic' (Aisher and Damodaran, 2016, p. 298).

This ontological turn encourages the rethinking and considering of other worlds, beyond merely those focused on the agency of humans (Archambault, 2016), and pays attention to everyday engagement with more than human beings and things. Furthermore, not only does ME explore the intertwined existence of multiple species, but also considers the land and oceanscapes as an assemblage of affective entities where these varied species interact (Aisher and Damodaran, 2016; Watson, 2016). In classical social scientific research, it was believed that one of the most distinguishing features of humanity was agency over the world and lives they share with non-humans, this meant 'strong tendencies remain (even among biologists) to evaluate diverse forms of life against an animal standard, and to regard non-animals as having diminished agency and capacities' (Hathaway, 2018, p. 48). Even in ethnographies dealing with interactions between humans and non-human animals, focus has been

The darkness and the shark, 2017

on the human dimension of the discourse. However, that idea is being challenged, and there is an emerging dialogue of a continuum of agency among humans and non-humans, even if the agency of the non-humans may be exercised in a different way than that of the humans (Caillon et al., 2017; Hathaway, 2018). ME traces the multidirectional interaction and

effect of multiple species on each other, and not just the effect of humans on other species (Pooley and et.al, 2016).

Human and animal lives are interconnected socially, politically, and geographically (Lloro-Bidart, 2018b; Kirksey and Helmreich, 2010). This also means that multispecies ethnographies are 'shifting from a species-specific to an interspecies frame of reference...reconceptualizing the focus of anthropological research' (Singer, 2014, p.1282). Indeed, even sociablitly is argued as intersubjective fields of action and co-equal society builders, where identities are created through interactions, and the tendency to socially interact and being socially available to each other. It requires understanding situations, and being able to adapt and engage in dialogue, symbolizing inter-subjective exchange. 'Studying animals helps us grasp what it means to be human; which mechanisms we avail of when interacting with others; how we build up social meanings...how we organise our lives; how we perceive our links with other living beings...' (Tedeschi, 2016, p. 164).

Kirksey and Helmreich, in their seminal article, 'The emergence of multispecies ethnography' (2010), pointed out that ME has its roots in the classical works of the late 19th century. They noted that scholars worked between boundaries of social and natural sciences; for example, in the works of Lewis Henry Morgan (Morgan, 1868), where he discussed the transmission of engineering knowledge among beavers, and drew similarities between human and beaver engineering knowledge. The 20th century saw interest in human cultures and their association with animals through hunting, husbandry, and animals as totem. Then, in the 21st century, natural scientists and anthropologists re-examined race, gender, genealogy, and human–animal divisions. This meant that by the early 2000s, there was a call for re-examining the human and non-human animal relationships and interconnectedness. The reflections of Tim Ingold on 'anthropology beyond the human' that aligned with the work of philosopher Dominique Lestel, examining hybrid communities with species sharing meaning among themselves, led the way in multispecies anthropological research (Ingold, 2013, from Fijn, 2019). Kirksey noted,

> New brands of animal anthropology twist the old, as more anthropologists have become curious about the lives of animals in labs, on farms, in agricultural production, as food, in rapidly changing ecosystems... [they] began to attend to the remaking of human nature, others began to follow related logics of remaking at work in nonhuman natures. (Kirksey and Helmreich, 2010, p. 551).

Currently, in more and more contemporary ethnographic investigations, human and non-human animal relations are an active part of anthropological investigations, and the lines between nature and culture are blurred to point out the emergence of human and non-human life through these multispecies relationships (Lloro-Bidart, 2018a). This new wave of human-animal research emphasizes the 'coexistence, conviviality and interactional encounters between humans and nonhumans [sic],' and explores the 'multispecies intersubjectivities in which all human lives and cultures are enmeshed' (Aisher and Damodaran, 2016, p. 294). Indeed, the contemporary explorations in ME aims to elevate the significance of experience, behaviour, and social life of multispecies beings, instead of treating all other species as non-agentive elements and objects in and around human life (Singer, 2014). In the same sense, the practice of the White shark cage diving in New Zealand is an endeavour, which is dependent on the intertwined agencies of the sharks, the wind, the ocean, and even the other material elements significant for the operation as much as the humans.

Here it is important to acknowledge, there might be an argument regarding the name applied to the type of research undertaken, as can increasingly occur in any research where researchers are primarily focusing on the relationship of human beings with one other species specifically; the best label may be 'bi-species' ethnographies as opposed to 'multispecies' ethnographies (although 'more than one' literally can be still considered to be 'multi'). However, presently, in most research focusing on human interaction with one major other species, be it primates, elephants, hyenas, crocodiles or even bees, the term 'multispecies' is used (Andrews, 2019; Fuentes, 2010; Radhakrishna, 2018; Locke, 2013; Wanderer, 2018; Baynes-Rock and Thomas, 2015; Fijn and Baynes-Rock, 2018; Baynes-Rock, 2019). Hence, to avoid any terminological confusion, I am using the term multispecies ethnography in this book, even though this book is primarily exploring the interaction and relationship between two species: humans and the white sharks.

ME follows within a broader conception of the anthropocene and posthumanism. The anthropocene is the proposition for the contemporary epoch where the magnitude of human activity is so grand that it is visible in the earth's ecology, atmospheric chemistry, and biodiversity. It is argued to be initiated around the 18th century with the development of the steam engine, where novel ecologies were created by humans, but not entirely controlled by them. This in effect has encouraged a post-human turn in social scientific investigation (Andrews, 2019; Lindgren and Öhman, 2019; Wolfe, 2003, 2009, 2010), where there has been a divergence from the human centred perspective, and a recentring of the agency of non-humans in shaping the world around us, be it material objects or non-human animals (Webber et al., 2017). Wang and Wang (2019) described posthumanism as 'Evolving from the crisis of humanism, meaning that the ubiquitous and powerful function of man in the humanist era has now come to an end, and man has entered into a "post-human" stage in which man is no longer regarded as the only species that has rational thought, nor does he play the leading role among all the species on the earth' (Wang and Wang, 2019, p. 4). Hence, post-humanists argue on a flat ontology, where anyone or anything are perceived from a similar platform of significant impact on each other (Cohen, 2019). It is pivotal in exploring nature-society relationships, where humans, non-human animals, and things create a complex socio-environmental relation (Rose et al., 2019).

ME aims at engaging the social beyond only humans, attuned with the ontological turn in social sciences in exploring the knowledge created through the conceptual breaking of strict barriers between nature and society or culture. ME benefits from conceptual development framing reciprocity patterns among varied species to displace humans from central positions in varied multispecies assemblages and, therefore, recognizes the needs to highlight on novel paths to navigate the anthropocene (Mason, 2016).

All the meetings of these varied species happens in a contact zone, cultural and physical spaces where they '...meet, clash and grapple with each other, often in contexts of highly asymmetrical relations of power, such as colonialism, slavery, or their aftermaths as they are lived out in many parts of the world today' (Pratt, 1991, p. 34). These are embodied sites where encounters take place; and more recently, this term has begun to be expanded to include multispecies ethnographies, where often violence and uneven power-relations are enacted, therefore 'more-than-human applications of the contact zone might include fuller, more serious considerations of nonhuman agency' (Isaacs, 2019, p. 735).

Wilson (2019) elaborated on being or becoming animals to now being concerned with people negotiating with creatures in such contact zones. Such contact zone negotiations are vastly different from their usual anthropocentric perspectives and norms, be it coexisting with coyote and cougars in Americas, to macaques of Singapore, or sharks in Australia, among others. These interactions and often emergent conflicts indicate to certain physical and conceptual boundaries, it is a struggle between 'us' and 'them' (Wilson, 2019). 'Many examples of human–animal encounters are about the breach of spatial and regulatory boundaries—home spaces, urban borderlands, safe swim zones, and so on...' (Wilson, 2019, p. 717).

The work of Donna Haraway (a prominent scholar of science and technology and feminist studies) has provided one key starting point for this 'species turn' in human–animal research. Haraway popularized this concept in human–non-human perspectives, and included that of non-human animals and humans interacting in contact zones; to her, these are spaces 'for multispecies entanglements and assemblages where the hardest lessons are learnt culminating in significant transformations' (Banerjee, 2013, p. 422). She noted, 'If we appreciate the foolishness of human exceptionalism, then we know that becoming is always becoming with—in a contact zone where the outcome, where who is in the world, is at stake' (2008, p. 244). These contact zones are where the lives of humans and other animals intersect in biological, cultural, and political intersections, creating mutual dependencies of the humans and non-humans. This 'species turn' within a contact zone reflects on the emergent hybrid ontologies arising from it, and also how the human is created from relationships, encounters, and entanglements with multiple species (Faier and Rofel, 2014, from Aiyadurai, 2016).

An example of such a contact zone can be found in the research of ethnoprimatology. Ethnoprimatology is a holistic perspective of understanding human-primate interaction and cohabitation, and has been one of the most influential methodological platforms to argue for the significance of multispecies ethnography (Fuentes, 2010; 2012; Jørgensen and Wirman, 2016). Such scholarly explorations analyze the worldmaking through multispecies entanglements in social and physical spaces where the divisions between nature, culture, human, and animal blur. They call for '...increased attention to the contact zones where the lives of humans and other species biologically, culturally and politically intersect, as a counterpoint to the dominant planetary perspective of earth systems and conservation science. They underline the importance of deep relational analyses of human interactions with other life forms, through renewed attention to multispecies histories, locality, and forms of knowledge rooted in place...' (Aisher and Damodaran, 2016, p. 294).

The dynamic contact zone of cage diving has not been considered in multispecies research. It is a contact zone created between the lives of the sharks and humans encountering each other in Foveaux, shaped by both their agency, coexistence, and sometimes, violence. Here, not only varied humans with different histories and agendas, but male and female sharks with diverse experience and ages meet. Even though the specific encounters around the cage facilitated through cage diving may be peaceful, the overall impact of the creation of this practice on the sharks may, however, be more asymmetrical.

The species turn in social scientific investigation has challenged the nature/nurture binaries and the humanistic epistemology of classical social scientific investigations. Roots of this can be found in animal rights theories, eastern philosophy, ecofeminist theories, and indigenous scholarships (Lloro-Bidart, 2018b). The species turn is also shaped, as discussed, by the tradition of posthumanism (Fijn, 2019; Lloro-Bidart, 2018b). The development of the multispecies approach in anthropology correlates with development in other related disciplines, be it in geography (Gillespie and Narayanan, 2020; Whatmore and Thorne, 1998), psychology (Adams, 2018), or archaeology (Boyd, 2017), arguing the significance of the human–non-human animal perspective to understand the human condition. ME has further influenced through other theories and scholarship, like actor network theory and philosophy; disability studies and queer studies; and feminism and gender studies (Lloro-Bidart, 2018).

There has been exploration in education research (Deckha, 2013, from Lloro-Bidart, 2018b, p. 263; Philo and Wilbert, 2000; Davies and Riach, 2019); work and labour research (Dashper, 2020), geography (Lloro-Bidart, 2018a; Whatmore and Thorne, 1998; Bear and Eden, 2011), psychological research (Adams, 2018), and in conservation science (Caillon et al., 2017).

With these varied disciplines working towards the same premise of understanding how we share our lives with other animals, the methodologies and methods for studying them holistically had to also evolve accordingly (Ulmer and Ulmer, 2017), often leading to multidisciplinary approaches. Researchers have shown interest in lived geographies of experience of the animals and the humans interacting with them, and emphasized the need for interdisciplinary investigation with tools and methods from disciplines like ecology to create a holistic picture of the interactions such as trackers, video recording (Hodgetts and Lorimer, 2015). Palmer et al., (2015) used a combination of ethnography and ethology to study orangutans' relationships with one another and caregivers' interpretations of orangutans' bonds in Auckland Zoo, New Zealand. They argued that qualitative and quantitative investigation methods used in ethnography could be useful for describing multispecies lives, by being able to compare data derived from the two methods, and create bridges among varied but co-dependent (in the long-term) disciplines.

To address the hurdles of disciplinary barriers in species conservation efforts, researchers dissected the meanings of biological science terms in comparison to anthropological terms within the same investigations with human–non-human animals, using the example of coexistence between people of the Nalu ethnic group and critically endangered Western chimpanzees at Cantanhez National Park in Guinea-Bissau (Parathian et al., 2018). They found, in the biological sciences, 'human–wildlife interactions are the actions resulting from people and wild animals sharing landscapes and resources, with outcomes ranging from being beneficial or harmful to one or both species. [Furthermore] humans are considered part of nature in an evolutionary sense but are traditionally viewed as separate from nature in an ecological sense' (p.749; 751). Whereas they found that social sciences '...have considered the multiple possible realities perceived by diverse human communities and individuals that are shaped by religious and cultural beliefs, historical and social backgrounds, and ontological reasoning' (p. 749; 751). In effect, an argument was made on the significance of ethnoprimatology as a platform of interdisciplinary methods.

One of the major barriers of enmeshing social sciences in conservation efforts and research is lack of understanding of subject specific methods and tools among other disciplines, and often, the lack of willingness to bridge this gap (Parathian et al., 2018). Furthermore, discipline-specific language and varied theories for understanding specific arguments and topics may be inaccessible to non-specialists of the particular disciplines. This is because, the language used by social scientists can be intentionally ambiguous, to reflect alternative worldviews of cultures that oppose

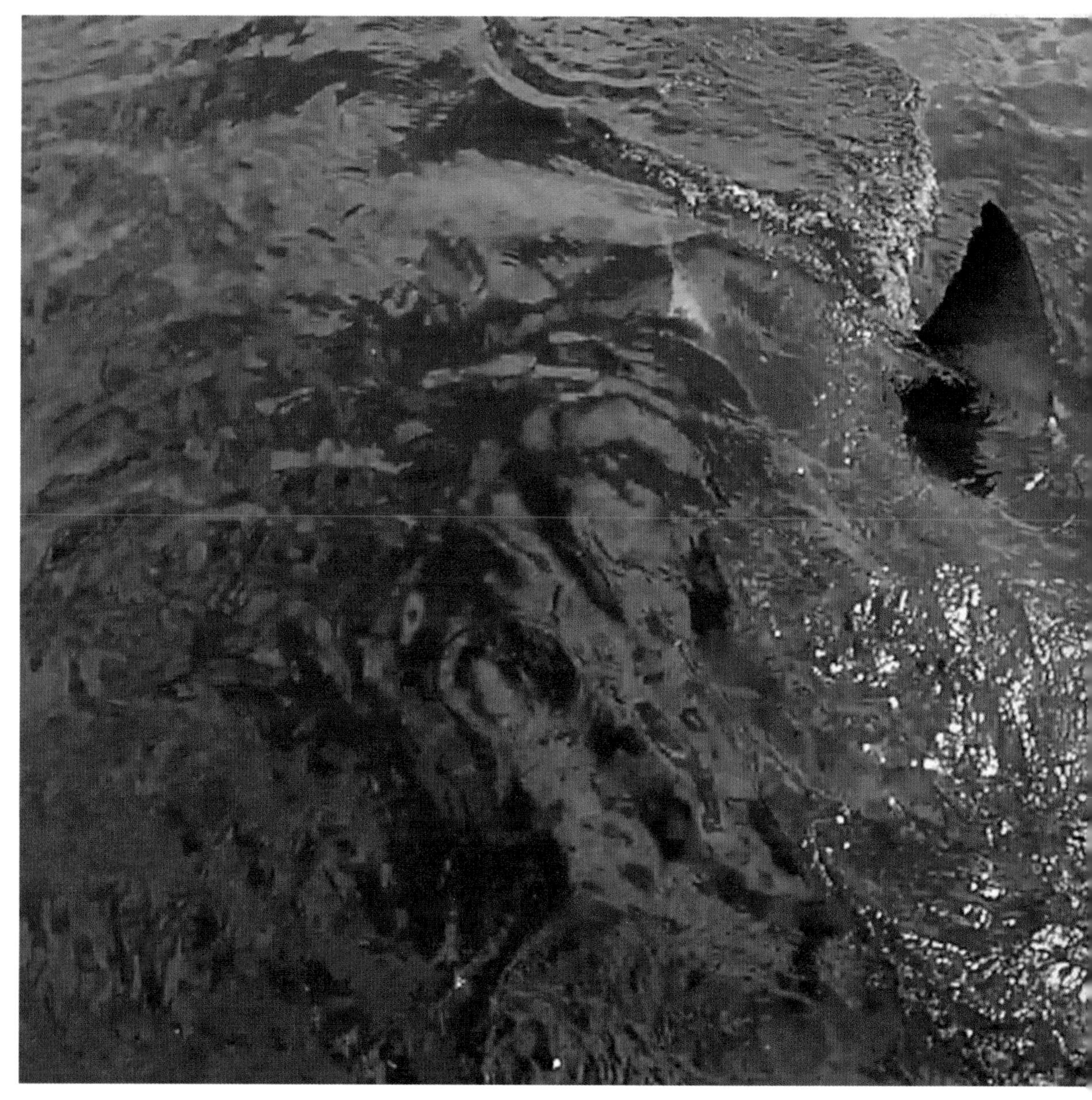

The fin slicing the water, 2016

Western dichotomized notions of nature (Descola, 2014; Kohn, 2007), and/ or to challenge preconceived ideas and assumptions about the world that characterize a Western scientific approach.

Multispecies ethnographers have explored interactions, encounters, and cohabitations with various species (Lowe, 2004; Petropavlovsky, 2014; from Kirksey and Helmreich, 2010; Davies and Riach, 2019). There has been investigations with agrarian and herd animals, like cows, buffaloes (Galvin,

2018; Lorimer, 2006; Fox, 2006; Johnston, 2008 from Bear and Eden, 2011; Hardy, 2019; McLoughlin, 2019), primates (Radhakrishna, 2018; Waters et al., 2019; Jost Robinson and Remis, 2018), and elephants (Barua, 2014; Locke, 2013; Kopnina, 2016; Erickson, 2017). There has been research on companion animals, such as dogs (Kohn, 2013; Sands, 2019; Satama and Huopalainen, 2019) and raccoons (Taylor and Pacini-Ketchabaw, 2017). Resaerchers investigated human connection with birds (Isaacs, 2019;

McClellan, 2015), plants (Archambault, 2016; Chao, 2018), amphibians like salamanders (Wanderer, 2018), microbial life (Greenhough, 2012; Ingram, 2011), fungi (Tsing, 2012; 2013), and even insects like bees or arachnids like spiders (Andrews, 2019; Engelmann, 2017; Ginn, 2013; Nxumalo and Pacini-Ketchabaw, 2017, Fijn and Baynes-Rock, 2018; Davies and Riach, 2019; Kelly and Lezaun, 2014), or other invertebrates (Lloro-Bidart, 2018, p. 28; Moore, 2015). Despite the variety of investigations with human life entanglement with other species, multispecies research with large predators, however, is comparatively limited, and they are only considered as groups, not as individuals with particular histories. Yet, recently, this has begun to change and researchers have already investigated lives shared with tigers, hyenas, bears, and even crocodiles (Aiyadurai, 2016; Baynes-Rock and Thomas, 2015; Boonman-Berson et al., 2016; Simon Pooley, 2016).

Not only on predators, multispecies research focusing on 'slimy' animals like fish is also limited. So, not only is there a gap in multispecies research in large predators, but also slimy fish, and the white shark is a large 'slimy predator' (Bear and Eden, 2011). Filling this gap, this book elaborates how the contact zone created among the humans and the sharks through cage diving is as much affected by the agency and will of the white sharks as much as that of the humans.

Alexander and the Sharks

Drunk and disoriented, my unruly feet fell on the cold Dunedin Street one after another. I needed a cigarette! I had quit smoking cigarettes officially three weeks back, but I had a couple of Guinness—I needed a smoke. Soosan had gone to drop Steve (Stephen Crawford) to his hotel, and I was left to my own devices for about 15 minutes. I found a liquor store, in I went, and was surprised to see packs of ITC Insignia, my favourite brand. I lit one and started walking down the path. Right by a field next to the University of Otago, I rested against a big tree, looking out at the traffic, and started smoking. First one, second, and then, the third, and as I was about to finish it, a shadow came and stood beside me. The voice asked, 'Hey bro, can I get a smoke?' 'Yeah sure.' The darkness was lit with my lighter, and we both stood still under the tree, smoking. Turns out, this young man was trying to get his law degree at the University of Otago, but kept on goofing up, because, he proclaimed, 'I get into a lot of fights.' He was complaining about how no one understood him and there was no way of expressing his energy. Interestingly enough, it turns out, he was somewhat of a martial artist. In a drunken mind, I asked what kind of techniques he used. And he started explaining the different styles he had practised over the years. Some time later, Soosan got back to the tab where she had left me, but after not finding me, started driving down the road to see if she could spot me. I was later told, she did find me, underneath a tree in the darkness, in a field, drunk, and teaching advanced pressure point techniques to a stranger with a green and pink mohawk.

White Sharks of Foveaux, 2017

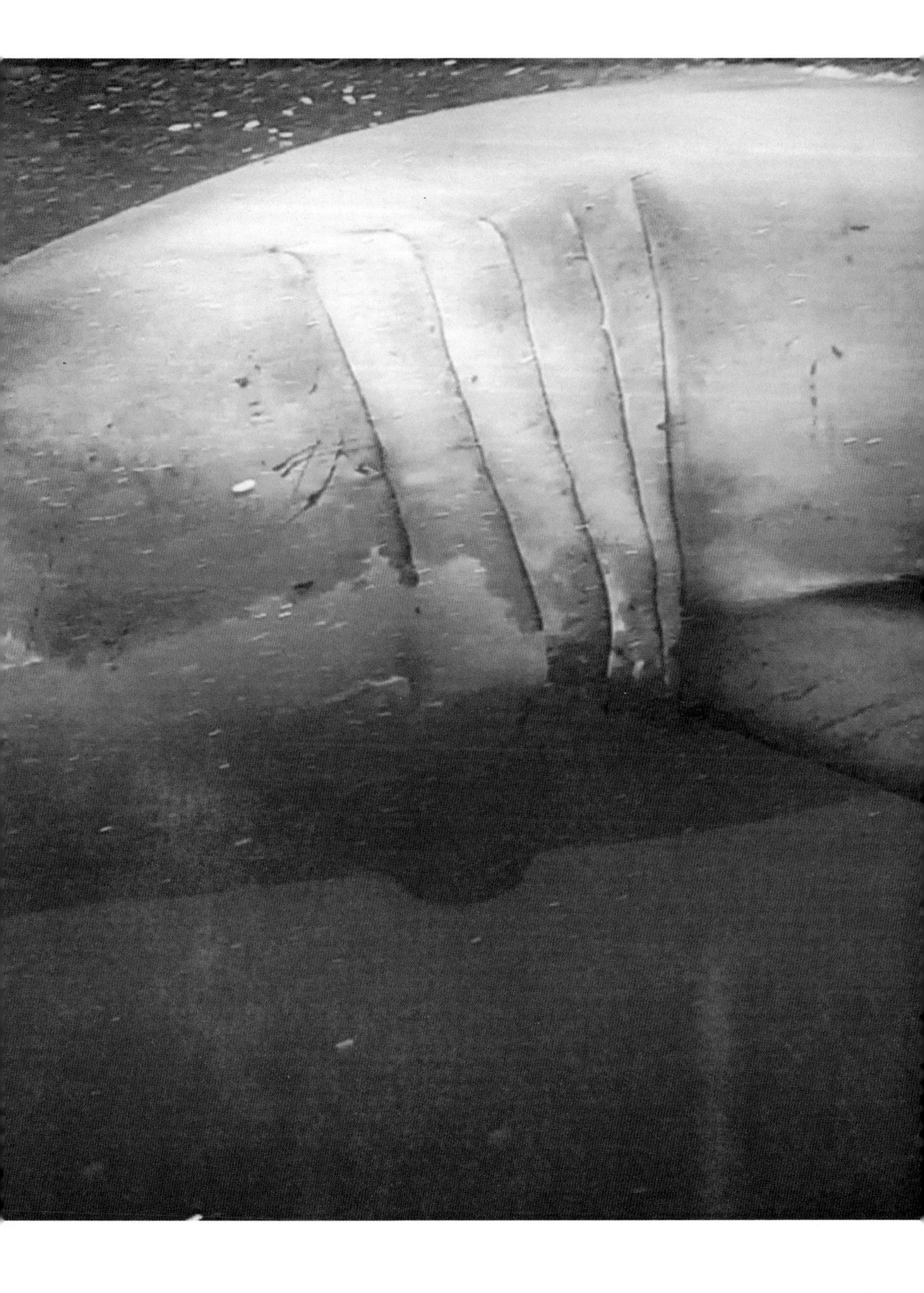

My faux pas was the result of an intense meeting with Associate Professor Stephen Crawford at the Eureka bar in Dunedin. Because New Zealand is a white shark hot spot, some national and international scientists conduct extensive research on them and their habitat. This meant, I was able to meet the national experts from the Department of Conservation (DOC) and the National Institute of Water and Atmospheric Research (NIWA), and learn about cage diving and the white sharks in New Zealand. Before I started my fieldwork, it came to my knowledge that there was a researcher from Canada who was conducting research on 'Local, Indigenous and Science Knowledge Systems' about white sharks in southern New Zealand. Eventually, Associate Professor Crawford of the Department of Integrative Biology at the University of Guelph in Guelph, Ontario, Canada, would publish his seminal work on the topic (https://www.whitepointer.cloud/). I requested a meeting (June 2016), and this meeting would have significant impact on the fate of my research.

Journal Entry
June 2016
3:16 am
Stewart Island, New Zealand

Soosan and I had never been to Stewart Island. Staying in Dunedin the night before, we planned to catch the ferry from Bluff to Stuart Island across Foveaux Strait at 7 am. We started off late at 4:45 am, racing through the foggy darkness of New Zealand rural wilderness, with me on my GPS directing Soosan at the helm of her BMW Z3. In the distant field, sparse houses looked like spaceships suspended mid-air, and sudden alien eyes lit up to our surprise—only to be deciphered as the confused gaze of the grazing sheep. I imagined a big dinosaur, or better yet, King Kong suddenly jumping out of the bush, looking at us in dismay due to this inappropriate hour and place to drive. Every 20 km or so, a car would pass us by cutting through the darkness, and here the only light is the light of our car, as I stare in awe at Soosan's focused driving, informed by the knowledge that she did 600 km just the night before. As we make up our lost time, minute by minute, I could see the distant horizon slowly being smudged with dashes of vermilion.

Thankfully, we caught our ferry just in time and reached the pristine shores of Stewart Island. Expensive houses, beautiful boats, and blue waters, one can imagine why the residents would want to maintain the perceived status quo of beauty, isolation, and safety. On reaching the shore, with some time in hand, we hired scooters and rode around the island, stopping intermittently, mesmerized by the beauty of the sun on the crystal-clear turquoise waters. Around the Half Moon Bay, we walked up to the sea, and looked out in the ocean for what it seemed like hours, naive as we were—mistaking every unruly ripple as a shark fin jutting out of the water.

But this was a particularly interesting time for this research, because I did not know how to pay for the next steps, but decided to have the meeting with Dr Crawford nevertheless. We went to his accommodation at the University of Otago field station for the meet, but it did not go how I had imagined it. Dominating, highly opiniated, and as sharp as a knife, for the next three and a half hours, I was grilled on everything from philosophy to scientific method, and knowledge about white pointers. When the meeting concluded, I was just relieved that it was over, and wanted to give up the research and go home to India, and further convinced about the legitimacy of my imposter syndrome about this work. This has been the case with every other meeting I've had with Steve over the following years, and at the same time, like forging of steel, each one of those meetings have always left me a better scientist, and dare I say a stronger man, than I was an hour ago.

Later in the evening, Steve invited Soosan and me for dinner, up in the old church (transformed into a restaurant). As we met, we started talking again, and as Soosan described, there developed an immediate bromance. Steve explained that he was testing me, if I was a serious scholar, and incidentally, I passed. While we were engulfed in our discussion, Soosan had stopped to talk to a random stranger. It turned out that this old fisherman had a lineage with the Waitaha people, and Soosan was compelled to talk to him when she saw him, because she identified deeply with the philosophy of the Waitaha too and was simply drawn to him.

After the dinner, we three walked down to the ocean's edge, sharing stories, frustrations, and laughter, and yes, knowledge about sharks. Steve explained to me that there is so much we don't know about the sharks, and he was not convinced about a lot of claims that many 'experts' make about their behaviour and ecology in New Zealand. He reminded me: 'Raji (his name for me), as a scientist, be a sceptic, but not a cynic. Always remain open to being convinced by the evidence.' Turning to Soosan, he said, 'You are a Viking Berserker, and I know you will go on to do great things; I don't have to worry about you. As for you (me), before I talked with you, I thought you were just another shark enthusiast trying to get his piece of the action. But now I know you are a rogue scholar, and I don't care if you have to beg, cry, or whatever—I want you to finish this project.' The next day, upon reaching Christchurch, I sold my only earthly possession, my Triumph Rocket III touring motorcycle, to pay for one more year of university fees, and the stranger Soosan met in the island was Stony (Kevin 'Stony' Bourke), who gave me my first giant white shark story.

The largest officially accepted white sharks are not more than 20–23 feet. The largest recorded surviving white shark has been named Deep Blue, first recorded in Mexico in 2014, and then in Hawaii in 2019 (Lampen, 2019). But every shark researcher at one point or the other comes across a giant white shark story, mine was from Stony. Potentially, any white shark beyond 18 feet is a 'giant' and rare indeed, but some stories, such as Stony's, speak of much larger individuals.

Dictaphone Recording Transcription
June 2016
Bluff

Raj: *Please tell me about the incident with the large white pointer.*
Stony: *We were heading south; at Cold River on the north side, there is a big rock; we say it is on roller skates, because it tends to move around. So, I'm looking for the rock, while all of a sudden, up ahead, holy hell, is the*

rock, and I'm on a collision course with it. I spun the wheel, but we didn't seem to be turning. I turned again very hard, then the boat went sideways, and I was waiting for the impact. My crew mate Tim came running up, and asked why the sudden stop? I said, 'There is something here in the water, and I think it is the Cold River rock.' Looking over the side, expecting it to hit underneath the boat, Tim said, 'Oh, that might be it!' What has happened was this thing was coming from the shore, and it turned, so I was unable to avoid it, but what I was seeing underwater, because it was so massive—could only be a rock. So, it turned again, and came to have a look at us. It swam down the other side, and I said to Tim, 'Let me know when the tail is gone past the stern.' I was at its nose when Tim shouted, 'The tail has gone past the stern!' 36 feet, by the size of the boat—12 paces, and the boat is 47 feet long.

Raj: *How did you know it was a white pointer?*
Stony: *'We had a 15-footer on the stern we got that morning that got caught in a net. There were basking sharks in the area, but a bask is an entirely different kettle of fish. This one—I looked at it—and I stepped back. Tim said, 'I will get a big hook,' and I said, 'What will I tie the line to?' It was like a 36-feet boat going past under the water...huge man... Huge!'*

Raj: *Did you see if the shark was looking at you?*
Stony: *No...because I didn't want to lean over and attract its attention. I know it won't climb over the boat, but the lethality of it, its potential! It was hardly moving—effortless! But I would have loved the skin of that and covered a catamaran. Can you imagine entering into one of those American races with a catamaran made out of two hollowed white pointers! Incredible thing, absolutely incredible!*

But this story is not in isolation; indeed, there are many stories of giant white sharks in the region, including the 20-feet-shark, named KZ7, off the Otago Coast, who was named so because he was as big as America's Cup yacht (KZ7 was 12 metre or 36 feet long—New Zealand's first entry in America's

Cup (1986/1987). I had collected one more story of a 20-feet-long shark from Carwyn (Mike Hein's partner at the time):

When we had the Candice, it was a catamaran [Mike Hein's previous boat]. So, it had the deck space and a platform on top. She was over on the west side and she was deep. Me, Ruby, and one more person on the boat were the only ones that saw her. We saw her getting under the boat, when the man saw her tail was still showing on one side, Ruby and I saw the head coming out on the other side, that's why she was at least 20-feet-long for sure.

The validity of the actual sizes of white sharks in these kinds of stories cannot be categorically contested, as there is ambiguity as to the apparent size of these animals, because in many of the instances, these measurements are in comparison to the sizes of the boats the fishermen are on. Rather, the tales of experiencing such large sharks have to be considered significant, because they are part of the contemporary myth of the giant predatory sharks in seagoing communities and contribute to the cultural narratives of these sentient, lethal monsters residing somewhere in the vast ocean, and as symbols of the awesomeness of 'nature vs man'. For example, the Black Demon shark, in the Mexican Baha coast, who was supposed to have scared fishermen out of the water, and then, the story of Submarine, the giant shark encountered by famous shark researcher Craig Ferreira (White Shark Video, 2015), at False Bay, South Africa (one of the comments in the video comment section was 'the Underwater Bigfoot', which is a great way to summarize the debates about the existence of these alleged animals).

The Southernmost Cage Diving in the World

I still feel the chill of Antarctica creating goose bumps in this hot night. The builders excavated the walls, and the stories eroded of their fetter. The house is two storeys, with two terraces on each floor. I sit on the first floor with windows agape. Once in a while, I look back at the feeling of being watched, but this destitute shadow reminds me that the only haunting here is of my own spirit, sitting at the corner there, by the scratched-up cement wall, and looking at me. With shaking feet, I wander into my library and my gym, next to my living room. The gym and the library in the same room was kind of kooky, but somehow made sense to me. The shadows of the open window penetrate through the skin of the rooms.

The lustre of the wall has been substituted with necro-textural chill. The gold paint has been peeled layer after layer in search of treasure. My iron free weights on one side, and book case on the other. A couple of 5-kg iron weights still lie on the ground, rusted and untouched, who knows for how long now, and in the middle, my old bench press in perpetual rigor mortis. I can vaguely see my study wall now filled with *hijibiji* (unruly, in Bengali) writings and drawings. The process started when I had to visualize a chunky statistical calculation during my last PhD, and there wasn't a long enough sheet of paper to draw it on, and 'Excel' was just not visualizing it correctly. I put on my headlamp, and like an archaeologist, I start dusting off inch to inch of the wall, looking for something—anything worth saving. There are ATM passwords, some lines of poetry, smudges of paint I think I was sampling, a lot of information about sharks, exercise, psychology, music

A large male white shark passing by the cage, 2017

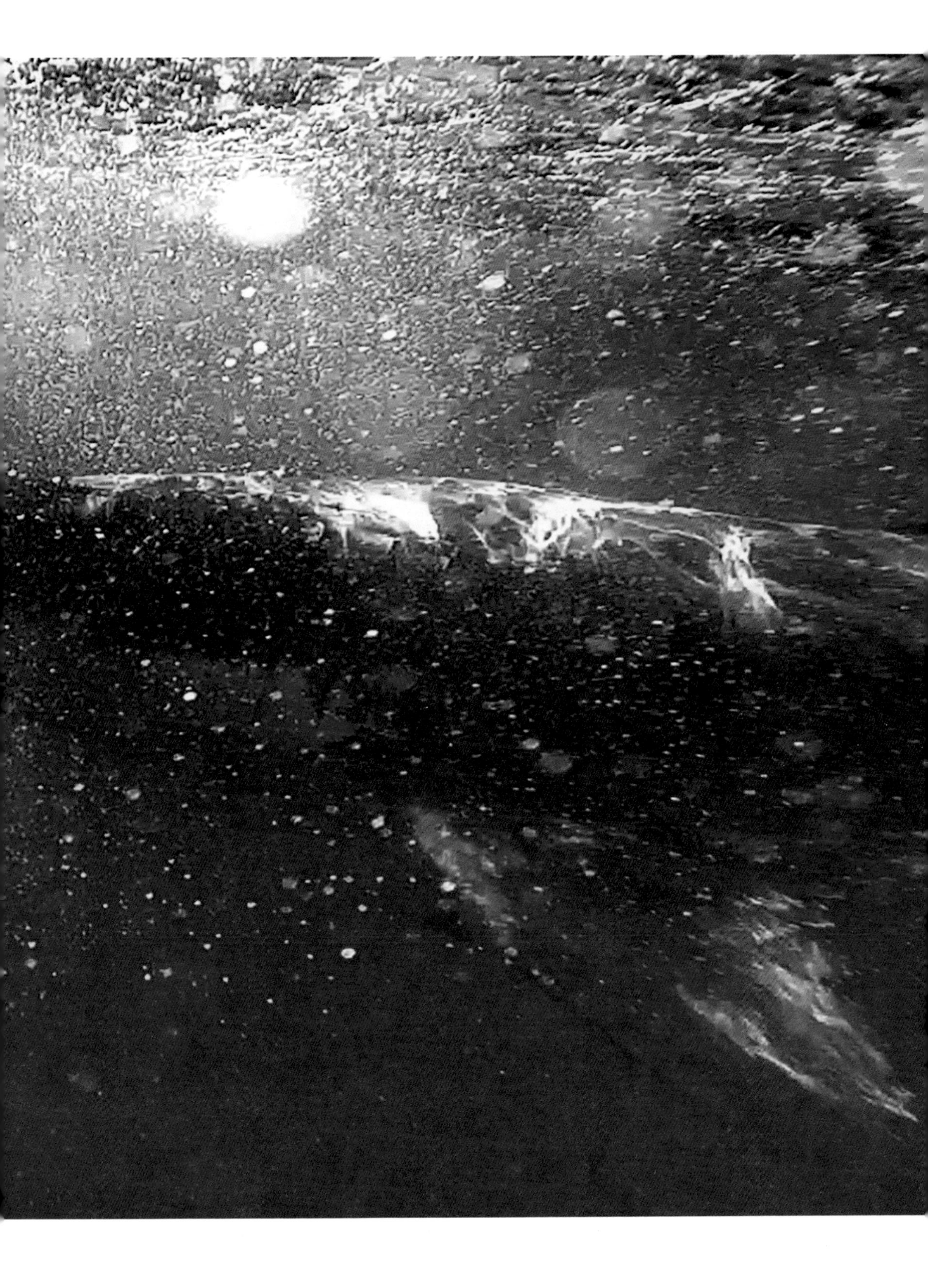

notations, and God knows what else, that seem like 10 different individuals had written on this old wall. I used the bench to look at the writing close to the ceiling, and in the dusty, mouldy, shattered left top end of the wall I see a writing in purple pastel... 'I too am vulnerable.'

It is true...beyond my irrational fear of them, I know, white sharks find their lives and oceanscapes in peril now. In a 2019 study (Queiroz et al., 2019), researchers from different parts of the world collaborated to create a report on overlap of pelagic shark movement and fishing hotspots. They used data gathered from 1,680 tagged sharks of 23 species and found substantial high sea fishing taking place in important global shark hotspots. They found that 24 per cent of the mean monthly space used by the pelagic sharks falls under the major footprints of pelagic longline fisheries. 'Areas of ocean that are frequented by protected species, such as great white sharks and porbeagle sharks, had an even higher overlap with longline fleets—around 64%' (Carne, 2019).

Sharks and rays of the open ocean, estuary, deep sea, and fresh water are in decline globally, and at least 67 species are Critically Endangered or Endangered (Simpfendorfer, Heupel, White, and Dulvy, 2011). Over-fishing, beach-netting, and anthropogenic ecological changes such as global warming and acidification of oceans (Camhi, et al., 1998) threaten their lives, but by far commercial fishing and by-catching are the most serious threats to their survival (Topelko and Dearden, 2005). 'These large marine predators, which include some of the latest-maturing and slowest-reproducing of all animals, are under greater threat than even the amphibians (formerly thought to be the world's most-threatened vertebrates) and mammals' (Ebert, Fowler, and Dando, 2015, p. 9). Ebert, Fowler, and Dando also note that large sharks are in more risk of being threatened than smaller ones, and the ones of shallow coastline, and open water habitats are at more risk than deep water ones (2015), and the white sharks are large, coastal, and open water sharks.

According to Evolutionarily Distinct and Globally Endangered or EDGE (EDGE of Existence is a global conservation initiative to focus specifically on threatened species that represent a significant amount of unique evolutionary history), 'This species [white sharks] is protected in parts of its range, such as Australia, South Africa, Namibia, Israel, Malta and the US, but lack of enforcement allows poaching to continue with demand in the black market. Species experts believe that an urgent change of perception of the species is needed to allow for more effective conservation' (Edge of existence, n.d.).

South Africa became the first country to declare white sharks fully protected in their waters (1991), and they are also protected under the Marine Living Resources Act (Act 18 of 1998) (Kwazulu-Natal Shark Board, n.d.). White sharks have been protected in California waters since 1994 and Atlantic waters since 1997 (Pallardy, n.d.). They are also protected under the Shark Conservation Act of 2010, and the Shark Fin Conservation Act of 2010 (Shark Stewards, n.d.). In Australia, they are considered to be under the vulnerable and migratory species and protected under the Environment Protection and Biodiversity Conservation Act, 1999 (Department of Agriculture, Water and the Environment, n.d.). The Memorandum of Understanding on the Conservation of Migratory Sharks was developed under the auspices of the Convention on the Conservation of Migratory Species of Wild Animals (CMS). Australia became the 14th country to sign the Sharks MoU, signing on 4 February 2011 (including white sharks). The Australian white shark recovery plan, originally drafted in 2002, was revised in 2013.

According to my personal correspondence with the Department of Conservation, New Zealand, the term conservation in the legal term means:

> The preservation and protection of natural and historic resources for the purpose of maintaining their intrinsic values, providing for their appreciation and recreational enjoyment by the public, and safeguarding the options of future generations. And in broad terms a species would probably be considered to have been conserved if it had a non-threatened threat classification and was occupying its full original range. (Personal email correspondence, 24 September 2018)

Considering that the great white shark is listed as 'vulnerable' globally by the International Union for Conservation of Nature (IUCN), and the deep cultural significance of white sharks in New Zealand, it comes as no surprise that New Zealand has played an active part in their conservation efforts.

White sharks are fully protected in the New Zealand waters under the Fisheries Act 1996 and Wildlife Act 1953 since 2007 (Department of Conservation, n.d.; New Zealand Herald, 2006). On 6 July 2015, New Zealand signed up an international agreement to help protect migratory sharks. Primary Industries Minister, Nathan Guy, and Conservation Minister, Maggie Barry, signed the memorandum of understanding on the conservation of migratory sharks, which helped save seven species vulnerable to exploitation, including the Basking Shark (*Cetorhinus maximus*), Great White Shark (*Carcharodon carcharias*), Longfin Mako Shark (*Isurus paucus*), Porbeagle (*Lamna nasus*), Shortfin Mako Shark (*Isurus oxyrinchus*), Spiny Dogfish (*Squalus acanthias*), and Whale Shark (*Rhincodon typus*). It was stated that the populations of some of these species of sharks, including the white shark, have been severely depleted by the global increase in unregulated fisheries, and markets targeting sharks and their fins, and this treaty was an active step against these atrocities. This international memorandum is the first global instrument of its kind, and they have 38 other countries as signatories (Radio New Zealand, 2015).

Great white shark population density has been decreasing drastically in the last few decades. In 2003, it was claimed that the white shark population had decreased by 79 per cent globally (Baum et al.). An article released in December 2018 (Roff, Brown, Priest, and Mumby, 2018) reported that the population of white sharks have decreased 92 per cent since 1960, along the Australian coast—an analysis based on shark catch in the region. Commonwealth Scientific and Industrial Research Organisation (CSIRO) estimates that there are about 2,000 adult white sharks off the coast of Australia. Fisheries and Oceans Canada notes that there seems to be a decrease of 63–80 per cent in north-west Atlantic. Studies from Guadalupe Island, Mexico, indicated about 142 individuals in 2012 (Nasby-Lucas and Domeier, 2012) and 219 in California in 2011 (Chapple Taylor K. et al., 2011).

Studies estimated a significantly higher number of sharks in South Africa with the superpopulation (a theoretical or hypothetical population) of 908 individuals in 2013 (Towner, Wcisel, Reisinger, Edwards, and Jewell, 2013).

There isn't enough collaborated data to estimate the total global number of white sharks, and what limited data we do have is indicative only. The IUCN website indicates that the global population of the said animal is 'unknown' (The IUCN Red List of Threatened Species, White Sharks, n.d.). The only thing we might conclude at this point of time is that there is a lack of reliable knowledge regarding the distribution or abundance of this species globally, but the data does indicate that their population is drastically low, despite white sharks being an apex species (Towner et al., 2013). The birth rate of the species cannot keep up with the rate at which they are being culled. White sharks mature at the age of 10 and are considered to have a two-year gestation period, producing only a few pups; so, their population cannot recuperate as easily as other fish. Sharks are hunted for their liver, skin, and their famous jaws, and as trophy to compensate the male ego. Consequently, white sharks are now considered to be a 'vulnerable' species by the IUCN.

Social and political scientists point out that though there is public interest in conservation of other marine megafaunas such as whales, which are considered to be closer to human sensibilities, there is not as much interest in shark conservation because of the negative image they have gained in public imagination in the last 40 years (Thompson, and Mintzes, 2002; Friedrich, Jefferson, and Glegg, 2014). This is especially relevant for the 'mindless eating machines', the white shark (O'Bryhim, 2011, p.16). Public policy researcher Christopher Neff addresses the 'predator policy paradox' and notes the complicated issues in creating a sound policy for safeguarding predators, as sharks generally evoke unfavourable attitude from within the wider population (Neff, 2014, 2015). The creation of these kinds of negative images in the public perception are sabotaging conservation initiatives, since the proliferation of such images harms the interest in legislation, which might help increase shark numbers in the oceans. Consequently, there needs to be efforts to challenge this negative image of sharks in public imagination, if legislation promoting the conservation of these animals are to be created and sustained.

> Creating positive attitudes toward sharks and their conservation is critical because attitudes have the ability to guide, influence, direct, shape, or predict an individual's potential behaviour toward a species and its conservation... Positive changes in the public's attitude toward an environmental issue can also lead to major shifts in policy pertaining to it. (O'Bryhim and Parsons, 2015, p. 44)

It seems that personal experience is an 'important influence on environmental attitude and motivation for personal engagement and conservation behaviour' (Friedrich et al., 2014) and direct experience has positive effect in creation of environmental values (Bögeholz, 2006; J.R. Miller, 2005, from Friedrich et al., 2014). One potential avenue of personal experience, and hence, demystification of sharks and challenging stereotypical negative image created of them by the public media, may be direct controlled encounters in their natural environment (O'Bryhim, 2011; Friedrich, Jefferson, and Glegg, 2014). In a study, Dobson (2007) found there was significant positive change in people's attitude towards sharks with some shark-focused dives. However, considering that there are currently no white sharks in captivity, the lethal potentiality of the white shark and our lack of understanding of their behaviour, the only controlled avenue of direct encounter with them, minimizing safety hazards for the fish and the humans alike, is through cage diving. (It is not necessary that other sharks be non-lethal, but there are groups who do cage-less diving with big predatory sharks like tiger sharks, but not so much with white sharks, due to, perhaps, a combination of multiple factors, like the sheer size of the animals, their strength, and our lack of knowledge of their behaviour. One curious thing that these divers do with bull sharks, and even tiger sharks, is to turn the sharks upside down and put them in the state of tonic immobility.)

Initially introduced as a scientific tool to observe shark behaviour by Jacques Cousteau (Townsend, 2011) while filming his documentary 'The Silent World' (Cousteau and Malle, 1956)—the commercial practice of taking tourists cage diving with sharks—has received significant attention in the last two–three decades. Though cage diving has been practised with different species of sharks (bull sharks, blue sharks, oceanic white tips), arguably, the most popular form of commercial shark cage diving is with the great white sharks. After Rodney Fox introduced cage diving in Australia in the 1970s (Kemper, 2017), cage diving tourism with white sharks is now practised in five major white shark hotspots of the world: the Guadalupe Islands in Mexico; Farallon Islands, California, in the USA; False Bay and Gansbaai, South Africa; Neptune Island, Australia; and Bluff in New Zealand (Kemper, 2017). Here, I need to point out, a few experts with good faith have told me that there has been a handful of white sharks sighting around the Indian subcontinent, such as around the Andaman and Nicobar Islands and Srilanka, although I have no evidence to back these claims.

As Rodney Fox, the father of white shark cage diving as we know it, notes: 'It's better to respect through understanding than kill out of fear' ('In our Blood,' n.d.). Furthermore, as Mary Curullo, the associate director of Friends of Casco Bay, in South Portland, Maine, notes: 'Scientists and photographers who have met a Great White Shark admire this awesome animal. But people who only know this shark by reputation fear it' (Cerullo, and Rotman 2000, page 1, para 1). As O'Bryhim (2011, p. 22) points out, participants often emerge from the cage diving experience with an admiration of the white sharks as the 'magnificent' animals they are in their 'natural' environment, instead of the *Jaws* image that may motivate them to do the dives in the first place. However, the empirical investigation of such claims exploring the effect of direct encounters with white sharks on the diver's attitude towards them are still quite limited.

Kirin Apps and her team found, 'Exposure to sharks, in combination with education and interpretation, reveals the potential for management and SCDOs [shark cage diving operations] to play a role in providing interpretive and educational initiatives which stimulate conservation, inspire appreciation, dispel myths, and influence attitudes to achieve a more balanced understanding of the plight of this apex predator' (Apps, Dimmock, Lloyd, and Huveneers, 2016). They also found that tourists' awareness, understanding, attitudes, and concern for sharks were positively enhanced, that there was an increase in declaration of participation in conservation-related behaviours and in actual conservation participation, and that their enhanced understanding of sharks was most strongly correlated with emotional engagement during the experience (Apps, Dimmock, and Huveneers, 2018S).

White shark cage diving in New Zealand is the southernmost cage diving operation in the world, and one of the newest ones dating from 2009 (when I started the book, there were two operations; before the completion of the book there was only one left). Always negotiating between the great white shark, the hyperreal trope of the 'man eating monster' and the white shark, the fish, as white shark cage diving in New Zealand facilitates the coming together of two sentient species whose lives and oceanscapes are intertwined in similar political, social, legal, economic, and ecological forces.

While discussing the real-ness of the American wilderness, Vidon, Rickly, and Knudsen (2018) argued—

> The choice to use American wilderness to illustrate the ways postmodern authenticity accommodates psychoanalytical authenticity has been a deliberate one...wilderness is often held up as the authentic landscape par excellence. It is, however, hypernatural, a simulacrum, and thus provides the perfect opportunity to explore postmodern and psychoanalytical authenticities working together. With a combination of fantasy and seduction, hyperreality and meaning-making, postmodern authenticity is manifest in the American wilderness of the Adirondack Park. (Vidon, Rickly, and Knudsen, 2018, p. 68–69)

Indeed, Leather and Gibson (2019, p. 98) argue that being outdoor, no matter how remote, is not the real experience of nature as we claim it to be, in the advanced capitalistic society we are all part of. On the same token, what is the 'real', 'natural' experience in the oceanic wilderness? Something that may be questioned while looking at it through the lens of hyperreality. Is it not an orchestrated operation where the setting is created artificially? And everyone including the materials, the humans and the sharks, playing a part?

Since 2009, two operators (Shark Experience Bluff/New Zealand and Shark Dive New Zealand) had been facilitating this encounter about 200 days in a year—20 km from Bluff (the last point of South Island, New Zealand) and 8 km from Stewart Island, in the turbulent Foveaux Strait, circumnavigating the small Edwards Island. Initially, Peter Scott, the owner and operator of Shark Dive New Zealand, made a cage, and used it for himself and friends to get in the water to see the white pointers in Foveaux Strait. In 2009, the Discovery Channel was making a 3D documentary on the sharks of Foveaux. They contacted Mike Haines to film with the sharks in the region (Great White Shark 3D, Cresswell, McNicholas, and Newton, 2013). Mike agreed to take them out; Peter Scott was a crew in the expedition. The team got a special cage made where they could put their large 3D camera. One risky incident happened when an 18-feet shark got inside the cage because of the large cage opening made for the camera and everyone had to escape the cage immediately.

This came at a time when Mike was no longer interested in taking tourists out for charter fishing, because of the greed he saw among the people and the waste of natural resources was demoralizing. This proposed a viable avenue for him to have a thriving economic endeavour; hence, Mike started his own cage diving business. Simultaneously, Peter Scott also started his business. Mike used to operate from Bluff, while Peter used to operate from Stewart Island. Around 2015, Peter Scott decided to start operating from Bluff—one of the reasons may be related to the alleged threats he received from some Stewart Island residents (as is known in the local community); since then, until the 2017–2018 season, both of them operated from Bluff.

This practice facilitates the coming together of two apex species, whose lives and oceanscapes are often intertwined in similar political, social, economic, and ecological assemblages. The cage diving encounters take

The decal of the shark on the Southern Isle, 2016

place in a marine contact zone as Barua notes: 'Encounters forge "contact zones", where people across cultures, with different histories come together into composition, interact, and intertwine. But encounters are also between beings of biologically different origins, equally vital to such histories and world-making' (Barua, 2015, p. 265).

Cage diving creates an avenue of controlled encounters among two species in the middle ground, on equal terms, where the sharks have as much significant agency as the humans, and both experience the subjectivity and intentionality of each other. It is a practice meticulously mediated by the protocols of the Department of Conservation (which looks after the welfare of the sharks), and the Ministry for Primary Industries (which is observant towards the industrial implication of the operation, and the safety of the passengers and the crew). Furthermore, over the years, it has created its own norms and protocols, for the ease of operating, safety, and increment of opportunity for having a successful encounter on a regular basis.

However, as of 4 September 2018, the practice of cage diving in New Zealand was announced as impermissible, by a ruling of the court of appeal based on an interpretation of Wildlife Act 1953, because 'disturbing, molesting and pursuing' of a protected species was an offence under the act. Shrouded in all this controversy, cage diving is a hotspot of social and political turmoil in South Island. The groups opposing cage diving (primarily residents of Stewart Island and paua divers of Foveaux Strait) often argue that cage diving increases the potentiality of 'shark attacks' on humans sharing the water space, by attracting sharks into regions where they were not there, making them more aggressive and training them to associate humans and boat with fish. Alternatively, the supporters of cage diving argued that the cage operators have as many rights to use the water for their benefit as the other locals, and the practice does not increase the chances of shark attacks in the region. Rather, it creates an avenue for demystifying the sharks and challenging the 'monster' image by altering public attitudes towards them, which is pertinent for their global conservation efforts, and providing a definitive service for many individuals from all over the world

Soosan and Raj, in their first cage diving expedition with Shark Experience, 31 December 2015

who have been fascinated by this animal for a long time. Furthermore, the supporters point to the significant shark knowledge emergent from this practice, about their migration and behaviour, and the economic and social benefit to the local community of Bluff.

Blue shark, mixed media painting, Raj Sekhar Aich, 2008

Rough and Tough, They come from Bluff

Matter is a composite of molecules, masquerades as intrinsically discrete, but physical decay deconstructs the bounded-ness of objects. In the object-world, things generally withstand the ambiguity which inheres in their composition by being situated in distinct locations: in place. But in ruins, things gradually lose resemblance to that which they once were, and the barrier between one thing and another evaporates. ... Dissembling material separates into parts, which stretch away from each other as joints weaken. In the slow processes of corrosion, things give up their solidity, their weight, become spectral forms bearing traces of their past. (Edensor, 2001, p. 46)

If you had happened to have a chance to witness the exhibits of the Southland Museum (currently closed) and Art Gallery Niho o te Taniwha, Invercargill (New Zealand), you would eventually stumble upon a giant poster depicting an image of a night sky illuminated by brilliant iridescent colours, hues ranging from green to purple amongst deep blackness. Among those colours, you would have seen a curious sign post with the direction and distance of a few major cities of the world, such as 'London: 18,958 km', 'Tokyo: 9,567 km', 'New York: 15,008 km', and 'Stewart Island: 35 km', even if it did not show the distance of 11,079 km to Kolkata, which used to be the centre of my universe for the majority of my life. If you pressed a small switch on the side of the poster, you would be enticed by a sound that can

BLUFF
NEW ZEALAND
LATITUDE
46°36 min 54 sec
SOUTH
AA
NEW ZEALAND
LONGITUDE
168°21 min 26 sec
EAST
LONDON
TOKYO
15008 km
HOBART 1680 km
KUMAGAYA

The famous Bluff sign board, showing directions and distance to some major global destinations, 2017

be best described as whale songs to the uninitiated, and one could mistake it as such, considering the giant baleen whale bones displayed nearby. That sound was actually the radio frequency of Aurora Australis, and the poster was of the Sterling Point in Bluff, New Zealand, in Māori, Tahu-Nui-a-Rangi or the great glowing sky.

Just about 28 km from Invercargill and 1,600 km from Antarctica, the small fishing town of Bluff sits at the very bottom of the South Island of New Zealand. Arguably, it is the oldest town in New Zealand at its original site. The only other mentionable settlement after Bluff and before Antarctica is the even smaller population of Oban in Stewart Island, about 35 km away (as displayed on the sign post), and the expanse of waters between these two pieces of land is Foveaux Strait. The opening paragraph of the book *Those Sheltering Hills* reads:

> The port of Bluff is on the south coast of the South Island of New Zealand, its geographical position is 47 south latitude. A range of hills running in a south-easterly direction for nine miles, from the New River heads to Look out point, Bluff, shelters Bluff harbour from the prevailing wind which blows from the south-west. This range of hills is a gigantic rock breakwater and its highest point is Bluff Hills 870 ft and the lowest and narrowest point is at oceanbeach [sic]. (Bremer, 1986, p. 3)

A few steps away from the whalebones is a large painting of a burning ship. It is a painting of the *Ocean Chief*, which was driven ashore at Tiwai Point, near the port of Bluff. She was later salvaged; however, due to some fateful turn of events, a few days later, she was set fire to by some crew members in the cover of night (Christiansen, 2015). In a manner, this painting is a depiction of the uncertainty involved with life at sea and the search of fortune. It may also symbolize the perilous travel that sailors had to take through Foveaux Strait. However, it is also a symbol of the presence of absence, that is often vividly apparent in museums around the world. It is the memorialization of matters and incidents which 'had been', but

The Bluff sign, end of highway one (New Zealand), 2017

are not anymore, at least in its original form, displayed in its singularity, often out of a larger context of jumbled up images and objects, curated on the whims and aesthetic choices and the historical understanding of the curators, simulating the real matter, or experience (Meyer, 2012)—much like the philosophical conundrum of the ship of Theseus, questioning the existence of identity over time, as all the elements of the self are shed and replaced. The Greek king, Theseus (who supposedly founded Athens), had his ship memorialized by the Athenians in his honour. Over years, rotten planks were removed one after the other to preserve the ship, until no original plank remained. The question that arises is: did it remain the same ship? On the same token, one might ask, in a museum situation, when the exhibits are exhibited in a sterile manner, often fixed to fix the specific geographical and architectural structure of the venue, are they able to retain any of their original identity? (Cohen, 2004; Worley, 2011).

'Memorial sites...are actively fashioned as experiential and textual devices to shape a particuiar affective response and emotional reaction from the visitor' (Simpson, Miller and Del Casino, 2018, p. 662). Some might further argue that the painting is a way of restructuring and representing

The Bluff War Memorial on the marine parade, 2017

the innate imperialistic motivations symbolized by the ship that led the Europeans to the shores of New Zealand in the first place. Goals, which led to many significant sociopolitical and ecological effects on the region, included the mass scale unregulated slaughter of the above-mentioned species of whales.

Before the Europeans arrived, Bluff was not a settlement. They came to the Bluff Harbour in 1813, and between 1816 and 1826, about a hundred sealers lived semi-permanently around Foveaux Strait. James Spencer, the founder of Bluff, arrived in 1824, set up a fishing station and store, and imported a herd of cattle. He was among the first to marry a Māori woman, and supposedly built the first European house in Southland (Bluff Museum posters; 'Bluff New Zealand History, Bluff,' n.d.; Bluff History Group, 2004). As the Bluff heritage trail pamphlet mentions:

> The first European name for the town was 'The Mount' followed by 'Old Man's Bluff'. The term 'old man' from the Celtic meaning High Rock. This was eventually dropped and became 'The Bluff'. Māori called it an island 'Motupohue' or the 'Island of Convolvulus'. In 1856 it was changed again to Campbelltown and then officially Bluff again on 1 March 1917 [sic]. (Bluff History Group, n.d.)

The heritage trail, running about 20 km across the town, displays some of the important institutions of the town, including the Greenhills Church, the Greenpoint Walkway and Ship's Graveyard, the Monica II in the Bluff Maritime Museum, the Radar Station, Old Bluff Cemetery, Post Office Club Hotel, Old Town Wharf, War Memorial, Bluff Gunpit Coastal Defense Camp, and finally, the Stirling Point.

It is generally accepted that whaling and sealing was the primary reason why the Europeans got interested in Southland, New Zealand. Historians argue that the genesis of this interest can be traced to Captain Cook's stories about the treasure trove of seals and whales in the New Zealand coast in 1775. In 1792, Captain William Raven was the first English mariner to arrive at Dusky, in his ship *Ritannia*, for sealing in the southern waters. Subsequently, explorers and hunters of many countries showed interest in the sealing opportunity of Southland and headed this way. Eventually, they discovered a small harbour guarded by hills, which they named the 'Old Man's Bluff' and made it into a sealing base. But because of the bad weather and heavy seas, the sealers spent a substantial amount of time inland. In the process, they realized that the timber in this region was of very high quality, of which the region eventually became a significant exporter. With deforested land, more crop plantation started to take place, including wheat, oats, and potatoes. Trade flourished, and many European men started marrying Māori women, and the settlement of 'Old Man's Bluff' thrived (Bremer, 1986).

The first whaling ship in the New Zealand waters was *William and Ann* in 1791 (McLintock, William Henry Dawbin, and Taonga, n.d.). 'Being situated at about 47° S., Bluff was right on the path of migrating whales and it was quite common during the whaling season to have several whaling ships in the port together' (Bremer, 1986, p. 29). The ships came to the harbour for supplies and processing the oil for their journey home. By 1835, whaling in Foveaux started to take shape into a viable industry. In 1844, the Harbour was named Bluff Harbour. By 1853, and the forming of the constitutional government, Southland had a significant European population; however, the whaling around Bluff ceased altogether by 1859, even if it had become an internationally famous whaling industry base (Bremer, 1986). In 1874, the first immigrant ship arrived, and by 1877, Bluff Harbour Board was established. More and more European emigrants started settling in Bluff, as there was at that time no custom charges in Bluff, and it was also the nearest port to Australia. As a popular history book of Bluff states:

> A new race of people had evolved; they were the children of two cultures whose fathers were seamen from countries of the northern hemisphere and their mothers were decedents of the great Polynesian navigators. When the boys were old enough, they worked with the whaling gangs becoming experts in the handling of whaleboats and the killing and treating of the whale carcass through the various stages till it was put into casks as whale oil.

The carcass barnacles, 2017

When the whaling seasons ended at about October each year they fished in the straits catching mostly blue cod which they salted and smoked. The sea was their livelihood; they had salt water in their veins. As the girls grew up, they learned from their mothers the ancient arts of weaving and basket making, and during the

whaling era some of the girls married Pakeha (white European) whalers and had children who were mixed race [sic]. (Bremer, 1986, p. 32)

Bluff developed with the development of the train line and the proliferation of the Bluff port due to the mail service to Melbourne in 1863. Many industries developed; the freezing work factory in particular. However, in 1991, the freezing works closed down and Bluff started losing Bluffies.

The current population of Bluff is about 1,800, and a sense of absence is glaringly present (particularly to visitors), and the scenario is quite different from that of the heydays. As you enter the town of Bluff from Invercargill, on the right-hand side of Highway One, you see a big hoarding of a helicopter service connecting Bluff to Stewart Island, like the metaphoric end of the fishing line, which Maui used to pull up the North Island with. Maui, the Polynesian demigod, would be stowed away in his brother's canoe, when they went fishing. He made a magical fish-hook from ancestral jawbone, and out in the open sea, he revealed himself to his brothers. Using his magical powers, he threw the hook into the ocean and eventually hooked a huge fish. This fish was the North Island; the South Island is considered to be his waka (canoe) and Stewart Island is his anchor stone. To the left of the road, there are the remains of one of the boats, which succumbed to the forces of Foveaux Strait, the *Olivia*. Nearby, opposite to the Greenpoint Cemetery, a stroll down

The scarred carcass, 2017

the board walk is the Greenpoint Ship's graveyard, where broken-down submerged carcasses of ships are 'displayed'. Invisible and visible at the whim of the ocean, these carcasses not only live in the liminality of the ocean and the sky, but that of death and being alive. For even if they are dead as ships, they are alive as symbols of the perils that awaits in Foveaux.

In comparing cemeteries and museums, Meyer and Woodthorpe (Woodthorpe, 2008; Meyer, 2012) note, 'We can "feel", "see", and "hear" absence. In museums, we are confronted with the absence of the "world

out there" and/or the "world that once was"...hence, do something *to* and something *with* the absent—transforming, freezing, materializing, evoking, delineating, enacting, performing, and remembering the absent.' The Bluff ship graveyard is both a cemetery and a museum. This creates a unique juxtaposition of memorialization of the non-existence of the humans and the ships, both of whose 'lives' are affected by Foveaux. The *Olivia* and other ships in the graveyard have lost the 'traits' that are important for a material object to be called a boat/ship, but still symbolizes the idea of one. Indeed,

An old ship in the ship graveyard, Green View, 2017

The memorial for the group lost in the sea, in Bluff wharf, Soosan Lucas, 2016

Details of the memorial for the group lost in the sea, in Bluff wharf, Soosan Lucas, 2016

these ideas of the ship are significant symbols of the identity of the region, and part of the 'heritage trail' of the Bluff walk. It may be argued that certain phenomena or ideas can have a significant effect on the world around it, often specifically because of its absence, understood as the paradox of 'the presence of absence' (Bille, Hastrup, and Sørensen, 2010, p. 4). There is a paradoxical 'presence of absence' (Bille, Hastrup, and Sørensen, 2010, p. 4) intrinsic to the architectural and social environment of Bluff. Presence of absence embedded in its everyday life, and its integral social, political, and material architecture is affected by presenting things and ideas, which are 'absent in one way or another' (Bille, Hastrup, and Sørensen, 2010, p. 18), as much as the ones which are present. One such visual artefact symbolizing the presence of absence and warning about the unpredictable Foveaux is a mural on the wharf, commemorating the *non-return* of the four German tourists in 2014, who were presumed to be lost in sea, due to an unexpected storm in Foveaux Strait. Their dead bodies were never recovered.

Foveaux Strait

Indeed, encountering the town of Bluff and the white sharks in the waters of Foveaux Strait is encountering Foveaux Strait herself. The Tasman

Leaving Bluff for Foveaux Strait, 2016

Sea and the Southern Ocean meets in Foveaux Strait and creates heavy currents, and it is often compared to a lake by long term sea people like Mike Haines, as the formation of waves happens very close to each other. The sea surface temperature of the coast of Southland seems to range from 11–15°C. 'Southland is both the most southerly and most westerly part of New Zealand and generally is the first to be influenced by weather systems moving onto the country from the west or south. It is well exposed to these systems... The region is in the latitudes of prevailing westerlies, and areas around Foveaux Strait frequently experience strong winds' (Macara, 2013, p. 6).

Journal Entry
19 April 2016
8:40 am
Out in Foveaux Strait

We are 9 km out in the ocean, returning to the shore, as the trip had to be cancelled midway. The waves are 10 to 15-feet-high with 40 knots winds, and the sky is again filled with dark ominous clouds. Mike took the decision that we needed to go back, for the safety of the passengers. He knows how the waters can get, and says that there are no rogue waves, only careless captains. One of the tourists has thrown up a few times and few others are resting their heads on the table; there is one other individual who has barricaded himself inside the bathroom. I stood on the deck for a few moments, holding on to whatever I could; I have never thrown up on the boat, but undoubtably have come close. Maca shouted out, 'Come in bro! I have been in the ocean for 25 years now, I still do not do that! One wave and you will be washed away.' Even though a part of me did not want to believe him and wanted to see if I was 'man enough', I came in. Just a few minutes after that, two large waves fell on the deck, and if I would not have been flown away, it would have been close. I can now see the lighthouse on Dog Island with a halo of light over it. Hard orange beams of light cut through the greyness over him, and if I did believe in heaven, this might as well be one of the times I could claim I was seeing

God's light shining through and bringing us home. A scene, I can imagine, inspiring classical painters to paint the sublime and the terrifying beauty and drama among the matal *(drunk) waves. The brilliant light, different from that in Europe and the tropics, ignores the grey wave and the grey sky, and lights the lighthouse. Now I sit on the floor of the boat and write this, with my cold fingers and taste the salt dripping down my lips.*

Two books by the Bluff history groups chronicle many of the disasters of vessels and lives lost in the waters around here. One book—*Hey-Day to May-Day* (Christiansen, 2009)—chronicles the loss of 106 boats registered in Bluff, drowning in the strait and surrounding regions from 1960 to 1985. The other, *Shipwrecks Bluff Area, 1845–1920*, published in 2015, goes further back and discusses about shipwrecks near the Bluff area from 1845 to 1920 (Christiansen, 2015). These accidents took place in varied locations of the strait, including the Long Point, Mokinui, Looking Glass Bay, Jacky Lee Island, Bauza Island, Cape Saunders, Cantre Island, Port Adventure, Rabbit Island, Bluff harbour, Solander Island, and the Edwards Islands. Over the centuries, these accidents have been related to many factors, such as a sudden north-easterly gale, collision with submerged logs, rocks, leaks in boat, engine failure, fire, breaking of stern, freak waves, broken pipes, torn mooring, search and rescue severely delayed by bad weather, rope tangling around propeller, sudden fog, unexplained explosions, and even treachery by the crew.

One account documented in the book exemplifies the trials and tribulations of such an incident quite vividly—

> On October the 10th, 1971, 6 miles west of Centre Island, the Seabird was staggered by a heavy sea, a second wave turned the vessel completely over, trapping the two crew members in the submerged wheelhouse. 'It was like a nightmare', was how 21-year-old Nick Hansen described the drama in which he and his crew mate Ross Beaton aged 17 were nearly drowned in Foveaux Strait. The wheelhouse door was jammed hard by tons of water as Mr. Hansen pulled at the door he heard the voice of

God's light comes shining through, 2018

> another skipper on the radio 'my God' they are gone, the boat is over'. The door finally dislodged, and the 2 crew fought their way to the surface where they were tossed in heavy seas next to the upturned hull of their boat. Dessmoore-Carter in the Seagold was the first boat to arrive 5 minutes after the capsize; the 2, cold, exhausted men were dragged on deck where they collapsed. The younger crewmember Ross Beaton could hardly swim yet recovered quickly. On arriving home, after being kissed by his mum, his first words were 'What's to eat Mum, I'm starving!' [sic] (Christiansen, 2009, p. 38)

Incidents have been accounted of fishermen chased by seals after being shipwrecked near Solander Island: stories about fishermen surviving on a cup of Milo, two oranges, and two apples for three days in deserted islands; fishermen living on paua (a large edible abalone, a kind of mollusc) for six days, making tents out of sails; a boat going down with its skipper jammed in the wheel house, who had to kick the door open and swim through thick bull kelp to reach the closest island; and finally, upon return on safe land, sailors have been heard saying things as 'I think I will go chicken farming' (Christiansen, 2009, p. 60), and 'there's got to be an easier way to earn a living' (Christiansen, 2009, p. 63).

Not unlike many isolated regions of the world, isolation is part of the identity of Bluff now. The documentary *St. Helena—A remote island in the Atlantic* (Denzel, 2017) mirrors a similar picture of the cultural relationship with isolation in the island of St. Helena, off the coast of South Africa. On this island, only accessible by a five days' journey by ship until quite recently, and where Napoleon was exiled, some cherish this isolation, while others despise it. The younger ones leave because of lack of opportunity and exposure, and there are some people who argue that more tourists should be encouraged to the island for economic benefits in the region, while others staunchly disagree, and love the land precisely because of its isolation and perceived safety in absence of the unknown others. On the other side of the world, Bluff is going through the same debate.

The Bluff Maritime Museum is a significant landmark in the town, symbolic of presence and longing for that which is absent. The walls show a way of life, which is about to go extinct, and the pride in the classical practices, including oystering, muttonbirding, and fishing. The presence of absence is palpable from the lack of people in the region, and the number of rundown homes. Many of the young women and men tend to leave Bluff and Invercargill once they are in the position to, due to lack of vocational opportunities, and other potential life experiences. Whenever I have asked tourists about how they felt on reaching Bluff, often I have come across the phrase 'the end of the world', which is not just the representation of the geographic placing of Bluff, but indeed a feeling of 'nothing', 'nowhere to go', and such other things—absence of not only materials, but also of a way of life. It seems true that the lack of *things to do* or *people* in Bluff may be unsettling for tourists, and even the younger generation of residents, but, to a significant portion of the long-term residents, it is something of pride and happiness, particularly for those people who have stronger economic stability, primarily through fishing and other ocean related endeavours. One could argue that there is almost a sense of Eden in the way their lives are portrayed, reminiscent of biblical stories, in the isolation from others, and a longing to return to the *original* state of *nature*, which seems to be not there anymore.

There is a longing for the absence—absence of the 'others', and the ill effects of globalization. Many residents of the region enjoyed their own privacy and their 'seclusion' from the rest of the world, even the rest of the country, which should be understood from the historical context that for a long time, Bluff was separated from Invercargill and effectively, the rest of the country, by a 30 km-long bog (Coote, 1994; Bremer, 1986). Accounts of many senior 'Bluffies' mention, how beautiful it was in the 'older' days. They romanticize how all the young ones at that time *played* in the *lap of nature*. There was regular music and dancing around; even if there was no money, there was never scarcity of food. The community always supported each other and provided food from the freezing works, the sea, and the farms. As the book *Those Sheltering Hills* (Bremer, 1986) discusses, everyone knew each other, the land and the residents provided safety and shelter to their neighbours, before all the 'strangers' started coming. However, one of my participants who had a long-time association in the region mentioned, 'Just remember, many of them like to be left alone, so that no one else can lay claim to resources, which they claim to be theirs.'

Food is a major currency and symbol of pride of the region, blue cod, marlin, sharks, cray fish, muttonbird, oysters, and kina. Muttonbirds are chicks of sooty shearwater, which the local Māori have harvested in the Southland region for hundreds of years, all across Foveaux islands or the Titi islands. They were one of the most important source of animal protein the Bluffies had. Now, even though there are plenty of other sources of protein available, the practice of harvesting the birds has become a family tradition for the locals and an important part of the cultural identity of the region. The only people allowed to collect the muttonbirds are the Rakiura Māori, and their immediate family. Originally, these birds were collected and packed in kelp bags and then, put into flax baskets. The equipment has become more sophisticated now, but the essence of the practice remains the same. Every year, the local Māori still go spend a few weeks in the Titi islands. I was lucky enough to accompany Mike to drop a father-and-son team of birders to one of the islands. We carried an inflatable dinghy, and all other equipment and amenities to last the weeks. There were generators, fuel, light, food, and other necessary amenities. Near the island, we deployed their inflatable dinghy and let them be on their way. A very interesting stretch of water with plenty of seals frolicking in the shallows, and plenty of sharks around too, swimming among the jagged rocks sticking out of the water.

Many years earlier, an incident happened when Mike had come to drop his friend to the island. It was late in the evening; they were unloading their small boat and just about when they were finished and Mike turned to his right, a huge shark jumped out of the water with its jaws open. Yes, they knew the sharks were there, and the birder mentioned that he could see, from his hut, the sharks swimming near his island during the day too, but it was not particularly something they were bothered with.

A few weeks later, we again picked them up from the island, and to show their gratitude, the birders presented Mike with two buckets full of birds, buckets which would fetch hundreds of dollars at the market, but this was an essence of the exchange that kept the region alive and the community close-knit.

There is one more kind of food, which is not just well sought after in the region or the country, but has made Bluff seafood a global attraction—Bluff oyster. Bluff's oyster industry turns over NZD 20–25 million each year, and it is an important symbol of the region. Around the 1800s, European sailors found a shallow bay at the east coast of Stewart Island, filled with oysters, and they named the bay 'Oyster Cove'. The oysters in the region became famous, and the interest grew worldwide. However, whenever a new spot of oysters was found, within a few years, all the oysters were wiped out due to overfishing, and in 1960, scientific management schemes of oystering were initiated in the region.

On 21 May 2016, Soosan and I had the opportunity to witness the Bluff oyster festival. The official website of the Bluff Oyster and Food Festival presents an appropriate picture of the region: 'The salty seaside township of Bluff is home to this festive winter event. Located in Bluff and brought to you by the locals of Bluff with a unique southland flavor' (see http://Bluffoysterfest.co.nz/). The quiet town of Bluff, of about 1,800 residents, changed into a crowded festival ground for a day. Six thousand people from all round the country and overseas congregated in the celebration of oysters, and other specialities of the region, including kina (sea urchin), muttonbirds, venison, and even specialized chocolate. The 'rustic' seaside town of Bluff celebrates the largest oyster festival in New Zealand, and one of the largest in Oceania, the Bluff Oyster, and Food Festival, the slogan of the festival being—'Unsophisticated and proud of it!' Grown slowly in the cold clean waters of Foveaux Strait, some consider the Bluff oyster to be one of the best oysters in the world, and a major global export of the Southland region of New Zealand. During the festival of 2016, 24,000 oysters were consumed in 4–5 hours. This festival provides a unique insight into the community and the culture of the region. The festival started with the 'odes to the oyster' and then diverted into varied events, including a tattoo display contest and an oyster opening contest. People walked around the festival with

The end of Highway 1, at the edge of Bluff, 2017

The oyster festival, Soosan Lucas, 2016

colourful hats and beer cups slung around their neck. By the end of the day, everyone was dancing on the tables, and people of different generations freely intermingled with each other, creating a safe avenue to break the normal subdued life and decorum of the town. The festival itself was the performance of the local community, and it was rustic in more ways than other. The ever-vacant streets of Bluff were overflowing with cars, ranging from old pickups to brand new flashy Jaguars.

The line of cars ran from the beginning of the town to the very end of it, the Sterling point, named after the influential settler and a manager of a whaling station—James Sterling. The last point of New Zealand mainland and the end of the 'Highway 1' is celebrated by the modern (and might I add, expensive) restaurant, Anchorage, overlooking Foveaux. Beyond that are the Bluff walks—the longest being the Ocean Beach Track, which takes you across the entire local coast of the bottom of the South Island through and by the abandoned Ocean Beach meat works, and leads to the Ocean Beach Road. The corner of this Ocean Beach Road is where I was based for the first part of my fieldwork. This cottage became my refuge for the next few months, as I embarked on the first part of my ethnographic explorations.

Uncertain Hypnogogic Liminality

Maybe I should not have left! I had a good life. A bachelor, in the city of Kolkata, reasonably famous—there was no reason to leave. I was safe, sheltered by the warmth of friends, family, and yes, this vivid body of brick and mortar. But I knew, something was missing. I felt—'I have to get out'. It was not for comfort or monetary fortunes, but I knew that I had to get out.

On the day of the trip, to locate myself to the field for the first time, Soosan, with inexhaustible energy, packed up all my belongings into the car, and took the responsibility of the expedition manager and my research assistant. Sitting at the porch, watching her, I could not help but smile at the humour of it all when I saw a 'Britisher, and a white memsahib' carrying and packing the belongings of the 'native' as he gets ready to go out to the untrodden land to do his anthropological exploration. Everything was packed up on the back of the *Saab* with the door that did not lock, including my punching bag and the white piano! We drove again the almost-600 km from Timaru to Bluff. Crossing over the Leith Valley River and passing by the vibrant student town of Dunedin, then deeper into Southland, we passed through small desolate factory towns, and through kilometre after kilometre of free and open undulating fields we travelled, without seeing a single house. Lisa, with the false teeth, the owner of the hotel we had stayed in, had a 100-year-old cottage for rent; she gave it to me for cheap, and I set up my field base.

The merging of shadows and leaves, 2018

I had been petrified of everything I had done since I had come to New Zealand. I was afraid when I started a PhD in anthropology, when I had not done a single paper on it, directly delving in from a PhD in psychology, riding a 2,300 cc motorcycle from my last one being 350 cc, to being a 33-year-old man who still had to sleep with a night lamp on, and the nights in Bluff were interesting to say the least.

Journal Entry
25 March 2016
1:03 am
Bluff

I have the habit of getting up at the middle of the night and going outside to look at the sky in the hope that I can see the Aurora Australis, but I have never been successful. For the last three days, there has been torrential rain, it's been cloudy and gloomy, and even the shark boat trips have been cancelled (which were generally not cancelled before winds of 35 knots). Since I do not have any form of transportation, I could not even go out to get food or other supplies. What's more, at one moment, it started raining so heavily today that I could not help myself but go out and take videos of it. One unruly gust and the door slammed shut and got locked. The unfathomable resource of my innate stupidity could not help but bring a loud laughter to my eminent blue lips. So, I was stranded outside in the rain for 20 minutes, in my shorts, with no phone, no keys, and no car. Finally, upon extensive search along the exterior of the throbbing wooden house, I discovered the hatch of the bathroom showing a centimetre-wide gap. I rummaged through the garbage and found a thick wire, and this I used to jimmy open the window lock and crawled in.

I am alone in a 100-year-old cottage, 1,600 km from Antarctica, an Indian man from Kolkata, where the present population is five million, as opposed to 1,800 in Bluff. Arguably, outside my comfort zone in many of the ways an anthropologist aspires to be, while doing their fieldwork. The only sound I hear is the rain outside, the cracking of the house around me, and the heavy winds looping through some loose electrical power lines just next to the house. I am scared of closing my eyes; today, I have a particularly severe attack of phobophobia. This isolation (and possibly the knowledge that this house may have had a long history of love, loss, and disillusion) is creating an extremely uneasy state of mind. I know, when I close my eyes, I will be haunted by ghosts and goblins and other phantasmal presence(s) (Gordon, 1997, p. 8), as I manoeuvre through uncertain hypnogogic liminality, and my psychological guards are down exposing my insecurities.

A stormy night in Bluff, with the harbour stopped in its steps, 2016

1:25 am

At the dead of night, the town surely seems lifeless. Every once in a while, a truck carrying oil from the oil refinery passes by, which makes me feel that I still am alive. The ever-tired port stays awake with its drooping eyes, hunched like a steam punk art work—cogs moving, the aesthetics ever-altering forms, and contours ever-metamorphosing in a repetitive continuation, until changed otherwise. It is not merely solitude; it is as if the town is dead, a town which I have been told was bustling with life at one time, but is not anymore. But then again, the people, who love the town, are in love with the solitude too; it is me who is unfit for her quiet and harsh beauty.

But at this moment, I am sure, I cannot stay in bed anymore; the silence of the house is more intimidating to me than the apparent darkness outside. I open the door; there are no lights on in any of the houses scattered here and there. Even the ever-awake docks have come to a halt. I can only see the chains on the big cranes move with the heavy wind and rain. My speakers play Springsteen's Stolen Car (or maybe, I imagined it to have been played). A million things run through my mind—healthy self-doubts materializing from my doubts on the virtues of solitude. Why the hell am I doing this? Who cares about it anyway? At the same time, boasting the audacity of reminding myself that I had heard even Cousteau asking himself the same questions in a documentary. But all these insecurities were intensified by one fact...

The grey sand, 2017

3 am

There have been no sharks for the last six days...

APEKSHA
The Wait

Shark Net

*** Storm warning in force**
** Westerly 35 knots to southwest 50 knots this afternoon
** Easing to southwest 35 knots Monday morning then rising to westerly 45 knots Monday afternoon
** Sea high at times
** South western swell rising to 5 meters for a time
** Poor visibility in occasional showers
(Met service reading of Foveaux Strait on a stormy day, 29 May 2016)

I did get out...but now, I have been stuck in India for almost two years... New Zealand has closed its doors for me, as if my life has been put on hold. I have finished the PhD, but have not got a position of work anywhere yet. I am back in the same place I fought so desperately to get out from. It is not that I feared being stuck, but have been afraid that I will give up the will to get out. A person who has left once, is destined to be homeless forever, never belonging anywhere. You cannot say, why you have to leave, you cannot explain to your family, why as a middle-aged man, who is supposed to be an 'expert' in your field, you do not have a job now, and you cannot explain to your friends why you feel like a failure. I belong nowhere, here in this dark night, my body lay, embraced by the home that once was, with bits of scraped paint falling from the ceiling, but in my mind, I am still playing among the waves.

Tying the boat to the bollard, 2017

Journal Entry
26 October 2016
6 am
Foveaux Strait

Our ship gashed through the turbulent waves of Foveaux Strait. If you fall in the waters, you have but a few minutes to survive before hyperthermia

sets in. She carries divers from all over the world, looking for one moment to come face to face with the fish of their dreams, and even nightmares, anchored about 500 feet away from an uninhabited island called Edwards; we lower our cage into the water and start 'chumming' with mashed up tuna. The weather has started getting really bad—windy, and with sharp rain. Cold lips sip on hot cups of soups and coffee, but eyes weigh in the distance for a fin in the water, or a shadow beneath. As the ship sways from side to side, my face and fingers feel frozen in the cold and my body nauseated from the back-breaking work on the boat. Hunger is making strange sounds in my stomach, but I stay up on the top deck, looking for the sharks, and keeping the spirits up of the tiring divers.

Suddenly, someone shouts—'Sharkkk!'...and out of the corner of my eye, a 3-feet-long fin slowly slices through the viridian blue water, 10 feet away from the boat. I jump down and get on top of the half-submerged cage, and sure enough, it is a 15-feet male white shark coming towards the bait. The captain gives me the bait line, as he runs to get more chum in the water, and I can see the body of a shark with its belly half the size of a big car, calmly and effortlessly coming towards me; for a moment, I feel a connection with this enormous strength and feel energized by his energy. We shout to the divers, 'Let's go! Let's go!' And it is the only time, where people jump in the water, when they hear there is a shark around. Hearts pumping, minds disoriented, and tongues dry, the divers jump into the cage, and try to hold on to it as hard as they can, as it sways heavily with the moody waves and waits for a feared entity to arrive in front of them. An entity who is still above us in the food chain, when we are underwater.

The practice of cage diving in Foveaux is a dynamic one. The Māori name of Foveaux is Te Ara a Kewa or 'The path of the southern right whales' (Taonga, 2006), because this is the path which the whales used to follow, making Bluff such a rich whaling station. The encounter that happens between humans and white sharks through cage diving, is mediated through the impact of the sharks, the humans, the environmental conditions, and the materiality involved. Instead of creating concrete boundaries between subjects and objects and their individual intentionality and agency, sociologists of science perceive them as co-mingled intertwined actors with relational attachment with each other (Jones and Boivin, 2010). All of these actors, through these relationships, create this social world that we are all part of. All the actors needed to act together for the successful completion of the human–shark interaction initiated by this practice. Much like Pickering's encounter (Pickering, 2010, p. 195), white shark cage diving in New Zealand, and the human–shark encounter facilitated by it, is a playing out of this 'dance of human and non-human agency' engaged in a dynamic relationship of cooperation, communication, and even conflict. The significant agents are varied, from individual humans to individual sharks, to sociopolitical groups, objects, and environmental factors, and a review of the daily practice produces a picture of the interconnectedness of these agents.

Social scientists Michel Callon, John Law, and Bruno Latour introduced the actor network theory (or ANT) as a theoretical method to understand the interconnectedness of agents in a dynamic web of relationships, which are ever-evolving, and the actors are ever-changing in hierarchy of prominence (Latour, 2005). The basic premise was that the world as we know is shaped by these relationships, and we need to investigate and follow the actors taking part in them to understand them. At the same time, he acknowledged that it is not a theory that explains why things happen, rather it tells 'stories about how relations assemble or don't' (p. 141). It is a way of understanding the networks when technology, nature, and the social are embedded in one another, and the actions and effects of one actor could be felt on all the others (Oppenheim, 2007).

In this way of looking at the world, all the actors become significant enough to consider having affective agency, not just among humans, but also non-humans, such as material matters, animals, and in this case, as I will demonstrate, environmental factors too. Like multispecies ethnography, ANT questions the human-centric view of the world and argues against the hard dualism between the natural actors and the social actors as they are classically known (Murdoch, 1997). Consequently, through applying ANT, we again challenge the perception that humans are the centre of such cultural discourse as cage diving. If any of the other actors does not 'cooperate' with the cage divers, the experience would be essentially unsuccessful, even if it was initiated by the humans. Furthermore, besides being actors themselves, all the agents act as mediators among themselves (Law, 2004; Law and Hassard, 1999; Murdoch, 1997; Tresch and Latour, 2013; Walsham, 1997).

While exploring the interaction with fish in angling, Bear and Eden noted angling is an assemblage of the fish, anglers, and technology, and the anglers attempt to think like fish, with consideration of their biology and behaviour, the environmental condition they are part of and even their individuality. They argued, 'the literature on "posthuman" and animal geographies has focused upon warm-blooded animals, paying little attention to fish and aquatic environments, and upon groups or categories of animals, such as species and herds, rather than differentiations within species or between individual organisms' (Bear and Eden, 2011, p. 336). This means, while many researchers are attempting to bring animals into the purview of social geography and ME, some species are less included than others, like fish. There are three main explanations for this lack of focus on fish: their cold (although even being fish, white sharks are warm blooded) and slimy bodily characteristics, the alien spaces of inhabitation, and that the majority of animal geographers focus on the relationships on or above earth. But in the case of human and fish interaction, the worlds are clearly demarcated by the reflection and even opaqueness of water; so, the anglers are travelling between the airy and watery environment.

'The recreational nature of angling still frequently involves an engagement with the rhythmicity of fish, developing an understanding of habits and patterns at varying scales, from individual to species (and even genus) and through years of fishing or of observing a particular fish for just a few hours' (Bear and Eden, 2011, p. 349). While there may not be the same level of intimacy in comparison to livestock, and other companion animals, the anglers' relationship with the fish, by recognizing and naming them, is indicative of some form of intimacy, even if the relationships and interactions are fleeting and not sustained. In times of flood, when anglers take risks to catch fish, the anglers try to match the habits of the fish to interact with them, beyond the perils of the environment.

When humans get in the water to have a close encounter with the sharks—at the same time, particularly in case of white sharks who spyhop—they raise their heads out of the water into the zone of the air to look at the humans, and even may interact with them. There was an incident with a shark called Ra, a 13-feet resident shark of Foveaux, who hoisted his entire body out of the water on to the back platform of the boat and started gaping (of which Mike thought that it was attempting at some form of communication with the humans) (Kock, n.d.). This happened when Mike and team were out during the off season in Foveaux to check out the shark demography [according to Mike Haines, it happened a few years back, exact dates are unknown].

There is an intimacy of relationship (be it fleeting) in case of the divers, and more so in the case of the operators who recognized individual sharks by their body and/or behaviour. There were sharks like Ra, which the team recognized because of a big black ring around his body (possibly because of some unwanted encounters with humans) and his interactions with the cage operators; Free Willy, because he had a crooked fin like that of the orca in the movie of the same name (Wincer, 1993); Tim Fin (named after the New Zealand musician, Tim Finn); and then a female shark I named Aurora (after Soosan's daughter) and one I named Rocky (after my own puppy), because of his excitement and inquisitiveness around the boat). The individuality of the sharks shaped by their own sex, age, history, size, experiences, and potentially social interaction within themselves may shape their interaction with the cage diving operation and the divers, instead of being non-agentive random beings. In the act of shark wrangling (enticing the shark with the bait to perform for the tourist), the humans have to be attuned to the individual character of the shark as much as the shark seems to try to assess the behavioural patterns of the humans handling the bait.

Mike and his team, operating these cage diving tours over the years had created their own norms, protocols, and practices. Some of these are mediated by bureaucratic and legislative bodies, while others arise from their own experiences in the field and with the sharks. Not only did Mike have to make sure that everyone got on the boat at the right time, but that every one of the tourists got to see the sharks at least once. He had to make sure everyone was comfortable and engaged with the trip. Last but not the least, he had to ensure everybody's safe return to the shore.

Once the operation started in 2009, multiple social and political actors became involved in the practice—parties who wanted to regulate it, ones who wanted to stop it all together, and others who wanted to promote it. There were bureaucratic organizations who were more interested in the safety of the sharks in the practice (DOC), and those who were interested in the safety of the humans (Maritime New Zealand). Once all these actors were at least in a transient homeostasis, there was consideration of the actors involved in the daily operations, which includes the human and the non-human agents, the environment, the materiality involved with the practice, and the sharks themselves. By the same token, the network could be easily broken down by the actions of the actors, causing the tourist not being able to encounter the sharks.

In sociological investigation of education, it had been noted that the relationships between the human (students, lecturers and technicians) and non-human (handbooks, tablet computers, textbooks and cameras) elements constitute educational curricula, and are used to consider the ways in which these elements are arranged across temporal, spatial and geographical boundaries... (Tummons et al., 2018, p. 1913). In a similar sense, MacLeod et al. (2019) noted in the example of connection among humans and equipment in medical education,

The boat ready to leave in the morning, 2017

…a mundane object, like a thermometer, does not find its own way and 'jump' into a person's mouth, and 'decide' whether that person has a fever. A fever is determined through multiple relationships between people and things: a person feeling unwell, knowledge of thermometer use and normal temperatures, digital technologies, batteries, and innumerable

other factors...ultimately no actor, human or non-human, could exist completely on its own... Thus, agency is conceptualized as something other than causal, uncoupled from concepts like intention, subjectivity, and free-will...[and there are] various social and material mechanisms at work to hold together a phenomenon. (p. 179)

Consequently, here, discussing the various non-living material actors and their affective relationship with the White shark cage diving in New Zealand operation, I consider their agency as a co-created relationship, beyond intention, but nevertheless possessing agentive abilities.

Environmental factors were among the most significant non-human actors shaping the human–shark encounter. These factors not only affected the comfort of the trip, also the visibility of the ocean affected the visibility of the sharks for the tourist, and even an entire day's trip could be cancelled by their influence. Often trips were cancelled in the morning itself because of the turbulent and unpredictable Foveaux Strait. Even if the boat was able to go out to Edwards, it may have had to take shelter behind the island, or other islands if the winds were too heavy. If the water was too choppy, then the equipment including the cage could not be put underwater; so, all that the operators could do was wait. The average time of the trip depended on the time of the shark encounter: if the sharks were there when the operators reached the spot and everyone had got in the water and seen the sharks enough, the trip could be done by lunch time. However, if the sharks were not there, then it could be the entire day.

Out in Foveaux, any kind of unexpected wind could be potentially problematic, but generally, the team did not go out when winds were heavier than 30 knots. The most problematic wind for the operation was the north-westerly; because of the structure of the Edwards Island, there was no shelter, and no matter which side you were, a heavy north-westerly could potentially stop the operation. If there were very heavy westerlies, it was hard coming out to the island, or returning. Northern winds were not generally stronger than 30 knots. Down here, they didn't get lots of southerly winds, because the southerly winds were cut off by Stewart Island and were felt more in Dunedin. That is why, Mike often had to cancel the trips at the last minute, because he could not take risks with the lives and well-being of the tourists who came out with him to see the sharks. Tourists got annoyed with operators when the operators cancelled the dive in the early morning, because they did not realize the danger it involved (as neither did I when the trip was cancelled on my first try). Consequently, while in the field, my sleep often broke in the middle of the nights, to check the weather, to know if we were going out the next morning.

Sometimes, even if the team was out, and the winds came from unfavourable directions, they had to take shelter around other islands, like Jacky Lee and North Island. The visibility of the water also was a deciding factor as to how much of the sharks people could see from the boat. If there was too much sunlight, it was hard to see them from the surface because of the glare, whereas it was a better viewing condition under water; if there was less glare, it could work the other way around. Visibility was also affected by suspended particles in the water on that day. White sharks would often come near the boat, interested by the burley trail and to make circles near the seabed, but never come to the surface. Especially in those situations, if the water was extremely murky, then the tourist in the cage would never be able to see the sharks. On the other hand, if the currents were too heavy, we could not put the cage in the water. Mike called the cage the 'washing machine', precisely because it could get quite turbulent in there. The weather could be stunning, with crystal clear and calm water, and lots of sun and even warm temperatures, and even if tourists had to wait a long time for sharks, they would be cheerful and in a positive mood. Alternatively, if it was gloomy, it didn't take a lot to create an environment of frustration and even despair, especially if the sharks had not been spotted for a few hours. Finally, the cold waters of Foveaux (averaging from 12–15 degree Celsius) were not the best for divers in a 5 mm wetsuit staying static inside the cage. For many divers, it was the chill of the water that got them out of the cage, and not necessarily the sharks. This was especially true for divers who had never dived in cold water, and certainly for new divers, the first shock that everyone got was the water, particularly on the bare skin around their face.

I present here incidents of one frustrating day (the day I had been out the longest on the boat) when the environmental factors (including the winds, the currents, the geographical positioning) and the sharks demonstratively dictated the practice, the lives, bodies, and plans of the tourist, the operators and the researcher. Furthermore, the boat, and other elements of the boat, including food, played a significant part in shaping the experience. This day gives a very direct look at the significance of agentic ability of the varied actors, and to do so, I draw upon my field journal entries.

14 January 2017
Foveaux Strait
8:15 am

After leaving Bluff Wharf at 7:30, we anchored around the eastern side of the Edwards. It is very windy, and we are having quite a hard time putting the cage in the water. Peter is also here with his boat and tourists.

2:30 pm
No sharks, and the weather got very bad. Mike and Peter think that it is probably because of the heavy currents that the sharks won't come up to the surface. These currents also disperse the tuna oil (from the chum) too much, which is another part of the problem. Morale is getting low; however, all of a sudden, one tourist realized that there were some of Mike's famous Afghani biscuits, tea, and some pies still left in the fridge, and everyone jumped inside the cabin to get some hot food.

2:40 pm
We have lifted anchor and come to the northern side of Edwards; it is still so choppy that we can't put the cage in and had to put the straps on the cage again to secure them to the boat, taking shelter from heavy winds between rocks. Mike says that he has to make the best of this, because he is not coming tomorrow or the day after; all the tourists encourage Mike by saying that they have nowhere to go, so they don't mind staying as long as there is light, and we have not seen the sharks. It had been raining and the weather was bloody cold, and to put that into perspective, even mighty Mike has been constantly shivering, and the water is so turbulent that he is having a hard time keeping the boat in the right place. We are really close to the island, and for the first time, I can see the details of big monoliths hanging from a side

of the rock face, with scrubs growing out of them, and the seals, a few feet away from us.

3:20 pm
We can't stay here any longer and have to go now and take shelter behind another island, possibly the north island (not to be confused by New Zealand's bigger North Island), a kilometre from here.

5:17 pm
We set off from the north island for the eastern side of Edwards; Peter has left, because if the weather gets any worse, he will have problems returning. In our boat, everyone is again excited and even adamant about seeing the sharks.

6:15 pm
It is quite windy in the east, so we move to the north-east side. The tourists are very eager and everyone looks to the horizon. One of them says, 'I keep thinking that I am seeing things, but definitely I am not.'

6:45 pm
Well, maybe he did see a shark, because at 6:30, we spot our first shark near our boat, but only for a couple of moments, and before anyone can get in the cage, he is gone. One tourist and I still get in the cage, but the shark is nowhere to be seen.

7:16 pm
We again come back to the east side of Edwards.

7:40 pm
Fortunately, there is a shark, and this time, it seems to stay around the boat, and the tourists could enjoy spending time with them in the cage. Even if the water is choppy, they can still get in, be it for even just a few minutes.

Taking shelter around the north island, about 3 km from Edwards Island, 2017

8:10 pm
We have to set off from Edwards now, as it is turning dark.

9:15 pm
This is the latest I have reached Bluff, as the Southern Isle enters the harbour, and I truly understand why the Bluff hills are called the sheltering hills (Bremer, 1986). The weather was supposed to be so bad, Mike had handed out sick bags before setting off. He mentioned that he has never had such strong winds in all the time he has been working here. Even if the tourists were scared, the captain couldn't show fear (maintaining composure); however, both the captains later mentioned that they were quite concerned. In this kind of weather, Peter's catamaran is in more trouble, because even if it is faster in normal weather, in high seas, it jumps from wave to wave and falls flat with a big bang, while Mike's boat can sway with the waves. Peter said that generally, while coming back, everyone sleeps; but on this day, everyone was awake and looking through the window, because the waves were coming up to the window, and Mike mentioned that this was the first time he had ever had to slow down in this boat while crossing Foveaux.

While out in Fouveaux, waiting for the sharks, Mike would make his spiel, where he talked about the lives of the sharks in the region. Here, he would first start talking about Edwards Island and its inhabiting wildlife, as much as of the sharks around it:

'Now, this island is part of a group of islands called the Muttonbird Islands, or Titi group of islands. It is a very small uninhabitable island, with jagged rocks and no fresh water supply. And I'm sure, at least one person in this group can tell us what they are (as he smiles towards one Māori gentleman in the group). So, what has been happening for hundreds of years is that the local Māoris have been coming to these islands to take chicks of sooty shearwater. The shearwaters migrate between here and the northern hemisphere in hundreds of thousands and they come to these islands to nest. They lay their eggs in burrows about 4 feet deep, around September–October, and remain with their mates till their dying days. Around April–May,

The wind and the rainbow, 2017

the local Māoris come to these islands—a tradition that is still going on today. When the chicks come out of the burrows, they're covered with down and at their fattest, that is when the Māoris gather them, pluck them and preserve them, to sustain them through the winter months, and they would also trade with the northern tribes. If anyone likes sardines or pilchards, one would like shearwaters; I have been brought up with the former and I love the latter. They're very high in Omega-3, very fatty, very oily, and very salty; so, they are quite an acquired taste. And we don't like to share them with too many people anyway. The other thing with these islands is that there are hundreds and hundreds of seals, and around July, the females give birth to the pups. Right about now (December–January), the pups would be on an average about eight months old. They will all be frolicking in the water, as the day goes on. See them swimming in the shallows there (as he points out to the shallows near the Edwards Island). They are all fat, juicy, and very naïve—just the way the sharks like them.'

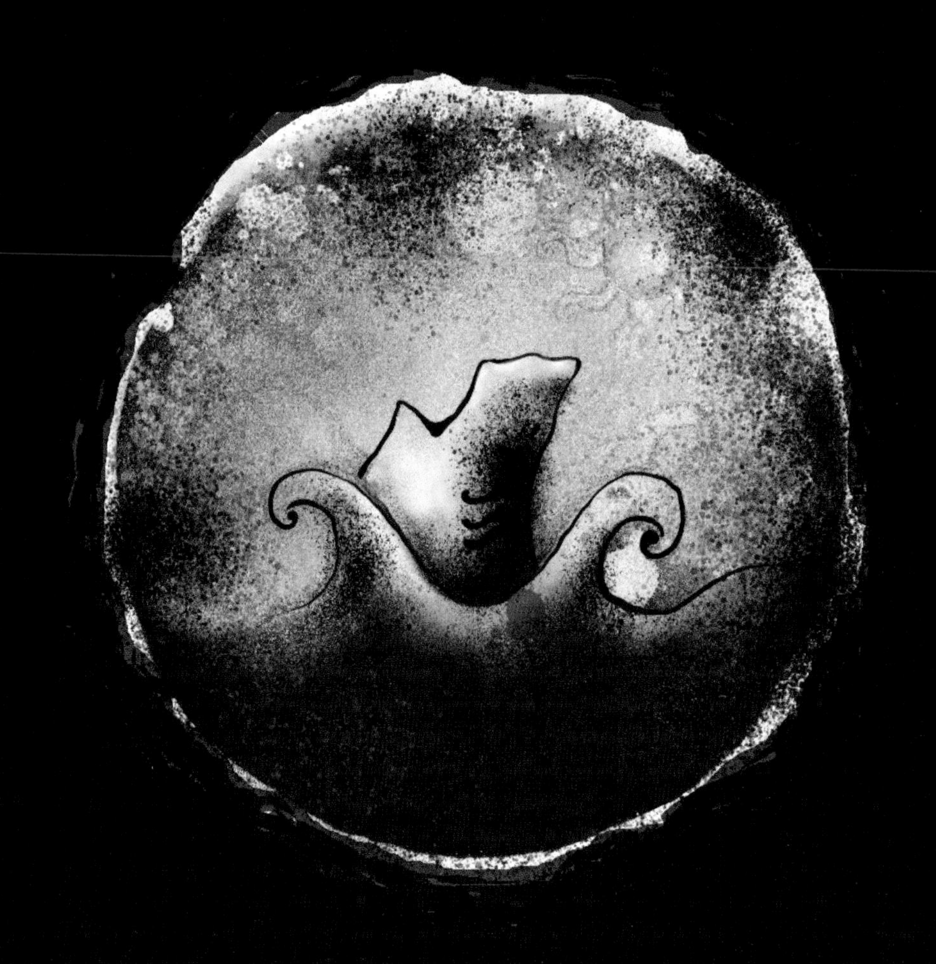

The shark and the wave, digital art, Raj Sekhar Aich, 2020

Reverie

Far away on the beach, there is a full moon party tonight, and our ship is about 10 km away here in the Cimmerian waves. I help put on the BCD (byuoyancy control device or byuoyancy compensator). Being eager as I always am, I jump into the inky darkness before anyone else. I switch on my torch, and aim it at the buoy line, suspended amid the ocean and travelling down the seabed. I look up at the boat, and I can still see your shadow hesitantly standing at the edge... I call you... And you do trust me, so you jump in with a big splash a few feet away from me. After you, one by one, the other three or four team members follow you into the water. It is our first night dive, last day of our advance scuba diving training. I grip your naked hand locked and strong, and slowly descend into the fathomless ink, only guided by the thin lines of light thrown away by our torch feet by feet. After about 60 odd feet, we feel the ground for the first time. We crouch down and sit on the seabed, me still holding onto your hand.

As instructed, now we switch off our torches and sit in absolute darkness. A minute goes by, then two, and three, and very gently, light starts coming back. We look around us, and the aquatic sky, and the abyssal roads are filled with gems of unutterable colour. These colours morph and evolve from coral, into jellyfish, then to fish, and into our dive instructor, mildly lapping fins and disperse into an underwater cloud. You and I start following the lines of light that comes off him. As we dance in this light in this unknown universe, I suddenly feel your hands pulling me, I can feel you rising up to the surface. Still holding your hand, I follow you up, and we break the surface.

A view of the shark approaching the cage

In the calm evening, we can see the dotted lights of the ship some few hundred feet away, camouflaging her with the dotted lights of the island a few kilometres away. As you open your mask, and take a shallow breath, I hold you and ask... 'What happened?' 'I can't breathe... I can't breathe... It is too claustrophobic.' I see another light coming up from the depth to us. 'You guys ok? What is going on?', 'It is fine, Skip, she just needs a breather!' 'Well, the ship is just a few hundred feet away. Why don't you come down and finish your dive, and she can swim to the ship?' As tempted as I was to go down, looking at you, I could not leave you, literally at the middle of the ocean, at night. 'It is okay, Skip, you guys continue; I will be with her.' Nodding in response, he went down.

'It is just a few hundred feet, maybe a kilometre or so to the boat; you inflate your BCD, and just relax; I got you.' Far away in the island, the party rages on, the echoes of the dances coming to us in the silence. The smoke from the barbeque lights up the horizon like the embrace of the Aurora. We follow the milky way to the ship, and like a lotus in a calm pond, you sleep gently on the wave as your skin glimmers on the deep maroon waves. We still have a few hundred feet to go, as I flap my fins and get us inch by inch closer to the ship.

My vigilant eyes look all around and to the ship. I realize, the waters below my fins moved a bit unruly. As I look down, a giant glowing green shadow with large fins swims by, inches below the surface. I cannot tell you what I just saw, I have to keep calm and get us to the ship. But as I look at the ship, I realize, we are actually further away from her than we were a few minutes ago. For a moment, I cannot understand what is going on, but in a flash, it dawns on me, the currents are taking us out. Still, I cannot let you know; my strokes start to get stronger and stronger. I make some progress, but still no one on the ship sees us. I take out my torch, and start strobing it to the ship, realizing that this can turn bad pretty fast.

Thankfully, now the ship spots us, but still we are a bit further away from the ladder. Someone has the right mind to throw us a rope. I hand it to you and ask you to hold it tight. Your tired arms, with all the equipment, are yet strong enough to pull yourself up. As I look up from the depth, you getting up to the ship, with a joyous smile on my face, I lay back on the ocean in relief... something very strong suddenly bumps my leg....

My eyes opened, the first thing I noticed is my bronze wristlet and your silver anklet dust-embraced somewhere in the corner, possibly for 10 years or so now. The walls have distorted sweat-laden impressions of limbs, hair, and hips. When you left for the last time, I was not here, but the only remnant I had of you was this terminal dream. Dreamscape is like a garden, and every moment I breathe, new flowers bloom in this garden. And every once in a while, in slumber, as I walk through this so precious garden of mine, I come across a smell I have not had for ages. In this garden, flowers young and old make love, share their colours, some perish, and some bloom anew.

My shark nightmares, all but stopped when I started my fieldwork with the sharks. In their place, I started to have dreams about how I am a 'shark anthropologist' and I have gone somewhere to work with the sharks. In one dream, I had come back home, and my mom had kept a big white shark in a shallow tank in her small garage, and she was telling me, 'He is so cute, I finally understand why you love them so much.' Upon deep self-analysis, I have come to the understanding that the image I have created of myself in the last few years has had a profound effect on my psyche, and this image is more significant than the fear itself. However, I am not sure if it indicates I am less afraid of them, at least at the conscious level.

Dreams transcend miles, landlocked or open ocean dwelling. Many similar themes of dreams were often shared among tourists on the boat about the sharks, indicating to the deep cognitive effect of sharks on our cultural and personal mindscapes. Although I will not attempt a dream analysis here of them, these dreams represent the varied images in which sharks come to us, no matter where we are from, be it any corner of the world. While the recounting of shark dreams to me as a researcher is admittedly anecdotal, it is fair to say that these dreams shape our perception of the sharks and are shaped by our perception of them; so, it is important to document them, and perhaps in later research, there will be valuable data that can be extracted from them. In particular, it will be interesting to know what kind of dreams will be had by people who are petrified of sharks and would never consider diving with them, as opposed to the ones who choose to interact with them in activities like cage diving.

The dreams were often about the fear of sharks, and a sense of helplessness, and they could be quite intimidating. A lot of the dreams were about boats sinking, a shark attacking family members and oneself. There was even a Canadian woman who dreamt that she fell in an icy lake and there was a shark underneath, but she could not get out. One person dreamt of a huge black shark walking up to the beach and devouring people. People were also afraid of the cage diving itself. Scott said, 'Yesterday, I was dreaming every type of scenario that could go wrong; the crew would be horrible, a shark would come up onto the boat, and the cage would detach and fall in the ocean.'

There were some who saw very vivid terrifying dreams and remembered them distinctly. A young doctor who was always obsessed with white sharks, but at the same time petrified of them, had flashes of them throughout the day: 'I can literally have day-mares about them, when I'm in the shower and I have bubbles in my eyes, I get scared, and even get freaked out in pools. The most vivid nightmare I have is the apartment I used to be in—there was water all of a sudden, and I was forced forward underwater. I cannot see the shark, but I knew it was a shark. It pulled me underneath the water. I tried to kick free. I woke up mid-dive and landed on my floor!'

A similar sense was felt by another American young woman, who could not see the shark. Interestingly, she actually welcomed these terrifying dreams: 'I used to have a lot of dreams of being attacked by sharks, but I always welcome those dreams, because I at least got to see them up close in those dreams. I could be in a pool or reef—I would not even see the shark, I would just feel the pain. I would see my blood cloud all around me, and all the pain and panic. The scariest thing is you don't know where the shark is! I don't have a lot of these dreams nowadays, unless I see *Jaws*.' It is interesting to observe that in the dreams, they could not see the entity that was attacking them, but they knew it was a shark, and even more specifically, the great white shark. This makes me wonder if it is the shark we are worried about, or the fear of the unknown, just beyond our vision, and sharks are just but a potent embodiment of this fear. Even though it is hard to say if this projection was present before *Jaws*, it is certain from interviews of the tourists from all over the world coming to this boat, and also the Bluffies, that it accentuated after people saw it.

Sometimes, these dreams were even peaceful, and the shark was a protector, said an Irish woman, Emma: 'I'm in the ocean, and alone, and a great white comes up, and swims peacefully beside me. It was huge, but I was really excited, not scared.' A Swiss male of 26 said, 'Sometimes, I grabbed the dorsal fin, and they were pulling me with them.' A Greek woman, Helen, said: 'I was walking on an island, and the shark was swimming near me. But it was more like a protection, or awareness, rather than danger.' And sometimes, it was more about inquisitiveness. Rose, a young New Zealand woman, recounted: 'I dream I'm in my office in Fonterra [the New Zealand multinational dairy cooperative], and in the afternoon, water would fill up in the office till about chest level. Then I would see sharks in the water—not scary, and I go from room to room looking for them. I walked into one room and I saw the top of its head; it was peeping out. But as soon as I got in the water, the water would vanish and so would the shark.'

Then there were the incidents I call para-real experiences, that is, experiences that are as real to the individual as are other aspects of reality to the general society of that time and place. And the real-ness of the experience to the individual is more relevant here than the scientific analysis of objective truth. An American woman and her husband had had a previous experience with cage diving before, in Port Lincoln (South Australia), and stayed in the boat that night, but when she was about to fall asleep, she could see sharks come in and swim all around her. In the morning, when she talked with the other tourists, all of them saw the same thing—'I saw the shark, really close to my face, not doing anything, but calmly swimming around.' And her husband added, 'I don't know what it was, but I did see a huge shadow passing by.'

Soosan also had some interesting experience with the sharks when she wanted to connect with them spiritually. This happened the first time we had gone to Bluff, but had missed seeing the sharks. She comes from a lineage of Vikings and Ogallala Sioux native Americans and has grown up with shamanic practices. It was early morning; she was sitting in front of the cottage we had rented for the night, and in her mind, she wanted to see the sharks and the sharks materialized in front of her. 'I prepared myself and as I sat and allowed my being to become relaxed and in a meditative state of higher consciousness, I called to the sharks—as opposed to whale calling, which I had practised and studied before. With no time to waste, a connection was made and the sharks were there, eager to connect. I immediately felt the greyness of the void, and yet also felt the vitality of "them". The mystery of the sharks and their voracity as apex predators scared me. A female came and was eager to talk; she said it had been a long absence of humans talking with them—I was calmed, and my curiosity overcame my fear. They said that they had much to say, many things to reveal, and would welcome connecting with me and the others. The group here in Foveaux was more removed from many other groups and although thought to be a singular animal, they did communicate and have a connected group. There were no words in this type of connection, but the whole being of myself and the animal communicated. I gave thanks and said I would connect another time and left and returned to my physical body sitting outside a motel on a hard deck chair on a cold, blustery day in Bluff.'

These experiences are beyond my experiential spectrum and explanation, it is irrelevant what I personally think of them or believe. What is important here is their significance and symbolism to personal myth, stories, and perceptions of the experiencers. Probably, they can best be understood and explained by alternate perspectives, which are beyond my own cognition and beliefs. As Stony pointed out—'I don't know where you stand with reincarnation, but these people were having dreams of sharks; could well be a memory they brought from the last life.'

Suicide Vest

I had the habit of getting up before sunrise. This happened after full nights of paintings, often preceded by deep travels in secret gardens. A body scarred with paint, scratches, dried up acrylic, remnants of varied aqueous essences, stumbled out of his den in search of a cup of coffee and pre-sun-cigarette. Thirty minutes of sword practice, and then I sat down for riyaz with my mother's old harmonium. My uncle's house was in the same property, where my uncle's family and my grandparents used to stay. Around 6 am, my grandmom would come to my house to pick flowers from the terrace for her puja and listen to me sing, and we would have tea and chatter about the day.

In Bluff, a typical day on a boat for me started off as I got up at 5:15 am. It was always a habit of mine to prepare everything the night before. Along with addressing my research objectives, the environment of Foveaux, my responsibilities as a boathand, and the confining physical environment of the boat, the fieldwork tools had to be carefully considered. Particularly because all the equipment had to be on my body at all times, because an important photographic chance or taking other forms of data, could be missed in an instant if not careful.

This is the equipment (as shown in the photograph), which I carried 'on my body' every morning while on the Southern Isle. I present them here, because my aim was to achieve maximum productivity with minimum physical footprint and independence of movement, and documentation of

Waiting for the tourist to arrive, Soosan Lucas, 2016

On-body field equipment, 2017

them may be helpful for future marine anthropological researchers. I used a green khaki fishing vest with multiple pockets to store and carry all my equipment, which incidentally looked like a suicide vest, particularly carried by a big brown bearded man, as Soosan rightfully pointed out.

1. The field bag—sturdy, waterproof, and just the right size to keep on all the time in a confined space, but could hold my laptop.
2. Camera clamp.
3. Mini collapsible camera-stick.
4. Large 5 ft action collapsible camera-stick (the largest professional camera sticks available in the market at the time)—to be able to video-record the sharks from the top of the cage, and the boat.
5. Laptop with ruggedized casing.
6. Waterproof, shockproof phone.
7. Waterproof diary.

8. GPS device to track the first locations of human and shark encounters daily.
9. Knife (as per the DOC handbook, there needed to be a knife handy, in case a shark got entangled in a rope).
10. Fishing vest—waterproof—and with lots of easily-accessible pockets.
11. Dictaphone for recording the interviews.
12. AEE action camera with special microphone accessory (used for primary video above water, mostly of the tourists).
13. Tally clicker, for measuring details about shark behaviour.
14. GoPro action camera (used for primary video underwater—of the sharks and the humans).
15. Action camera body harness, for shooting first-hand perspectives of the fieldwork.
16. Clip board to have a stable platform for the survey in the turbulent environment.
17. And then, there was equipment not shown in the photograph.
    - Olympus TG 2 all-weather camera (back-up video camera, and primary photographic camera).
    - Pens.
    - AEE camera remote (necessary for documentary footage, where I, as a researcher, needed to be in the frame and the camera was placed out of reach).

There was also a laser distance metre that I used to measure the distance of the sharks on each pass around the boat; however, the reflection on the water caused disruption with collecting reliable data. Even if this tool did not produce particularly reliable data in this expedition, this may be something to consider for future research, given the right circumstances.

Due to the saline water, the equipment would occasionally break, or lock up. Tools which are apparently benign, like a screw or a nut were realized to be crucial throughout the entire field day. That is why I learnt to keep extra fittings and equipment, and sometimes, a kind tourist would donate some accessories, which would keep me going. At other times, I had to be creative and improvise, like tying my camera with a rope on my stick, when the screw broke. Or even using hair bands to keep the water out of camera equipment after an accidental exposure (one of the perks of having long hair). One day I forgot to pack my long camera handle, so I tied my camera to the back of a broom stick to take videos of passing sharks.

Besides the tools for the documentation of the research, other important tools to consider were actually the dressing articles. These included the waterproof trousers, jackets, gum boots, inner garments, and a pair of sturdy half gloves, which allowed the articulation of my fingers. I have to admit that I needed to be more careful as I had not been in such an environment before and was unsure how my body would react to the cold, and the fact I would be wet for a considerable part of the day. At least for my initial fieldwork, my clothes were inadequate; so, for my second time in the field, I aimed at getting the right ones (which I did over the months I was back in Christchurch between the two field trips). Considering my limited resources, some going out of business sales were extremely fortuitous.

There were a few factors that I had to consider, including ruggedization, waterproofing, and portability. There was also the cost factor, which every research must consider, especially in projects that are resource-heavy as this. Any average good sea-going film-making camera runs in thousands of dollars alone. And in this project, understandably, I had to make a lot of compromises, because I had no financial backing. But still I did the best I could, given the circumstances.

Not only did I have to be aware of the turbulent waves, but also of the sudden and intense rain in Foveaux. The weather could change at any time, and often, along with the other crew members, I had to bear through the rain for hours on deck, baiting and waiting for the sharks to come. Most of my important digital equipment had to be ruggedized (a common term; however, the levels have more specific technical terms, like IP64, and so on, which are US military ruggedization specifications), including my phone, GPS equipment, laptop, and journals. The phone, GPS, and watch were purchased specifically, which were already ruggedized and water proofed, which generally meant that the devices were covered in some sort of rubberized shock resistant materials, and for waterproofing, generally the charging ports, and in some cases, headphone ports, were covered by protective flaps. As for the laptop, it was not waterproofed, but I had to purchase an aftermarket ruggedized case. Still, as for waterproofing, now there is new technology, which nano-coats the circuits to save them from water; however, in most of the cases, this nanocoating is not suitable for saline water, which should be considered before acquiring such devices; good old solid rubber covers still work the best, at least in my opinion, even if it adds a bit more bulk to the devices. I learned it first-hand and I have multiple dents on my laptop to prove it—a sudden wave hit me and along with my laptop, I flew from one side of the galley to another and hit the table edge.

Then there was also the need to consider the continuous working with heavy equipment, like the cage, the crane, and ladders. There was also the effect of water on journal entries that were already in my illegible writing, written amidst high waves, with cold shaking hands. Then there was the chance of sudden malfunctioning of an equipment, like if my camera got sea

A day of writing field notes on the boat before starting the use of waterproof paper, 2016

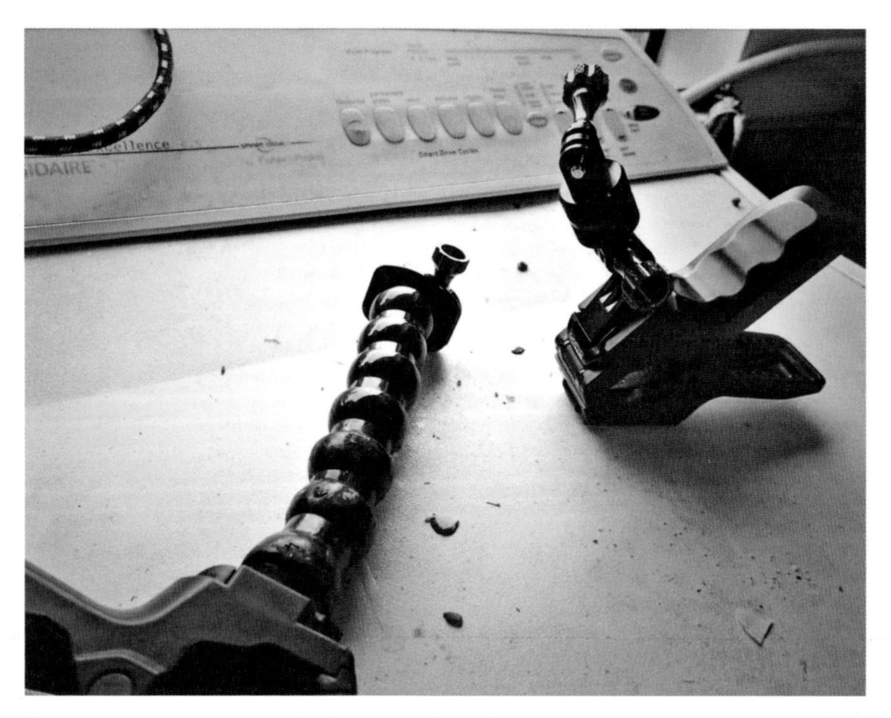

Cheap camera mounts, which often broke in the field, and super glue always needed to be in hand to fix them, 2017

water inside. The other factor while considering the tools for the fieldwork was portability. Due to the close quarters and my other duties on the boat, which needed free hands and agility, the equipment had to have a minimum physical footprint. Particularly while completing the surveys, I had to have the skills and tools to deploy and execute the task efficiently, considering the limited time I had and the tourists felt nauseated while writing on a boat.

So, once kitted, I would be out of the door by 6:15 am. In the first field visit, I briskly walked to the dock where the boat was kept and reported at exactly 6:30 am. Mike waits for no one. But the second time, Carwyn arranged for me to stay in a house next to theirs. Up and ready, I walked to Mike's house and waited in front of the car, helped with the gear, and we were off to the boat.

A view of the ocean while waiting for the sharks

View from the Cage Top

I have been considerably good at many things (except school, of course). When I painted, I was good enough, had a few national and international shows; when I acted, I was the lead of a daily soap in Kolkata; when I did bodybuilding, my instructor wanted to put me in the national competition; and God knows what else. And though I was good at all those things, I would have never been extraordinary, or indeed add to the development of the art; if anything, I was a celebration of mediocracy. But for the first time, I found something where all my knowledge could be put to use. All my identities converging into one—my mental, physical, emotional and sexual energy converging together like tributaries and creating a force of nature.

On that little boat out in Foveaux, everyone was performing their own varied identities. Goffman's *The Presentation of Self in Everyday Life* (1959) becomes an excellent platform to understand them.

Goffman (1959) developed the 'dramaturgical' analysis of social interactions (Fawkes, 2015), where he considered humans in social situations to be like actors (using the analogy of theatre), playing certain parts or acts in front of others on a stage to represent a desired image or impression of themselves consciously or subconsciously. At the same note, the audience is constantly analyzing the activities of the actors as information to create or organize their perception of the actor to new images, or in many cases, preconstructed schemas in their mind (Weigert, 2003; Van Dijck, 2013; Schultze, 2014; Squirrell, n.d.). Our 'self' is a 'reflective, co-creative process that relies not only on our own careful manipulations

but also on the way others understand and interpret our conveyed image... The performance, to be conveyed, requires an audience, be it an individual, a team, or a group of some other kind' (Zavattaro, 2013, p.512, 514).

These performances took place on a stage or the front. And while creating fronts, 'The actor employs props and performs in settings...creating a believable role that signifies the actor's intended meaning.... This act culminates in a "dramatic realization"...of a performance in which the staging is favourably received by the audience' (Hoxsey, 2008, p. 4). Much like in arguments about creation of hyperreal images (Wolny, 2017), elements relevant to creation of fronts are: the settings—physical elements or material construction of the situation, including the props, scenery and locations where the social interaction is taking place; appearance—elements which follow the performers no matter what the context, and are supposed to display the actor's social status and role, particularly elements of dressing and props, and other accessories on their body; and the manners—how the actor is playing the role attitudes or behaviour of the performance in a specific situation, including matters like facial expression, and general demeanour (Thompson, 2016; Zavattaro, 2013). Furthermore, actors often use identity pegs, 'on which to hang his self-narrative' (Down and Reveley, 2009, p. 380). Often these pegs are clothing and other behavioural mannerisms (Chriss, 2015). Being the vessel to carry the people across Foveaux, the Southern Isles (the boat of

The emerald chalice, 2017

the Shark Experience team) was such a dynamic stage, where everyone was performing identities with the crew (and the researcher) playing the actors, and the tourists were the audience.

There are three stages of such performances: the front stage, the back stage, and then the offstage (Benford and Hare, 2015). The actions and images of the actors are regionally segregated in these stages, differentiated by accessibility of the audience to the actors and the presentation they create for the performance (Tewksbury, 1994). 'Front-stage performances are those designed and presented for consumption by either a specified or a generalized audience [in other words, it is in full view of the audience]' (Goffman, 1959, p. 112). They are an integral part of the performance involved in the image creation, defining the context and situation of the performance to the audience (Goffman, 1959). The front stage was created on the boat with the decals of the sharks on the body and the nautical memorabilia in the cabin. The interior was given a lounge feeling, so that the tourists could sit with each other and face each other. The warmth in the cabin gave a sense of comfort and safety after been drenched in the southern wind and waters. The sense of safety and confidence in the professionalism of the operation was further emphasized with the cleanliness and continuous upkeep of the boat, and the equipment, especially the cage. Furthermore, the cage and the deck also became the front stages at times, especially when looking for or viewing the sharks.

The backstage is 'a place, relative to a given performance, where the impression fostered by the performance is knowingly contradicted' (Goffman, 1959, p.112). This region often remained hidden from the audience (Zavattaro, 2013). The captain's cockpit, and the galley, and the hull (where all the equipment was stored) were the backstage on the boat. The tourists generally did not venture here, and personal discussion about the operations would take place in the cockpit, often during the journey to and back from Edwards. The third stage is the offstage, which are '...those locations and occasions that are physically, temporally, and/or behaviourally separate and distinct from a given identity; the off stage is the interactional realm that is completely removed from the theatre of a given performance' (Tewksbury, 1994, p. 3). In the context of Shark Experience New Zealand offstage was outside the boat.

Mike was the captain and unquestioned leader of the boat; he established that during introducing himself while leaving for Edwards, when he mentioned his most important job was to make sure that everyone arrived back safely. His usual calm demeanour all through the trip helped the tourists to have confidence in him and the operation, even when the sharks did not arrive for hours, or the weather was bad. He took the decision where to anchor the boat, and furthermore, this creation of his social position in the operation was especially visible when he was insistent that every one of the divers would see the sharks at least once, even when some of them had a hard time learning to breathe underwater initially. Mike was authoritative and his directive tone ensured that everyone finally learned to breathe underwater and saw the sharks. Carwyn, as a mother figure on the boat, made everyone feel warm and taken care off; she presented the feminine side of the entire experience and she was aware of it. Maca, the dive instructor and the boat hand, created a happy and relaxed environment for the tourists. The material that acted as identity pegs for the crew were generally the T-shirts with the brand logo of Shark Experience, and gumboots. For myself, my most significant pegs (Hoxsey, 2008) were the on-body equipment, and stickers on me. There was also my knowledge about the sharks as much as the equipment on me as an onboard researcher and potentially a conservationist, such as my camera, note books, etc.; another significant identity peg was my Megalodon tooth, which is considered to be an iconic artefact among shark lovers.

Foveaux Strait was as much the setting for white shark cage diving in New Zealand, the play, as the boat. The play and its presentation of the front stage started at about 7:15 am, as soon as the tourists arrived at the boat and were greeted by me or Maca, and the play finished when the tourists got off the boat.

The backstage work started as we all converged on the boat by 6:30 pm. Cap (i.e., Captain—my name for Mike), Maca, and myself would go to the boat, while Carwyn and/or Maddy (Mike's daughter, who studies psychology at the university of Otago and came to be on the boat in the holidays) used to open up the information centre and wait for the customers to come in, check their diving credentials, and got them registered. Cap went into the galley to start the engine, as Macca and I would untie the ropes that tied the boat to the new wharf, where we were anchored. I got up on top of the

Taking notes on the boat, 2017

cage to untie the big orange rope from the boardwalk bollard, and we were on our way to the old wharf, where the tourists would be boarding. Here lay one of my most dreaded parts of the day—unroping and roping the boat, a part of the practice, which honestly, I never got good at. The ropes would get tangled with ropes of other boats; I would put improper knots, and due to low tide, heavy tension on the ropes meant they could not be opened, and had to be cut, this in effect would often get Mike frustrated at me. My soft Bengali

academic hands did not seem strong enough to serve the purpose anyway. In one instance, I was trying to set it free as the boat was leaving and Carwyn ran up to me and shouted to move away—'The boat would yank it off the bollard. Don't you know how many people died being hit by the rope!'

It was a two-minute voyage from the new wharf to the old wharf. Here we waited to pick up Carwyn and the passengers. This was the fuelling spot too, where we fuelled up about 1,000 l of the possible 5,000 l she could hold. The boat used about 300 l of fuel per day; so, we refuelled every three to four days. We checked the hoses and the air pressure: if it was below 3,000 psi, we started the compressor. Sea water was pumped into the chum barrel and was flushed out to clean it, and fresh water was filled up, as and when the water tanks were empty—necessary for drinking, showering, and making hot beverages. Carwyn would be back with the food for the sharks (tuna sourced from the west coast) and that for the humans. Rations like cheese, salad, cold meat, tea, coffee, bread, and Mike's famous chocolate Afghan biscuits would be stowed in the galley, and other clean-up operations were completed, and we were all set for the arrival of the tourists.

When the tourists started to board, Maca and I were ever-present on the deck with a smiling face and a cold wet hand, ready to greet the tourists who had often been driving all night to be here, parking in the wharf parking lot, and some who had spent the night before in Bluff. Helping them onboard, often carrying large amounts of baggage as they had been travelling, a big smile always helped to calm things down. Unfortunately, again, my lack of foresight implied that I ended up getting a tattoo on my right hand the day before my fieldwork began and I needed to put antiseptic cream on it. So, after a couple of embarrassing soggy handshakes, I had to enlist the help of an awkward left-hand shake for the first few weeks.

Once everyone was inside and seated, I would unrope the boat again, detach the metal stairs, and secure every movable object on the deck, including the stairs themselves, the buckets, and other utensils with rope, and the boat would set sail from the wharf. Upon embarkment from Bluff, Mike would make a small introduction of the crew, the safety and comfort considerations of the boat, and the journey to Edwards Island. He mentioned Macca, the main dive instructor of the day, and Carwyn, who was in charge of all the food and refreshment for the day (and she was also an instructor), and finally himself, the captain, whose job was to attract the sharks and make sure that everyone got to see them, but most importantly make sure everyone returned safely back to Bluff. He then mentioned me, a shark scientist, and for a long time, Mike had no idea what I was exactly working on. Following this, a video would be shown on the big-screen TV, describing the layout of the boat, including the bathroom, the life vest, life raft, and what to do if feeling seasick. Then the video would go on to 'safety in the cage', including information about putting the feet inside the ropes and never to put arms or feet outside the cage.

Just after the video finished, in the first part of the fieldwork, I had already taken my position outside in the deck in front of the cage, when the weather was all right, and would refrain from engaging in anything particular with the tourists. I wanted to make myself visible and 'perform' my role as the 'shark researcher'; more often than none, people would come up to me and start talking. However, in the second part of the fieldwork, I had decided I needed to gather statistical data of the pre-dive and post-dive tourist attitude towards white sharks; so, I had to initiate immediate rapport with the tourists. The sooner I could start talking and connecting with the tourists, the better it was; often, one hour seemed to be not enough time to create rapport and take the data from most of the tourists, and also arranging with them to collect another set of data on the way back. This was especially hard if the waters were choppy, and over the months, I had figured out the best way to do it. Inside my suicide vest, I had my laptop, pens, dictaphone, GPS, click counter, and for the surveys, I had single survey sheets, set on two A5 size waterproof clip boards, with attached pens, which I could fit in my vest's main compartment, and could deploy immediately. I made a disclaimer about my research, and asked them if they would like to participate, which most people gladly did, except the ones who felt particularly seasick. After the initial data collection, I would go around the boat and talk to everyone, in a sense to allow things to happen more organically and talk to the individuals who seemed to show interest in me, my work, and/or sharks in general.

Seeing me play the part of the 'expert' on the boat helped the tourists to open up. I had to be always listening, noting down, and observing what was going on, and had to be serious about shark conservation, look the part of an adventurer with all his tools and kits of 'researching'. Many times, the tourists ended up mentioning, 'It is so great to see someone so enthusiastic about their work.'

As part of the performance and image maintenance, the tourists showed that they were interested not just to actually see the sharks, but in their ecology, and behaviour, and one of the most common accessories they carried was their camera, to capture images. They often asked questions about the sharks in the region, where we were going, and how much chance we had of actually seeing them; most of them came with realistic expectations about shark sighting. The knowledge that they had a scientist onboard also possibly made them feel that they were part of something bigger, and not just a touristic endeavour; hence, they wanted to help as much as they could.

While leaving for Edwards Island, on many a day Mike said, in reference to the waves, 'Once we cross the hills here, it should be a fairly smooth ride.' When the weather was fine, the sun was shining, and the sea was somewhat flat, everyone came out to the deck, and talked to each other, sharing nervous laughs. However, when the weather was bad, the waves crashed on top of each other, tourists often stayed inside the cabin, the faces of some of them having turned blue due to sea sickness. They held on to the tables to save themselves from falling over, and, at times, even went to sleep. Mike always made sure that at all times, the doors to the deck were open to allow fresh air through. And if the tourists were too seasick, they were instructed to go and sit on the seats out on the deck to get fresh air and encouraged to look at the horizon. But often, nothing would help, and they would start vomiting, sometimes in the bathroom, sometimes over the side of the moving boat, and sometimes on the deck itself—which Maca and I had the pleasure of cleaning up. This is where the Vicks lemon menthol, tucked in my vest, helped. Then there were the ones who would always be out on the deck no matter what the condition, moving with the heavy waves, drenched with the splashes and rain. I would be also generally out, and every time there was an unruly wave or splash which would get us wet, we would burst out laughing.

As we got near the island, all the tourists would come out to see it. It would look isolated and even primitive, and the seals would be a sign that is the domain of the white shark. The setting itself created a great backdrop of a once-in-a-lifetime adventure experience. Mike would decide which side he wanted to anchor on, depending on the tides, winds, and the experience of shark sighting from the days before. As soon as we anchored, Macca and I

Tuna being used as bait, 2017

Raj working on the boat, Soosan Lucas, 2016

Leaving Bluff when the weather is not great, 2017

Leaving Bluff in the morning, 2016

The humans on their way to meet the sharks, 2016

would start getting the cage ready for deployment. The cage was secured on the platform behind the boat, attached with a heavy belt passing through and tied with another rope attached to the railing. First, I would open the ratchet, with which the belt was attached. Once done, the primary rope attached to the cage would be untied from the railings and re-attached to the crane hook; this was crucial, as otherwise, the cage could fall in the

water. Once it did fall in the water, and Carwyn had to dive to attach the cage with the crane line, for it to be lifted. This came as a shock to the tourists watching; as Carwyn tells, the tourists were mortified to watch Mike insisting 'a woman' to get in the 'shark-infested' waters to retrieve the cage.

Mike's daily spiel was a great source of knowledge for the tourists to learn about global and local shark population and biology. 'Worldwide, there

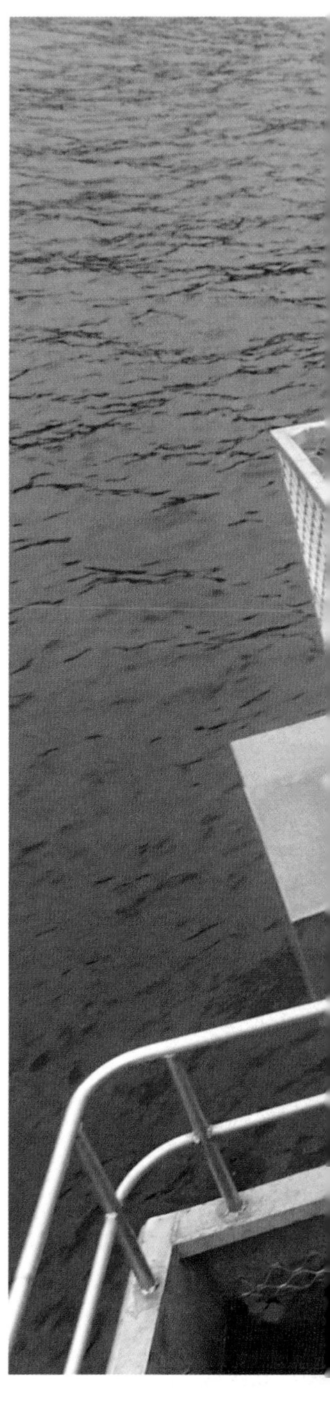

Above: Setting the camera underwater, 2016
Below: The cage being attached with bolts to the back platform, Soosan Lucas, 2016;
Facing page: The cage being lowered into the water, Soosan Lucas, 2016

are only 3.5 thousand [adult] sharks, so there's not a lot of them. Those figures are compiled in places like South Africa, Australia, the US, and New Zealand [he included Mexico later]. We're pretty sure of the number of sharks in New Zealand; we know that in these waters, each year, there are 120 to

The camera under the water, 2017

140 sharks, which is a lot of sharks in a very small area. Everywhere else in the world, they are solitary animals; so, there is a plus or minus 10 per cent than what we expect the numbers of sharks globally to be. The shark liver forms one-third of its weight; in fact, till they were protected in New Zealand, they were hunted for three purposes: for their liver oil, fins, and jaws. Here, for a four-metre [13 feet] shark, the jaw would trade for NZD 10,000–12,000,

but you couldn't eat the meat, because being our top predator, the sharks were very high in mercury.'

He would continue saying, 'The numbers are declining very gradually each year—that is one of the sad things about it, because what we can't afford to do is lose an apex predator. Thank goodness, they are protected, but unfortunately, there still are some local Stewart Islanders, who have been fishing them for years and still think it's fun to kill a shark and hang the jaw in the garage or at home.

The sharks have large black eyes; they primarily hunt through visual means; so, if they see a silhouette on the surface that could be prey for them, how do they know that it is edible? (Opens the question to the tourists) Bite them? Yes, through biting it; they don't have hands, and so they test by biting. That's quite recognizable; some new sharks we have never seen before, and sharks which have probably never seen cages before, come and attack the cage, and even the boat. That's why our boat gets smaller each year [everyone laughed]. But once they understand it's not edible, that's the end of that, and they don't try it again. They hit their prey at such a high speed, that they leave the surface of the water and breach, and the prey dies immediately. The last thing a shark wants to do is to fight their prey, because there's a chance of injuring themselves. So, once it has killed its prey, it eats it. Sharks love high-fat diet, if they are, for example, eating a carcass of a whale, they keep on eating consistently, and they can eat their own body weight. They then break that fat down into liver oil.'

Once the cage was attached to the crane, it was lowered into the water, off the platform, and then was attached to the platform with bolts. Hence, it was considered to be 'part of the boat'. The next part of setting the cage can be frustrating if the waters are choppy. The boat moves at its own speed, as does the cage, and it can be a delicate process of getting both the bolts right into the socket, considering how heavy the aluminium cage is. Even after doing it for months, I still used to get anxious with one leg on the half-submerged platform, and the other on the loose cage, trying to fasten it, when I knew there probably were big white sharks swimming below us at that very moment.

Next, I would put the generic blue and white dive flag on the top of the boat. Mike had asked the Ministry of Primary Industries to provide them with their own flag, blue and white with a shark on it, but they never got back to Mike and the team, and so, he used the normal diving flag. I had to deploy and set the underwater camera linked to the TV in the cabin. The TV screen became the window to the underwater realm, primarily important to spot

A human-feeding frenzy, 2017

The platform at the back of the boat at Southern Isle, and the cage is considered to be part of the boat, as it is attached to the platform with nuts and bolts, 2016

the sharks, and Mike had plans of having high definition cameras, to get footage that he could sell to the Discovery Channel. So now, the underwater space in front of the camera was another stage for the act to occur, and here, there was only one actor of interest—the white shark.

Jim Corbett, the famous British hunter and naturalist (Corbett, 1944, 1948; 1959) had been known to have spent thousands of hours on top of his treetop hut to observe and even kill man-eating tigers in India. My treetop observation platform, which was sometimes too close for comfort, was the back platform of the boat and the cage top. While being on top of the submerged cage (and the cage platform often kept me busy, because I had to attend to the tourists getting in and out of the cage), it did provide me a strategic position to observe the sharks and the human–shark interaction. Any research that incorporates situations where the observer has to be in close proximity of large predators, always brings its own challenges, particularly in case of a researcher who is not a confident swimmer, working inches away from 12–18-feet-long great white sharks and is petrified of them to begin with.

I had to remind myself of the perils of complacency, when often I would put my hand inside the water and outside the cage to take a perfect photograph of a passing shark, forgetting the most important dictum in the cage diving practice: 'It is not the shark that you can see is what you should be worried about, but the shark you are not seeing.' A rogue wave, a bump from a tourist, or even a bump from a shark can very easily drop you over the side into the water. As a matter of fact, I was told by the Shark Experience team that it did happen on the boat of the other company (exact date not known, as no media reporting could be found), when a tourist fell over the side and there were sharks around in the water, but thankfully, he was pulled in immediately. Then there is also the off chance that a shark actually bites onto my camera handle submerged in the water; it never happened to me, but I did have two incidents when sharks showed interest towards it and came in to investigate. Thankfully, I was able to lift it out of the water immediately, as I was not willing to test out my strength against the white shark in a GoPro tug of war.

To create objective data of the daily environmental conditions and human–shark encounters, every morning upon reaching, I took the GPS positioning of the first encounter with the sharks, anchor dropping, weighing time, and water temperature, because when the fieldwork started, for the first two months, there were almost no sharks in the region, and there was a belief that it might be related to dead sharks in the area, producing ammonia, and in effect, keeping other sharks away. For the first two months of the fieldwork, I recorded basic chemical indicators of marine water health (i.e., nitrate, nitrite, and ammonia) to document any observable indications of such claims with rudimentary chemical kits. Furthermore, I took regular photographic and video data to record the varied states of Foveaux Strait. To analyze the GPS data, the area around Edwards Island was divided into four quadrants and the coordinates of the first shark encounter everyday were plotted to explore any observable pattern of the presence and/or movement of the sharks in any particular quadrant of the boat.

On the boat (De Leon and Cohen, 2005; 'Ethnographic Interview,' 2017; Gobo, 2008; Nielsen, 2014; Schensul, Schensul, LeCompte, and Ph.D, 1999; Spradley, 1979), while interviewing the tourists, immediate rapport needed to be created upon embarking on the journey to Edwards Island, which was about 20 km away; going about 23 knots, it took us about 45 minutes to an hour to reach. Three elements really helped in facilitating the rapport, making sure I was present at the exact time at the beginning of the trip, when Mike introduced all the crew to the tourists, the eye contacts made, smiles, and the acknowledgement that I am the 'expert' was important. As Mike mentioned with a smirk, 'If *yous* people have a more complicated question about the sharks, just ask Raj!' Furthermore, Vicks honey lemon cough drops helped some of the tourists with seasickness. Finally, once I established that a tourist was deeply passionate about sharks, and was willing to interact with me, I showed the megalodon tooth I always carried, and the smile on their faces meant that I was successful in creating a strong bond, be it transient.

Besides the interview, I conducted surveys to quantitatively explore the effect of cage diving on tourist attitude (I shall discuss the major findings in later chapters) towards white sharks after encountering them through cage diving. Furthermore, the survey also explored any relation of attitude towards white sharks with scuba diving and academic education level, and other demographic factors. The initial survey was introduced before the encounter, at the time of the outgoing travel, as once the boat reached the destination, I had to attend to my duties of the daily operation, being a participant observer and one of the crew.

I created some empirical studies for observing the effect of cage diving on the behaviour and ecology of the sharks; however, it is pertinent to note, the reliability of the ecological data may not be the strongest, because of a lack of research tools, the inability to continuously focus on the behaviour and the short scale movement of the sharks due to the engagement in other responsibilities of the boat, and lack of experience in such methods of

The shark, Rocky, around the cage, 2017

data collection. However, methodologically, these attempts are significant to note for creating future direction in research, where social scientists can benefit in engaging with ecological and ethological methods, while delving with human–animal and human–shark research. Furthermore, it notes that any social scientist who wishes to conduct such interdisciplinary methods should then ideally take up basic training in ecology and its research methods from experts of the discipline.

To create demographic data of the sharks coming to the boat while I was in the field, I used the ecological research method of photo identification (Buray, Mourier, Planes, and Claus, 2009; Gore, Frey, Ormond, Allan, and Gilkes, 2016; Graham and Roberts, 2007; McKinney et al., 2017). Not only is this method useful to assess the regularity of individual sharks in a region, but also for assessing general short scale movements of species in an

Photo identification sheet for the male sub-adult shark named Rocky, 2017

area, and for assessing the effect of human activities, such as cage diving, on the movement and behaviour of sharks.

This method has some advantage over tagging, which is the other major form of shark identification method. Primarily, and most importantly, photo identification is safer for the fish. There is a debate (National Geographic, 2014; Shark divers, 2014; Shark Research, n.d.; Welovesharks, 2017) about the potential risks that tagging in general may pose for the lives of the sharks, let alone certain operations where, in undertaking tagging, sensationalized optics are created of hoisting sharks out from the water onto a ship platform (White Shark Video, n.d.). Indeed, there was a rumour (among the cage operators) that a team of New Zealand scientists may have caused the death of a shark due to this kind of an invasive procedure (this research has no evidence of this either way).

Scars on skin of the sharks, which is not always the best point of reference to identify sharks, because their skin heals quite easily, 2017

Photo identification is also safer for humans, especially if dealing with highly predatory species such as white pointers. Furthermore, tagging operations are extremely expensive and resource-heavy endeavours. Finally, in tagging, the extraneous variables are quite hard to predict and control, like the environmental factors, and the behaviour of the sharks, who have to come close enough for the harpooning of the tags on their fins. Photo identification, on the other hand, is non-invasive, comparably cheaper, the sharks can be much farther away, safer for the sharks as well as the humans, and it can enlist help of local scientists who capture photos of sharks for creating a global database.

This method has been exceptionally useful for sharks and had been used to create a data base for varied species, such as Lemon sharks

(Buray, Mourier, Planes, and Clua, 2009), Basking sharks (Gore et al., 2016), Grey Nurse (Barker and Williamson, 2010), and Whale sharks (Andrzejaczek et al., 2016; Arzoumanian Z., Holmberg J., and Norman B., 2005; Brooks, Rowat, Pierce, Jouannet, and Vely, 2010; Graham and Roberts, 2007; Holmberg, Norman, and Arzoumanian, 2009; McKinney et al., 2017). Photo identification has been extensively used in white shark research. Researchers have used the method to investigate the movement and long term population trends of white sharks in California (Anderson, Box, and Goldman, 1996; Anderson, Chapple, Jorgensen, Klimley, and Block, 2011), for concordance of genetic and fin photo identification, and population estimates in South Africa (Towner, Wcisel, Reisinger, Edwards, and Jewell, 2013; Gubili et al., 2009; Hewitt, Kock, Booth, and Griffiths, 2018), for annual re-sighting at Guadalupe, Mexico (Domeier and Nasby-Lucas, 2007), and have even attempted at creating a large global photo identification data base (Andreotti et al., 2014).

Photo identification has its own challenges. To begin with, there is an assumption that all species of sharks have distinct identification patterns among individuals, which is not true in all cases. There is a chance of re-identification of individuals, because shark skin has a great healing capacity and scars and other identified marks may change dramatically over time (Marshall and Pierce, 2012). Also, high quality photographic instruments are quite expensive. Furthermore, as opposed to land animals, the only time you can properly know the position and movement of the shark is if they are about 50 feet in front (approx.) or 30 feet under the cage (because of visibility underwater, and glare on the surface). Even in the case of ocean mammals, they come to the surface—be it after a long interval—to breathe, but for sharks, there is no guarantee that they will surface. The shark might be 50 feet away, but underwater, and from a boat deck, one would have no way of knowing where they are and what they are doing (to note if the visibility of the water is good, drones could work, otherwise it is of no use; on the same token I desperately wanted to try out one, but could not afford it for the project). Furthermore, due to the camouflaged skin and shape of the shark, they may be just under the surface, but we may not be able to see them. This is a problem when we are dealing with a method that relies on visual awareness of the animal in question.

I used underwater photography to create a visual demography of the sharks, which I created while standing on top of the submerged cage and the cage platform using a 5-foot camera stick and action camera attached. As mentioned, during the second part of my fieldwork, my primary duty in the boat was helping the divers into the cage, which meant securing the air hoses, commencing final check-up of the diving equipment, and directing them to their position inside the cage. As much as this helped me to be close to the water while photographing the sharks, the opportunity was intermittent while taking care of my duties. Another significant lesson learned during this experience is the necessity of a display available to the photographer showing the image the lens is capturing, as they can then see where the sharks are and where the camera is facing. Often great photographic opportunities were missed due to the inappropriate angle of the camera.

Even though the investigation of cage diving's effect on white shark behaviour are limited, Bromilow did a very interesting bachelor's research on the feeding behaviour of white sharks around the cage (Bromilow, 2014 University of Michigan, Department of Ecology and Evolutionary Biology). She categorized shark behaviour with bait in the water into five phases, and made her observations accordingly. Phase 1, Investigation: Slow Swimming; Phase 2, Pursuit, Jump: Horizontal Lunge, Vertical Lunge; Phase 3, Prey Capture: Horizontal Surface Grasp, Vertical Surface Grasp, Lateral Snap; Phase 4, Prey Handling: Carrying, Thrashing; Phase 5: Feeding; and Phase 6: Release. Inspired by her method, I used similar tactics to document the behaviour of certain sharks around the cage. For a few days, I was able to document the behaviour of one juvenile shark, whom I called Rocky, and see how his behaviour changed over time as he got accustomed to the cage and the bait (un-published data). I discuss my findings in later chapters.

The Diving of Cage Diving

Once we were set up, it was time for the diving part of the cage diving. There were many sensory stimuli that worked on a diver when they got in the cage, especially on new divers. Not only did they have to learn how to breathe underwater for the first time, but endure the cold water, habituate with the dynamics of the cage space, and also control their fear of the shark that was just outside the cage. Admittedly, this could be properly described as a hell of a place to learn underwater diving for the first time; however, I would argue that the adrenaline in their body would have made the cold more bearable, at least in the beginning.

There were three forms of registration fares to the shark experience expeditions (as of 2018). First, sightseeing (NZD 299), for the tourists who did not want to get in the cage and just wanted to watch the shark and the environment from the top, was the least expensive. Second, there was the more expensive registration for trained divers to go in the cage (NZD 499), and third, there was the most expensive registration for tourists who wanted to dive, but did not have scuba diving certification (NZD 568). Before anyone could get in the cage, the non-certified divers had to be trained in basic underwater breathing. Technically, they were not scuba diving, because they were not 'self-contained'; rather, it was assisted diving, with air supplied to them through hoses attached to the compressor in the galley.

Mike had to make sure that all the wet suits and other peripherals were suited to the tourists. Generally, we had suits from the previous day on the

Divers being trained to breathe underwater, 2017

deck; if he needed new ones, he got them out of the galley. All the crew were dive instructors; so, one or the other got in the cage with the tourists to help them through the process, generally though it was Maca who performed the duty of the dive instructor. The process started with him having a 10-minute theory class with the trainees. Maca explained the basics of underwater breathing, and also the few exercises, which the divers had to perform, like clearing the mask, regulator, and ears. Once the theory was done, the tourists were taken to the deck, where they were fitted with the masks, and shown how to use the regulator. Finally, after the tourists got suited up, Maca would get into his dry suit (because he would be in the water for a long time) and get in the cage. I would be standing on top of the cage and help the tourists into the cage—one at a time. Over the months, I had my own system to make sure the hoses attached to the regulators did not tangle, because the regulators could easily come off from someone's mouth; furthermore, if I had to communicate with the divers by tugging on the hoses, that would create confusion.

Mario, my dive instructor, used to tell me that women are better at scuba diving, because they can be calmer, they generally have less muscle mass, so the air lasts longer, and (with a wink in his eyes) they look better than men underwater. Mike mentioned a similar idea that women seem to be better at diving too. However, in the months I have been in the cage, I noticed that for some reason, there were more women who had trouble to get underwater than men. I have no explanation for that, except possibly the stereotypical 'masculine' gender role in Western society may have prompted the men to keep going even if they were petrified. When new divers had trouble breathing underwater, and the instructor in the cage could not handle it, Mike stepped in. Being a long-time dive instructor, he was always calm, but firm; he didn't rush anyone, but was very precise in what he wanted people to do, and reminded them that Maca, or the instructor, can't be in the water the entire day for them. 'Step into the cage, put the mask on, put face in the water, put the regulator in your mouth, make a firm seal with the lips, take deep breaths; once you're accustomed, just walk right in,' he would say. Often, certain divers would get out of the water in high stress and exclaim that they could not breathe. As Soosan (with no diving experience) explained, 'Once I got in the water, I felt I was drowning...and was about to get up. Carwyn was really helpful, and she calmed me down. I told myself, I'm a Viking Brit! I can do it! And I got underwater, and slowly calmed down.' No matter what, I never saw anyone not complete the dive, because Mike made sure all divers saw the sharks at least once. Mike knew if the divers saw the sharks at least once (if the sharks were around), they would feel that the trip was worth it; that was great for the divers—hence, great for Mike's business.

Even if the water was cold, most people did not mind it, at least for a short span; however, when they came out of the water and faced the southern winds, that was when the cold set in. Because of this agonizing cold, the white faces change colour to blue (purple for others), and deep body shakes would set in. Mike did not let tourists stay underwater for a long time when there were no sharks, because it could have led their core body temperature to drop, which in the worst of cases may lead to hyperthermia. One particular day, I was about to have such a fate myself.

Date...not known
8 pm
Bluff

I wanted to make it to one hour. But as I got up and tried to take the suit off, I realized my fingers had no feeling, neither did my lips. I changed in whatever wet clothes I had (not being prepared for the dive that day). I drank a hot cup of tea, but could not get warm, all I was hearing my mind say, 'I am cold, I am cold', while I saw the cup in my hand shaking. After we anchored at the wharf, I jogged back to the cottage to get warm, but it did not help. As I got in, I got out of the wet clothes. I was still shaking. I got under the covers and passed out, and my last thoughts were, 'I am cold, I am cold...'.

Divers getting ready to dive, 2017

Looking from the cage to the world above, 2017

The Cage

All the material equipment involved in the diving played a significant part in the success or failure of the operation. The significance of the cage is evident from the name of the practice, which is commonly called 'cage diving', and less so 'white shark diving'. Furthermore, the inclusion of the term 'white shark' is also important to note here. What would happen if it was called 'fish cage diving'? Would the name have such an emotional charge as 'great white shark cage diving'? The cage is the contact zone, where the humans and the sharks have embodied encounters with each other. As I was helping the divers set up the regulators on their suit, it was also my responsibility to instruct them how to enter and exit the cage properly in a fast and safe manner.

Not only is it a distinctive geographical existence that is located in a particular area of water, but a psychological one too. As a researcher, every time I was in the cage, the windows were as much a physical attribute as they were screens into the 'outside' world—the world that is potentially dangerous for me as a human being underwater, surrounded by white sharks. It was a psychological barrier between the worlds of the

The cage, 2017

sharks and the worlds of the humans, no matter how intertwined it was. So much so that a mere finger out of the cage is strictly prohibited, the divers were reminded by the blue sign board in the cage, 'Keep all body parts inside cage at all times', and often when they got on the boat, they would look at the cage minutely, and touch it, even take a photograph of the sign. This sign was a symbol of the 'danger' involved in the activity they were taking part in and had bragging rights. Interestingly enough, the sign is not red (in which generally all danger signs are) but in blue, possibly because Mike did not want it to be too alarming, or even remind people of the colour of blood.

The environment inside the cage is to an extent devoid of the unpredictability of the environment outside. It was cold and hard to touch, with jagged edges. In this embodied experience, this geographical space—12 ft wide and 5 ft tall is 'safe' and suspends the potentiality of humans being under the sharks in the food chain, if the humans adhere to the rules of the sharks and the cage. The cage is attached by the bolts to the boat and, as the big volume of the boat moves with the waves, so does the cage. The only way it seemed we could stop this was by attaching a rope from the bottom of the cage to the hull; however, it was risky for the sharks, as they could get tangled in that space.

Both: The cage ladder and rope loops at the bottom for diver to put their legs in, 2017

In the 2016 season, Mike increased the size of his cage, so that five tourists could fit in it (as opposed to four before) at a time. This large size of the cage, first, allowed more divers to see the sharks (because the appearance of the sharks can be fleeting), and second, this set up allowed families and friends travelling together to get in the cage at the same time and have such an intense sensory experience together. The spacing between the rods of the cage is also very important, especially the windows. Too small a space and the tourist may feel disconnected from the environment, and not able to take photographs of the 'free' sharks (curiously enough, when the tourists are the ones in the cage). Alternatively, if there is too large a space, it may cause injury to the sharks or the people if the shark accidently wedged inside the cage.

The railing on the top of the cage in Mike's boat is also a very interesting apparatus. It served two primary purposes. First, the sharks could not accidently enter the cage from the top, and second, it ensured the divers could not fall over the cage while getting inside it. It also provided opportunities for the divers to stand on top of the cage, hold the railing, and look at the sharks from inches apart—safely, which they could not do from the boat, or if the railing was not there. This also provided the avenue of close observation and video documentation of the sharks for a researcher.

Diving Equipment

The equipment essential for creating a temporary sustainable habitat inside the cage are the 'on-body' diving adaptors. I use the term 'adaptor', as this equipment helps the humans to 'adapt' to the marine environment for sustained periods of time. I divide these adaptors into two major categories: the breathing adaptors and the sensory adaptors.

As I mentioned, technically speaking, the diving involved in the cage diving operation is not scuba, because it is not self-contained; rather, the appropriate term is 'surface supply'. The compressor in the hull of the boat creates compressed air, which is supplied through the hoses, and then to the regulator to the divers. If the compressor is fully filled, there is enough air to last one person more than two days underwater. The hoses also play

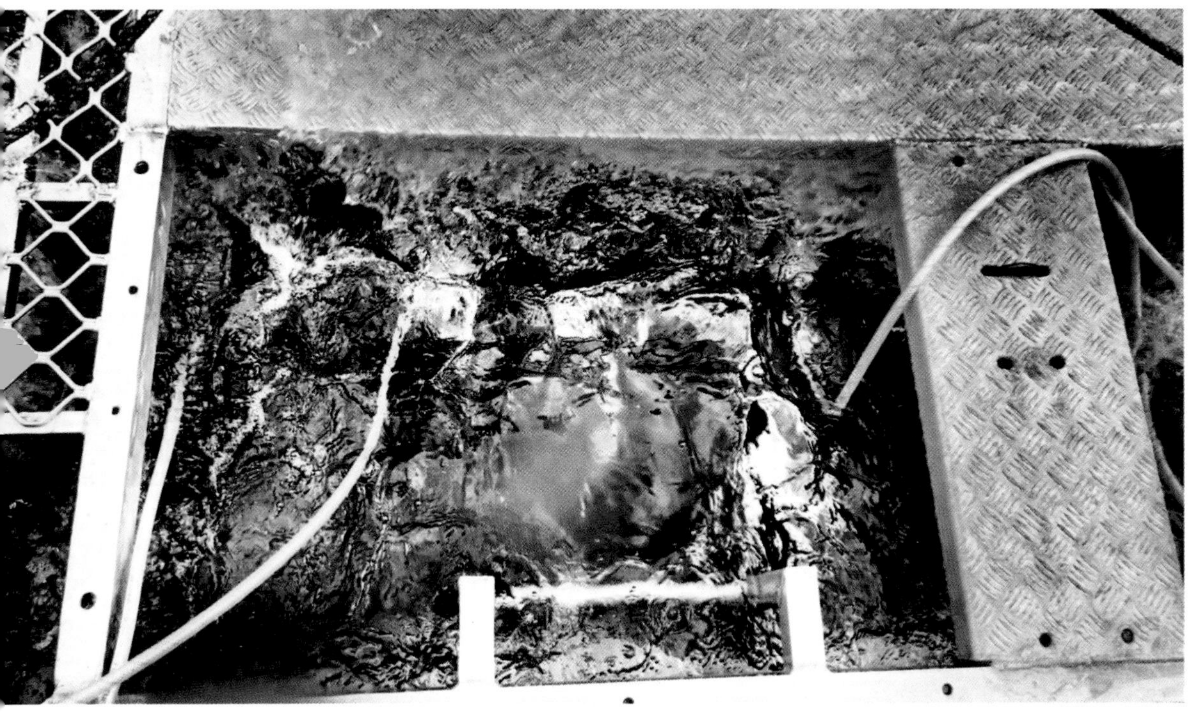

Divers in the cage, 2017

another important role, as a communicator. Traditionally, in surface supply operations, the breathing hose is part of the umbilical, which is also used for communication done through pulls and bells (short tugs), and a lot of detailed communication could be done through it. However, in the boat, the primary message I used to communicate with the divers was 'getting out of the cage', which was done with strong pulls.

The sensory adaptors were the masks/goggles, the wetsuit, and gloves and booties. They played a very important part in the successful encounter, especially for non-experienced divers. If the goggles did not fit properly, and there was a leakage, new divers often tended to panic and resurface. And in many instances, this happened at crucial moments—when the sharks were around. The same can happen if there is fogging in the glass, because the diver is not being able to defog them properly. The wetsuit is an essential part of the equation too, protecting the diver from the cold, and from

(Left) Wet suits hanging to dry, 2017; (Right) Goggles and gloves stored away, 2017

Maca the shark wrangler throwing the bait in the water, which helps to attract the sharks close to the cage, 2017

scratches and bumps in the cage as it moved around in the water. When divers got out of the water and would be shaking because of the freezing cold, one trick always worked to warm them up—that was emptying the booty of the sea water and pouring hot water in and putting it back on.

One final part of the equation were the weight belts; the weight harness was always more stable to control, as it did not fall off the waist, but it was limited. However, no matter how much weight was put on, the divers would always bounce around the cage if they did not put their legs through the ropes attached to the bottom of the cage, and crouch down, a fact, which was often overlooked even by experienced divers and paid for by bumped skulls.

Attracting the Sharks

There were two parts to attracting the sharks, so that the tourists could see them. It was to bring the sharks in the close proximity of the boat, and then bring them close enough to the cage, so that the divers could actually see them. There was no benefit if the sharks were around the boat, but the tourist could not see them from the cage. Therefore, attracting the sharks was done through chumming and baiting respectively, and the food source used in both the cases was tuna. As Mike said, 'It has a very high oil content, and it is a natural food source for the sharks. We mix it up with salt water in the barrel and try to secrete the oil from it. We tip it over the side, and what we're trying to do is to send an oil slick out with the currents. If you look at the back, you can see a shiny flow of the oil on the surface. We are trying to send it 2 or 3 km out into the ocean, because the sharks don't just sit here; they are travelling to the deep waters and the island all through the day and are bored to tears, and what they're waiting for is someone to entertain them like us. So, the first thing you should do when you see a shark is to get divers in the water, for two purposes: one, to entertain the sharks, and the other, to utilize the sharks while they are here, because you don't know how long they might stick around.' So, in that case, entertaining the shark may be considered a performance by the shark wrangler to entertain their audience, the shark.

Realistically, the best bait for attracting white sharks is probably by the flesh and blood of seals, as they are the preferred preys of white sharks due to their high fat content. However, hunting seals is banned in New Zealand, and most other places, for that matter; and the visuals of that may be alarming to the tourists—the tourists wouldn't necessarily see the hunting, but they would see the seal body parts (even though I don't really think they would mind if they got to see the sharks, which I elaborate on in a later chapter). Interestingly enough, Mike and team did try with other meat like beef, but it never seemed to work effectively. And allegedly, people had tried every different kind of meat, so much so, there are stories that an entire carcass of venison was put in the water to attract the sharks, but it did not work. This is important to note, because it directly contradicts the common notion that sharks are attracted by human blood or the blood of other animals that are not part of their regular diet.

The New Zealand laws are the strictest in the world regarding bait restriction, and the operators were allowed to use only one piece of bait in a day, if the bait was lost and the shark had taken it, the operation could not use any more bait for the day (Bruce, 2015). To make sure the

Grinding the tuna in the chum barrel, which is drained in the water as chum for attracting the sharks to the general area, 2016

The bait, ready to be deployed when the shark is in close proximity, 2017

operators stuck to the rules, the Department of Conservation often sent secret observers to the operation in guise of tourists on the boat to check. Furthermore, operators were not allowed to use decoys of seals as they do in places like South Africa, which creates the spectacle of white shark breaching, because the material of the decoys may be harmful to the sharks, and it adds to the fear that the sharks will be trained by this activity and be inadvertently harmful to the humans in the water. For most of the days, in the morning we would load two to three big tunas on the boat, which we used to get from special fish suppliers from Invercargill. Most of the tuna would actually be mashed up in the mincer, mixed with sea water,

Carwyn wrangles the sharks as the tourists look on, 2016

and dumped outside the boat as chum. The heads were usually used as bait, the wrangler would tie a rope around it and throw it overboard, trying to entice the shark to come close to the cage.

Sometimes, when watching from the surface, one may just see a silver streak passing below the surface, which is only shown if the shark is onto its side, showing its belly. But this streak can be—and often was—confused with a passing barracuda. Most cage diving operators and shark spotters

recommend that polarized sunglasses are to be worn while looking for the sharks to decrease the sun glare, and for better sub-surface visibility, but personally, for me, the darkness of the sunglasses seemed to make the spotting harder.

The upper deck was restricted to the tourists when the boat was in movement and there was no smoking on the lower deck, but the upper deck was the best spot for shark spotting and for a smoke. Spotting a 15-feet shark seems like a straight enough task in theory, but in practice it is quite different. The eyes should be observant to everything happening on both the surface and sub-surface in four directions—near and afar. White sharks are not white on the top, indeed anything from blue-grey to almost black. Particularly around the island, because of the kelp floor, the sharks were practically invisible. Tourists often asked, when the first time you see the sharks, do you see a fin like *Jaws*? Or do you see them moving underwater? The truth of the matter is, there is no specific way to know when the shark is nearby. There are a few elements, which have to be considered together. Even if the sharks are near the boat, chances are that we may never see them if they are swimming deep.

Alternatively, all of a sudden, we could see a fin in the distance, slicing through the water. This created a further opportunity for illusion, as when the sun sparkled, the ocean ripples mimicked the shape of shark fins. One also had to keep one's eyes open for the surface movement of birds, and when watching underwater videos, be a keen observer of the other smaller fish around, as an absence of these fish may indicate the presence of a shark. But actually seeing the sharks for the first time is an awe-inspiring experience, especially if it is a big shark. Sometimes, because of the sheer

Spotting a white shark underwater can be tricky, considering their dark top colour, and is dependent on the weather, 2017

size, I have had tourists asking me, 'Is that a whale?' Indeed, it is not the length of a pointer that catches most tourists off guard, it is the girth. The rate of increase of the girth of a white shark increases with age than the rate of increase in length; so, a 13-feet shark and a 16-feet shark would look substantially different.

People would get ready and suited up, with camera in hand, waiting for the sharks. The sharks can be there when we reach the diving spot, and alternatively, hours would go by and there would be no sharks. The tourists, half tired, may have sat with a cup of coffee and suddenly, there would be a big eye looking at them through the lens of the boat's underwater camera; this would be followed by a loud cry...'Shark!'...and everyone would rush out to see them. Initially, the trained divers would jump in the water. I would make sure that everyone got in the right place, so I could pull them up to put the next ones in. Once the trained ones were done, then it would be the turn of the untrained divers. This must be done swiftly; the main aim, as mentioned, was that everyone got to have the experience at least once. Once everyone had seen the sharks at least once, and the sharks were still around, then the people could take as much time they wanted to be underwater.

Anglers in the UK noted that the smaller fish were sprinters, like teenagers, and took chances with every bait, while the matured ones were more careful (Bear and Eden, 2011). Such was the case with the white sharks too. The younger ones did seem more aggressive and went after the bait with more voracity than the adult ones. According to the team, white pointers behaved differently with age; the matured females were even more careful while dealing with the bait and coming up to the surface. They would have an entourage of young males with them, and often the males would come up first to 'check out' the bait, and then the females, and the others would then give her space. The younger ones were just like human teenagers, more active and less fearful, and came closer to the cage and took more bites at the bait. Larger sharks seemed to be more conscious of their actions, more aware of the divers in the cage, and everything that was going on, undertaking deliberate actions. They seemed to go deep and then swim up to get the bait. If they were not able to get the bait, they swam back deep. All this added to the delicate embodied experience that is shark wrangling, where the body of the sharks and the humans merge together through the bait and its line, but at the same time, the worlds and physicality of the two species are carefully kept apart. The wrangler tries in all his ability to 'think like [a] fish' (Bear and Eden, 2011, p. 336), so that he can make the fish perform as he wants to. It is frustrating and requires a lot of patience, and the individuality of the sharks is as significant in it as much as the individuality of the wrangler; so, the shark and the wrangler affect each other's behaviour and tactics of interaction.

Markuksela and Valtonen (2019), examined match fishing (i.e. competitive, timebound, fishing competitions) in Finland, and argued that non-human encounters in such practices could be characterized as partners engaged in multispecies agentive dance. 'Match fishing is a dance between a fish and an angler that takes place in a dance hall of waterbodies... In the 'dance-like' practice of fishing, the weather acts as a dance orchestra, providing music for the dancers. The rhythm of the music, in turn, orchestrates the circumstances of the dance hall, that is, the waterbody. It moulds the context in which both fish and anglers saunter... The weather above the waterline also shapes the weather below the waterline' (Markuksela and Valtonen, 2019, p. 355, 362). The weather and the waterscape were similarly affective actors in shark wrangling, as the wrangler had to dance with the rhythm of the sharks and the weather. Often, for hours, Maca (the primary shark wrangler) would stand in the stern of the boat with rope in his hand and wait, no matter if it was raining or no one had any idea if the sharks were around. It was particularly a careful undertaking in New Zealand, because as I mentioned, even if it is the primary method to attract the sharks close to the cage, the number of baits allowed to be lost was one, so if the shark took the bait, that was it for the day. And this could happen after hours of waiting: the shark would suddenly appear and go for the bait in the blink of an eye. This begs the question, why would a shark spend so much energy at getting a bait, which is not really a lot of food for them, sometimes half a fish head? Do they not understand it is not a live prey? This is unlikely, considering all their sophisticated senses. So why would they chase after it? Can I use the forbidden word when concerning with fish—'play'?

'When fishing, the angler does not necessarily need to know his or her dance partner beforehand. However, to "hook" with a partner in this dance, he or she must first be properly introduced to the prospective dancing partner. [At the same time] The angler can sense the fish, as an individual, and connect this sensation to the fish's species-specific attributes. In sensing like a fish, the angler is, in a way, becoming an animal...in terms of the fish's practical understanding...[on the other hand] the fish has the prospect of becoming aware of the angler's intentions and practices, of learning how a man acts. Thus, the fish is acting not on but with human anglers...[for example] it can choose whether to strike the lures offered. Thus, the fish has pivotal agency' (Markuksela and Valtonen, 2019, p. 358–363).

In the same sense, the shark showed pivotal agency in the dance that was shark wrangling. Peter Scott, the owner and operator of Shark Dive New Zealand, was considered to be able to make the sharks 'play' with the bait. Often, both the boats (that is, Shark Experience New Zealand and Shark Dive New Zealand) would be anchored close together, and they operated alongside, to increase the chances of both the operations spotting the sharks (I should note here, that I had contacted them to allow me to be a participant observer in their operations too, but they were not interested. Hence, even though I did observe them most of the days from a distance, I make no comment about their operation logistics, besides the methods which were distinctively apparent to everyone observing). I would be on top of the boat observing the sharks as they interacted with both the boats and the humans in them. For quite a lot of the time, the same shark who acted calm around our boat seemed to act more 'excited' in front of Peter's boat; hence, if there was empirical investigation of the close proximal behaviour of the sharks around the two boats, chances were that the boarders of the two boats would get somewhat different results.

The reason seemed to be that Peter left the bait in the water till the last moment where the sharks aimed at grabbing it, and it indeed seemed like it caused Peter to lose his bait more times. Mike, on the other hand, took a different approach; he wanted the sharks close to the cage, but did not want to agitate them. Hence, he instructed all his crew who handled the bait to take it out of the water well before the shark had the opportunity to lunge for it. Here, it is important to consider the term 'excited' as opposed to aggressive behaviour. Noted primatologist Jane Goodall stated, 'Animal behaviour research will never be a hard science. Science can be arrogant—it pushes aside a lot of fascinating things because it does not have the tools to study them' (Klein, 2015, p. 64). Such is the case with white sharks here, because we don't understand the behaviour of the sharks; basing on the aesthetics of the situation, we make an assumption of their intent. Goodall recalled an important learning experience, 'Naïve me, not knowing anything about science, I'd written a piece for my thesis that said that Fifi, a chimpanzee, was jealous of her sibling. My doctoral advisor said, "You can't say that because you can't prove it." He suggested that I say, "Fifi behaved in such a way that had she been a human child..."' (p. 68). If let us say the same happened with our cat at home, we would say we are making them more excited when they jump out to catch the string, as opposed to them getting aggressive (on the other hand, a house cat generally doesn't predate on a human). And the term we use while explaining their behaviour with the string is 'playing' rather than 'attacking'. Based on my observations with the shark wranglers, the wrangling strategies, and their effect on shark behaviour, I would like to note that the level of excitement of a shark around a bait in cage diving practice is directly related to the suddenness of the action of bait retrieval and the proximity of bait retrieval from the body of the shark.

However, even Mike lost baits at times. This was primarily because white sharks are unpredictably fast, even if you know they are coming for the bait. In any case, it is certainly evident that in wrangling, not only the inclination of the human wrangling affects the behaviour of the sharks, but how an individual shark is by age and gender, and how they are on a particular day also affects how the wrangler wrangles. Mike believed that it was not even the bait that the shark was most interested in, when he or she was in the vicinity, but rather the activities in the cage, and it were the bubbles and movement in the cage that attracted the sharks and kept them around. Often, when the sharks were accustomed to the boat and comfortable, they 'hung around' and kept on making circles around the boat, even if there were no bait around.

For me, wrangling was an especially exhilarating proposition; any time I did it, I was stressed and excited at the same time. To see a 15-feet white shark coming towards half a tuna head hanging from a short rope at the end of which I am standing is a sensation like none other. When you are in the cage, and observing them most of the time, the connection can be quite distorted, not so much when you are holding the bait line. You can see the focus in her movement which is being affected by the flicker of your wrist, and in effect, the position of the bait. As your body is tensing up, relaxing, pivoting, and crouching down to get balance, the effect on the shark can be seen demonstrated by a small flicker of her giant fin.

It is empowering and, at the same time, intimidating, knowing the capability of the being who is communicating to you through her movement, and I am communicating with her by the flick of my wrist—simultaneously, avoiding any closer physical and tactual communication at all cost, making sure the rope is not too tight around my wrist, so I can let it go if the shark happens to get hold of it. It is an affective embodied experience between two beings, who are attuned to the minute physical movement of each other and is affecting the other and being affected by the other, a mutual inter-species communication—the effect and affect going both ways. As she passes by, you know that she is looking at you, and no one else. In moments like this, I had felt a deep sense of existential inferiority, by invariantly comparing my own feeble physical form, as opposed to her, whose body has been fine-tuned for 400 million years to be the best suited for her environment.

The season of 2018–2019, Shark Experience Bluff, New Zealand, was the only surviving cage diving operation. Also, in this season, still awaiting the final decision of the Supreme Court, the operators continued with 'No Incentive Bait Policy' (meaning cage divers were not allowed to use bait while attracting sharks to the boat) according to the changes to the interpretation of the Wildlife Act of 1953. However, personal correspondence with Mike Haines revealed that the chum was still acting to bring the sharks to the general area (which it originally did), and the sharks were, in most of the instances, within 16–20 feet of the cage, and sometimes, even closer, so the tourists could still see them.

The journey back after a day's work was usually quieter. Most people were sleeping, partners and friends keeping each other warm. There would always be someone on the deck, taking in the air (or the water for that matter, if it was raining). As soon as Mike called it a day, we lifted the cage out of the water, placed it carefully on the platform, ran the strap through the cage, and used the ratchet to tighten it up. The primary ropes were tied to the railing, and the cage was safe. The crane was closed, and the cover was put on, the underwater camera was pulled up, buckets tied to the railing, the 'do not enter' sign tied to a chain was again put over the step to the upper deck, and we were ready to head back. Immediately, at this time, I started to collect the remaining interviews or survey data that I needed, particularly the post-dive attitude towards the sharks. Thankfully, since I had informed

Returning to the Bluff wharf at the end of the day, 2016

the tourists before that I might take some more data from them, most of them did not mind, and were even happy to help. The longer I waited to take the data at this point, the less chance I would have of getting anything at all, because a majority of the tourists would fall asleep, especially if it had been a turbulent day.

Now, I would be sitting at the end of the boat, just thinking about the day, moving with the waves, sometimes dozing off, as everyone else rested in the seats. There would be the hum of soothing smiles, secret stories, and soft kisses, with only Captain Mike wide awake, taking us back home.

Furniture of Whale Vertebra

A few years before my *dadu* died, I knew that his time was coming, and so did he. He did not have any particular ailments, but told me, 'Well, something has to go wrong, I have to die some way.' I used to sit down with him and collect as many stories as possible, stories of his journeys, his fish, and his home. I don't remember my mother's face; she died when I was very young and all I remember is the night we cremated her in a bamboo forest near our house. Few months before he died, I got an invitation to present a lecture on empirical aesthetics in Bangladesh, while at the same time, my father had an art exhibition. I knew this was my chance to go look for our ancestral home. I asked *dadu* about the details of the place; he said, 'I will tell you, but how would you go there?' I said, 'I'll figure it out.'

When I reached, I realized that my father had also got excited about this proposal and had made arrangements for us to find it. So, my father, myself, and a few friends went on the journey to find it. Across the bright and green Bangladeshi landscape, we travelled until we reached the Meher Kali Bari (Kali temple) in Shahrasti; it is here that we were supposed to get information about our home.

What happened next is a story for another day, but we did find the house, which no one from our dynasty had visited for the last 70 years, and while returning, I got a bit of soil back with me. I presented it to my *dadu*, and that was the one and only time I saw tears rolling down his cheeks as he pressed the soil to his forehead.

The little shark and the ocean

In the calm town of Bluff, the soil, the water, and the animals connect everyone, but much like other documented interviews with old Bluffies (which is how the residents of Bluff self-describe) (Bremer, 1986), my interviews found that a lot of the old Bluffies echoed a sense of loss, a loss of camaraderie, of a vibrant lifestyle and a unique identity of being from Bluff. However, they still liked the isolation the place provided and were dissatisfied with the increase of foreigners who did not belong here and felt that white shark cage diving in New Zealand brings more problems than benefits to the community. I got a chance to talk to Mary, a true native of Bluff, as her great-great-grandfather was James Spencer, arguably the first European to settle in Bluff in 1823. She had been in Bluff for 71 years, and as I walked into her house, she was sitting on a stool made out of sperm whale vertebra, chatting with her friend Peter about a mako shark jaw that hung on her garden wall.

Mary told me the story of a place, which had seen its heydays, had now declined, and a place, which always had had sharks around, but not necessarily white pointers. 'The wharf was really big, and I saw the new wharf being built. We had a big rabbit factory, wool sheds, trucking business, taxis and buses, four pubs, and we even had maternity homes.' Peter, her friend, added, 'The main street had two butcher shops, two barbershops, two shoe shops, a veggie shop, and a bakery—she was a big little town. Ocean Beach was going, fishing was going, oystering was going—it was big, it really was. Rough and tough, they come from Bluff. The freezing works (meat processing factory) used to bring all the boats, and it used to get all the seamen ashore. Before, in the pubs there would be so many people, you had to fight your way in; now, there is hardly anyone. In the 60s, I saw 16 ships in the old wharf at one time; (the closure of) Ocean Beach (freezing works in 1991) killed Bluff.' There was a pride in the community, people helping each other, and ever available jobs, if one was willing to work. 'Everyone shared their gardens, when we went to church, someone would hand over lettuce or carrots, and Dad would come home from fishing and give it to everybody. We used to go down to the pub, there was a guitar and singing, and it was all good—that is no more there now. They had a big shed in the old wharf, and a town hall that's where people used to dance; it's all closed now. There were some real rich people in Bluff—millionaires, but they worked really hard for it. If you were willing and if you're able, you would always have work in Bluff.'

The economic downfall though is now glaringly visible in the community, 'Now, a lot of foreigners, all your relatives [Indians], and Chinese who come by have dairies [I smiled]. It is not the same, I could walk from here to the main street and not talk to anyone. All my kids have been doing the same job all these years, but they don't like it, and the young ones are smoking "P" (meth). They work, they drink, they smoke, and that's it; they have no outlook on life, and nothing to do.' At the same time, her friend pointed out, 'A lot of people like Bluff now, because it is peaceful. Where else can you get a cheap house by the sea. There's a lot of people from Christchurch down here since the earthquake.'

Despite having salt water in their veins, most non-fishermen have never encountered the largest predator around, no matter how long they have been in the area. As Carwyn mentioned, 'We had a dive club call Foveaux Explorer; we have dived here for 10 years. I have done over 200, and Mike over a 1,000, but we never saw a white shark. The only time we see them is when we use tuna to attract them.' A tourist asked, 'Wonder how many times the sharks saw you?' 'Well, that's a good question, that's why I never snorkel; even for paua diving, only dive,' Carwyn chuckled (because when one is snorkelling, they are on the surface, and the awareness of the immediate marine environment is less 'immersive' than when you are underwater; furthermore, it is generally considered that when white sharks actually attack, they do a vertical lunge from near the ocean floor and usually in a surprise move; that is why divers are most careful while in the process of surfacing). Alternatively, fishermen in the area have seen white pointers from a long time back, evident by Steve's transcripts in *White Pointer Chronicles*, and also in some of my interviews with locals, and generally, they were considered to be 'just sharks'. But something happened to this ocean-going community after *Jaws*. To my astonishment, even for them, the shark was turned into a monster. Particularly for the children at the time, for many who had never encountered them, or did not even know they were there in the region. Needless to say, the relationship of the town with the cage diving operations had also been complicated.

The interview with Mary revealed that the freezing works always brought sharks to Bluff, and how *Jaws* effected the perception of white pointers. 'There have always been offshoot going into the sea from the meat works. Did it bring sharks? All the blood would go into the ocean through pipes, and there would be heaps of sharks, there are probably sharks still there. Sharks even came up to the harbour, when we were kids, we used to swim in the sea, and we were always told to watch out for sharks. My father got a big Grey Nurse shark down there in the 50s, I have a photograph somewhere. It was about 18 feet long, they had it on display in the main street for about three weeks.' Peter said, 'They were "just a shark"; even as a kid, people used to tell us, don't get your feet nipped by a shark; it wasn't a big deal.' I asked if Mary had ever seen white pointers in all the years she had been here, any in the harbour. 'No, there were no pointers in the harbour, and I have never seen one, and I go to the Muttonbird Islands every year.' So, the obvious question was if sharks were such an everyday part of life, why are people so afraid of them now? 'Because they watched *Jaws*! Before that, we didn't really care. I think *Jaws* changed our ideas, even me, before when I used to go to the Muttonbird Island in our dinghy, I never used to bother, but now I think twice. I used to put my hand in the water, Daddy used to slap us and tell us not to do it. Barracuda or shark can come up and grab your hand, he would say and we used to laugh at it, but now I realize he was right. All the kids became afraid, and we used to have fun with each other—"Look out, *Jaws* is coming."' So, all that changed after *Jaws*? 'I reckon so! But I think my father's generation knew it was just a movie, it was more our generation that was really effected.' Peter exclaimed, 'So they were already there, but we never thought about them?!'

So, what is the community's thoughts on the cage diving today? 'They don't like it, it is wrong. They said that a shark followed the boat up to the new wharf, where the kids swim; someone is going to get hurt one day. Before, we had never heard of white pointers; we had heard of other ones, but not white pointers. Near the Muttonbird Island, the boys would be in the houses upon the hill, and they used to see fins passing by occasionally. I think they see more of them now, because of them fiddling with the white pointers, and feeding them.' I asked what she thought of the money coming to Bluff from cage diving and she replied, 'I can't see that; it goes only to the owner's pockets. Tourists are not going to spend all that money in cage diving and spend heaps of money in Bluff too.'

Mary is not the only person who felt that cage diving did not bring enough revenue in the region to justify its existence, Lisa, the hotelier, shared similar emotions. 'The cage divers are a small part of my business. Most of the people here, are to visit, or travelling to Stewart Island. As a matter of fact, the shark diving is more stressful to me. Because of the weather, they don't go out, and the diving is cancelled frequently. So, the tourist cancels the rooms too, and it is more pain than any good. I don't think it brings in a lot of money to the community, because the people who are coming for

A Sea lion feasting on octopus in the Bluff wharf, 2017

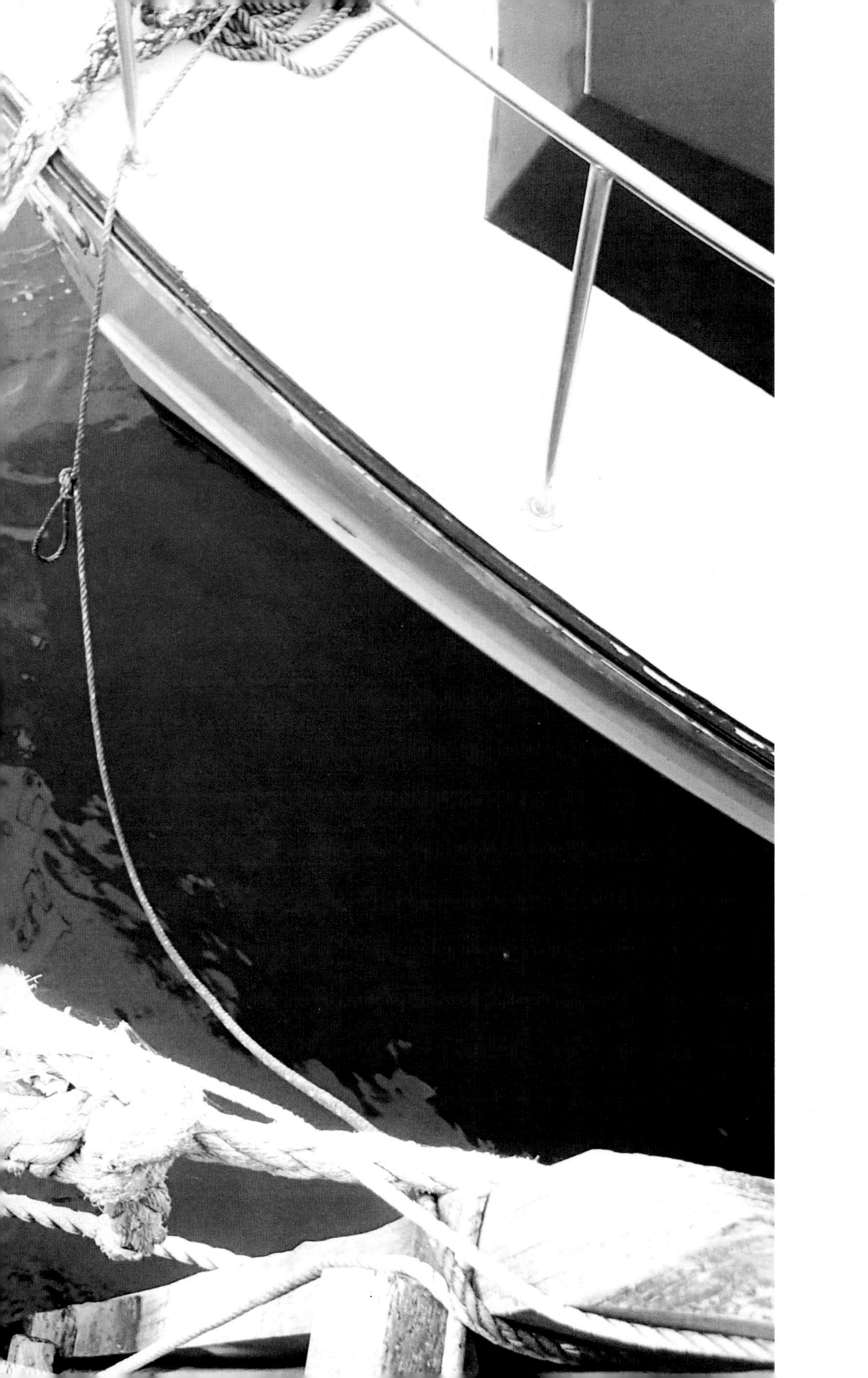

the dive, come in late from other places like Queenstown, and they leave the next day in the morning; so, they don't spend a lot of money here.'

The tourists who came to Bluff also felt the tension in the town about the operations. On the day of my first dive with the sharks, from the car park, Soosan and I were looking for the cage diving boat and went into the ferry terminal (the ferry to Stewart Island). But the reception we got was not warm to say the least; rather, we felt unwanted. For a long time, I thought that was just me, but over the course of my fieldwork, at least three other tourists told me that they experienced the same. Most could not understand why Bluff does not 'embrace this', and 'everything is not about money' and asked since this is already here, and it's not doing any damage to the environment like fishing, why the community was not embracing it? They explained, 'The hostel we are staying is horrible, if they had good places to stay, good restaurants to eat in, that would have been great.' The Bluff library building shares the same building and floor as the Bluff post office. The staff are all very friendly, and being such a small community, everyone knows each other. The workers had stopped for their lunch, when a conversation about the news report regarding houses catching fire came up, and one lady mentioned, 'Yeah, it gives me the shivers, much like the thoughts of the sharks there.' I could not help but feel her making that comment while I was around (being the shark researcher) was more intentional than not.

The Vikings are Coming

The last day of the year 2015, Soosan and I came down to Bluff to dive with the sharks, out in Foveaux Strait. Yet again, staying in Dunedin the night before, we drove through the twilight wilderness to reach Bluff just in time, as Mike was about to set off to the ocean. Soosan wore her red jacket, braved the waves, and finally met the sharks, although that story deserves a chapter for itself. Many times after that, she came on the daily trips, always bringing a little bit more light to the mornings. At the same time, she was then pursuing her degree in anthropology and art history at Canterbury University; so, often for a 40-minute class, she had to drive up for two hours, take the class, get back to her car and drive back two hours for her young minion. Her commitment to my work was astounding, even sacrificing her own resources and time. I am still baffled why she did that to be honest, although one day she told me, 'When I saw you that day in the art gallery, somehow I felt I was looking at myself.' She is a force of nature, and I have to admit, she is the only person, who has tackled me down, and has a louder voice than me. She brought me food, cleaned my cottage with me, and even got rid of spirits, which I felt were haunting me. And we have had so many adventures beyond that, climbing wooded hills, to dining on $300-steaks, and literally standing shoulder to shoulder in multiple fist fights.

When she came down to see me in Bluff, sometimes, we would drive up to the bustling city life that was Invercargill. Invercargill is the southernmost and westernmost city of the country, and one of the southernmost cities in the world. It is the trade centre of the Southland region. Life has formed

The Viking, 2015

around two large streets, named after Scottish rivers. The small city was filled with posh cafes and restaurants; for me though, the most exciting establishment was the world's southernmost Triumph motorcycle dealership. Invercargill had a deep love of motorcycle, especially since this was the stomping grounds of the legendary Burt Munro, who at the age of 68 set a below 1,000 cc land speed record. It was in Oreti beach in Invercargill where he practised, incidentally a beach which was known for it's white pointers to have 'nipped' at the surfers! This town had another interesting store for me, that was an antique store, and Soosan even bought me an antique glass nib fountain pen from it once.

One evening, Soosan and I walked into an art gallery and realized that it was an exhibition opening. Growing up in the art business, this environment was like second home to me, and if you do find me in one of these things, you will find a proficient art snob, with a certain amount of warranted cynicism about the condition of amateur attempts towards greatness. Soosan always being the centre of attraction was whisked away by the curators and others, while I quietly looked at what they called paintings. A lady, in her late 60s, approached me and asked, 'Excuse me! I don't mean to bother you, but I paint brown people, and you are a beautiful brown man. Will it be ok if I took some photographs of you?' This prompted a quick dismayed smile from me, and I said, 'Sure... Go ahead.' Her honesty was so nice, I could not say anything else. When Soosan came back and heard this, she wanted to decapitate her, and I had to coax her out of the gallery.

There was one more shop which had a unique experience in store for us. While walking down a quiet street, we noticed an 'Indian' store. As we entered, a nice young couple welcomed us, and while rummaging around their frozen section, we stumbled upon some *para* (Indian sweets). Getting it in our hand, we rushed to Soosan's little Z3 that we had named the 'Sharky'; inside, we pumped the heater to the fullest, and yet, like threads of silk, cool air seeped through the minute cracks of the soft top, and Soosan put on her red barrette to keep warm. We bit into the frozen sweets, and warm deep soft plump drops of tears started rolling down both of our eyes, it opened up two time and space portals for us: one which took me back to Kolkata, as now the memory of that taste again takes me to the town of Invercargill; the other one opened a doorway to London, and Soosan's days with her friends who opened her up to Indian cuisine. In milliseconds, we travelled time and spaces, and when we opened our eyes, two friends were sitting in a 2 x 4 ft cabin, smiling at each other. One day in the field, I realized that I needed a lot of printed copies of surveys and material from the university back in Christchurch. Soosan, suddenly having an unexpected time gap in her class at the university, decided to come and visit me and bring along the much-needed supplies. Setting off from Christchurch in the afternoon, for eight hours she drove through torrential rain and fallen trees to finally reach Bluff around 12 pm. After having the beans I had prepared for her, around 1 am, I packed all the equipment and we walked up to the Starling point. Our intention was to walk the 'Bluff hill walk', which goes around the entire hill that protects Bluff, and see the sunrise on the other side. To increase the intensity of the ambience, we decided to play classical Indian flutes in our headphones.

As we took the first steps towards the dark abyss that seemed to be a mere two-three hour walk, Pd. Hariprasad Chaurasia[1] played his Sivanjali raga, and accompanying was the shattering of the waves on the cliff below. It being a moonless night, we could only see to the 40–50 ft up front lit by our headlamps, and the illumination of our mp3 players in our individual pockets. I have had a lot of experience of travelling in the night and in the darkness, but this was different; it was almost musty and tactual. This

1 An Indian music director and classical flautist.

Soosan the fish chummer, 2016

darkness as a terrestrial experience was as immersive as the dark waters. We are not nocturnal animals, neither are we naturally aquatic species; in both the cases, we can find ourselves disoriented, and fearful, much like Trout's argument about the fear of darkness possibly related to the fear of us being predated by predators, where the darkness of the land due to the timing of the day, or the absence of light due to the increase of depth, and/or suspended particles in water, our fear of the unknown can be piqued. The moment we crossed the lights for the 'Oyster Cove' signboard (the last and poshest restaurant at the end of highway one), it was as if we got sucked into an intense isolation. Even if I had the firm knowledge that Soosan was walking only inches away from me, I was isolated from her and the rest of the universe in a flicker of an eye lid. As more and more we kept taking steps away from the lights, and into the walkway, I felt more distant from my home than ever before, even if I was in New Zealand for three years now. The road started to get woody with the overhanging trees, the timid stars were barely visible, and where was the Aurora when most desired? A few more minutes, and suddenly, my heightened sensations told me that I could actually smell the dead whale carcass dragged to these shores—a sensed memory of the place perhaps. I stopped, silent in the road. I uttered, 'We should go back.' Soosan said, 'Okay.' Quite a few times during this fieldwork I had such sensory experiences—experiences which cannot be always explained by positivistic paradigms, that Western academia is accustomed with. Rationality, and objective analysis of a physiological and psychological reaction in certain experiences such as this, may take a back seat to immersion in the experience itself. Maybe, these experiences were created from memories that the researcher holds deep within their own psyche even unavailable to their own conscious self, as much they are about the shared memories and the cultural experience of a place.

Married for 60 years, in this house, the day my grandpa died, he knew he was going, while everyone was rushing around him, trying to take him to the hospital—he laid on his bed put his hand on Mumma's head, gently stroking her, and saying, 'Don't worry, I have taken care of everything, you will be safe.' He left with nothing, but that little bit of land I had got for him from Bangladesh, pressed against his chest by Mumma. After he passed away, my Mumma's plump little face withered away like the banks of an old river, in a year or two. Now as I look in the night at the dark hole of a window of her room from here, seven years has passed, and she has full blown Alzheimer. She does not know anyone, the synaptic connections in her brain do not work and comes slow death of neurons in her limbic system, I wonder where have all the flowers gone. Still, some were deep at a basic cognitive level, once in a sudden instance, some connection is made, some flower blooms for a moment, and reminds her of a place where she belongs, and is safe, and her children are playing around in the dust; so, in the middle of the night, she tries to run away to this house. Maybe, it is the *runanubandha*, memory of the body (not to be confused by muscle memory). Deep memories that your body collects along the multitudes of lives lived, and it is said, getting rid of *runanubandha* or these unnecessary bonds, is the path to freedom.

Turbulent Waters

After meeting Steve in Stewart Island, while having a coffee before catching the ferry back to Bluff, an interesting poster came to my attention.

This poster depicted unequivocally the underlying tension among the cage diving operators and the residents of Stewart Island, and the 'community of water lovers' and the cage diving 'business'. It claimed that the operations were 'training' sharks to associate humans and boat with food. The poster raised the idea of danger to swimmers, kayakers, sailors, boaties, water-skiers, and divers (Paua and recreational). Hence, the use of the term 'water lovers', and its relation to the 'businesses of cage diving'. Even though this conflict is not the focus of my research and the human community I was primarily focused on was that of Bluff, considering the attention it has received in the New Zealand public psyche (often exemplified by newspaper reports from the last few years), and the relation of Stewart Island with Bluff, I have to make some comments on it, as I saw it as a participant observer of white shark cage diving in New Zealand.

Significant sections of the Stewart Island community strongly held the belief that cage diving was increasing their chances of being attacked, because white sharks were being attracted here—where they were not as prevalent before. It was also claimed that white sharks were getting more aggressive, and associating boats and humans with food. I am not a biologist or an ecologist, and hence, cannot authoritatively state on the reality of these beliefs and claims either way. Although, as the Commission on Environmental, Economic and Social Policy (CEESP) notes, biological

Welcome to Stewart Island!

Do your plans here include

SHARK CAGE DIVING?

If so, please consider the impact this industry is having on the Stewart Island community. The shark cage dive industry is not welcome or appreciated by many people who live here.

The cage boats are operating too close to where we live.

The methods they use to attract sharks are training great white sharks to associate humans and boats with meals, and since they began operating here, great white sharks have become increasingly aggressive toward boats.

Stewart Island is a community of water lovers, and we are horrified by the daily baiting of sharks a few miles from here. We are asking our government to stop them.
In the meantime, YOU can help us.

As a tourist, you have choices. If you choose to go shark cage diving you are supporting their business and contributing to our problem. We ask you to please think about the impact your choice might have on us. Ask yourself: would you want anyone teasing apex predators with food every day in *your* front yard? That is what the shark cage dive companies are doing here: teasing sharks with food every day next to humans in cages.

Thank you for your consideration.
We sincerely hope you enjoy your visit here.

For more information visit the Facebook page
Stop Shark Cage Diving Near Stewart Island

A poster handed out in Stewart Island, 2016

sciences are not sufficiently equipped to delve into these kinds of human-animal conflicts, and social sciences and interdisciplinary approaches are needed to play a significant part in them, exploring the sociohistorical reasoning for these conflicts (Madden, 2010). Hence, rather briefly, I contextualize the timeline of this conflict and explore the major social dimensions that may play a part in the emergence of this conflict. I have to admit, to properly delve into such a topic, I needed to spend as much time with the islanders as I did with the cage operators in Bluff, and it would have been a different research all together. But my research was not focused on this. Furthermore, because I was working with the cage divers, I would have probably not been welcomed in Stewart Island. So here, more than any authoritative dictation about the conflict, I hope to create a platform for future social scientific research on it.

The CEESP defines human–wildlife conflict as that which 'occurs when the needs and behaviour of wildlife impact negatively on the goals of humans or when the goals of humans negatively impact the needs of wildlife' (Madden, 2010, p. 248). At the same time, it is also generally acknowledged that human–animal conflict often concerns human–human conflict (Dickman, 2010; Pooley et al., 2017). 'Human–wildlife conflict, however, frequently involves an equally important conflict between people who have different goals, attitudes, values, feelings, levels of empowerment, and wealth. Conflict with wildlife may be rooted in struggles among people over empowerment and access to resources or needs for survival. The conflict may also stem from people who have different needs or levels of need, different perspectives on the world in which they live, and questions of who should have access to resources or control over them' (Madden, 2010, p. 250). The conflict related to white shark cage diving in New Zealand is as much a conflict between the humans of different groups, as much it is between humans and sharks.

Shark diving is a broad term, where you may or may not use scuba equipment, with or without cage, and be underwater or on the surface, and with varied species of sharks (Dive Worldwide, n.d.; Eco and Conservation Dive Liveaboard, n.d.; sharkcagediving, n.d.; MakeMyTrip Blog, n.d.; Scuba

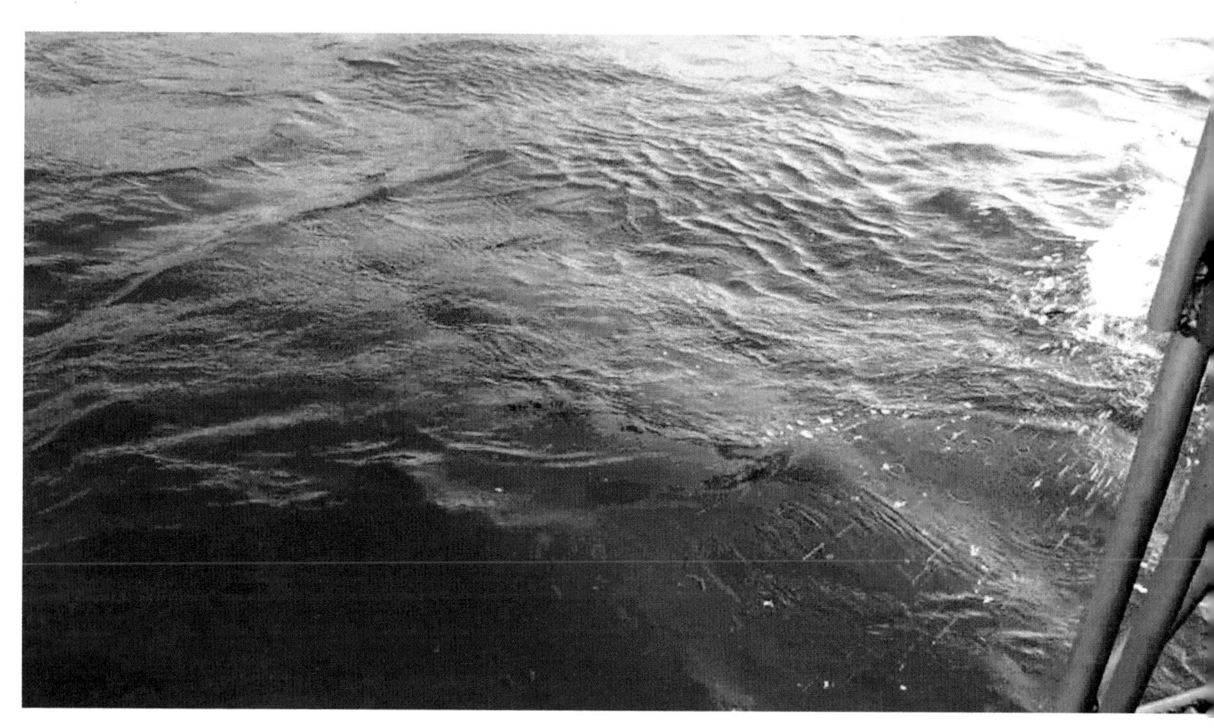

The blue shadow, 2016

Diving, n.d.; Shark Diving Tours, Honolulu, Oahu, n.d.), and commercial shark diving is certainly not new in New Zealand. There was cage diving with some different species of sharks off the coast of Kaikoura, New Zealand, that began in 1998 and closed around 2001 allegedly due to the lack of sharks. The dives attracted Mako and blue sharks, and was a rudimentary operation to say the least. However, for some reason it seems, the fact that there were and 'are' sharks in the Kaikoura region has been erased from the town's history, because it has become a surfing, dolphin diving, and whale watching town, and accounts of the shark diving and sharks are limited (Baker, 2001). Even if you ask people in the area if there are sharks there, you get a stern reply—'No'. It is as if there is an intentional cultural amnesia about sharks and cage diving in Kaikoura. One cannot help but draw the connection between this phenomenon and Kaikoura's other tourism attractions like diving with dolphins and other non-threatening species. The Kelly Tarlton's Sea Life Aquarium of Auckland has cage diving with sharks, with Wobbegong School, and Broadnose Sevengill Sharks (Kelly Tarlton's Sea Life Aquarium, n.d.). Furthermore, the national aquarium of New Zealand in Napier also had snorkelling and diving with sharks (National Aquarium of New Zealand, n.d.) like school sharks, rig, sevengiller, carpet sharks, and swell sharks. However, no operation has gained so much global and local attention and been the centre of controversy in the last decade as the white shark cage diving off the coast of Bluff.

In 2009, Mike Haines started cage diving operations. Around 2010, Paua Mac5 declared that they wanted shark cage diving regulated, and around 2013, the residents of Stewart Islands started protesting about cage diving too. In November 2014, the DOC introduced cage diving operation guidelines for the operators. In December 2014, Maritime New Zealand produced safety guidelines for commercial shark cage diving. Again, in December 2015, the DOC produced a revised version of the cage diving guidelines. 2016 brought new challenges when a judge ruled that the DOC does not have the right to give out the permits. By the end of 2016, the DOC gave the operators only a temporary permit because of the uncertainty about the legal arguments. In September 2018, the Court of Appeal ruled that cage diving cannot be permitted. Following that, Peter Scott closed his operations (Shark Dive New Zealand) permanently. However, Mike Haines (Shark Experience Bluff, New Zealand) still aimed to continue. Shark Experience completed it's 2018–2019 season, with no bait policy, and are awaiting the verdict of the case from the Supreme Court, the future remains uncertain.

After its beginning in 2009, around 2010, Paua Mac5 (the chief representative body of Paua fishermen in Southland) wanted white shark cage diving in New Zealand regulated, and when the DOC finally stepped in to create the permits for the operators, a mutual agreement arose to work around Edwards Island. Also, the area around the island is quite shallow, about 8–12 m, so even if the sharks are near the bottom, there is the possibility of observing them, once the tourists are in a cage.

In 2016, 786 people, headed by Helen Cave, one of the most influential business women on the island, and owner of the South Sea Hotel (Roy, 2016) (who was also distributing the pamphlet to stop cage diving) signed a petition to stop the cage diving operations around Stewart Island 'permanently'. She claimed that 'almost all Stewart Islanders oppose the operations despite the financial benefits they may bring to the local economy' (Petition, 2014/16 of Helen Cave) (it is an interesting point to note, after Peter Scott shifted his cage

PETITION 2014/16 OF HELEN CAVE

Petition 2014/16 of Helen Cave

Recommendation

The Local Government and Environment Committee has considered Petition 2014/16 of Helen Cave and recommends that the House take note of its report.

The petition requests

> That the House note that 768 people have signed a petition calling for shark cage diving near Stewart Island to be stopped immediately and permanently.

Shark cage diving operations near Stewart Island

Shark cage diving has been operating in the northeastern waters of Stewart Island since 2008. These operations were unregulated until 2014 when the Department of Conservation introduced a permit system to reduce the "risk of harm to great white sharks". Great white sharks are a protected species under the Wildlife Act 1953 and classified by the department as in decline.

In December 2014, the department granted permits to two shark cage diving operations. The permits limit the use of baits and regulate the construction of cages. They are due to expire in August 2016.

Petitioner's concerns

Reason for the petition

Ms Cave's reasons for petitioning Parliament are to increase awareness of the issues faced by Stewart Island residents and avoid future shark attacks around Stewart Island. She is seeking legislation to make great white shark cage diving illegal in New Zealand waters.

The petitioner is concerned about the effects of shark cage diving operations off the coast of Stewart Island on the environment, the sharks, and the local community. She notes that almost all Stewart Islanders oppose the operations despite the financial benefits they may bring to the local economy.

Clayton Mitchell MP, who spoke to us in support of the petition, said that the effects of shark cage diving operations need to be studied. He supports an immediate moratorium on this activity until such a study is completed.

Waters unsafe for recreation

Ken McAnergney, who spoke to us on the petitioner's behalf in December 2015, told us that he and his children are no longer able to safely enter the sea around Stewart Island to collect kai moana, as was once possible. We were also told that it is no longer safe for local children to have kayak lessons where they once did.

The Stewart Island petition

diving operation, Shark Dive New Zealand, to Bluff, whatever limited economic benefit of cage diving Stewart Island was receiving must have decreased drastically anyway). The petitioners stated their concern '...about the effects of shark cage diving operations off the coast of Stewart Island [8–10 km away] on the environment, the sharks, and the local community' (Petition, 2014/16 of Helen Cave). An argument was made that cage diving was bringing sharks into areas near Stewart Island, where there had never been a large population of white sharks and that also the white sharks were getting more aggressive, in effect increasing human and white shark encounters and the potentiality of a 'shark attack'. The community was enraged with the practice, and even though the initial urge by Paua Mac5 was to regulate it, the situation was elevated to such a deeply emotionally charged topic that the Stewart Island community pleaded to ban the operations all together.

Since then, the nation was made more and more aware of the perceived risks and possible future atrocities happening around Stewart Island in the name of tourism by various news media features, often with headlines like: 'Fatal shark attack likely if Stewart Island cage diving not stopped', 'It was like *Jaws*', and 'No longer safe to enter the sea'. Residents made emotive statements which often mentioned *Jaws*, and before cage diving, there was a peaceful coexistence between the man and the 'beast', but now it is an attack waiting to happen: 'In the old days, great white sharks and the residents of Oban coexisted peacefully in the New Zealand waters off Stewart Island, but nowadays, it's more like a real-life version of Amity Island from the movie *Jaws*' (Strege, 2016). Finally, 'Does it have to take someone gonna get eaten? [sic]' (Pennington, 2017). Political parties and ministers also joined in the discourse, party leaders of New Zealand First, Winston Peters and MP Clayton Mitchell, attended a meeting with the islanders on 21 April 2015, to discuss the matter (ZB, n.d.). In the meeting, the islanders wanted the permits to be revoked.

Speaking after the meeting, Peters said that he would question Conservation Minister Maggie Barry and the government on just how those permits were issued. 'I don't know how they could have made this decision,' Peters said. He suggested that pleas from the Stewart Island community had fallen on deaf ears. 'If I saw a hall this size and I was in government, I would listen to what they had to say,' he said. 'It was a pretty enthusiastic crowd.' Southland District Councillor, representing Stewart Island, Bruce Ford, said that he was 'sitting on the fence', but attended the meeting. 'I think the reason they went there is they're concerned about safety and someone would be eaten,' he said. When asked if Peters and Mitchell made any promises to Island residents, he said, 'All politicians do.'

On the 27 April, it was reported that Clayton Mitchell had called for a moratorium on shark diving. Mitchell laid the question to Maggie Barry in parliament, as reported on 15 June 2018.

> CLAYTON MITCHELL (New Zealand First) to the Minister of Conservation: What action will she take given that 768 people have signed a petition calling for shark cage diving near Stewart Island to be stopped immediately and permanently? Hon MAGGIE BARRY (Minister of Conservation): As the Minister of Conservation, I will not be taking any action as a result of the petition, but what I will be doing, as I have said from the start, is paying very close attention to the review that the Department of Conservation now has under way. We have said from the beginning that at the end of the season we will evaluate it. That process is under way now. I will be very interested to see what has been happening. There have been no formal complaints. There may well be changes or not as a result of that review. Any changes will be widely circulated and discussed before the start of the next season in December. (Question Time, 2018)

As of 4 September 2018, cage diving was impermissible in New Zealand waters, by a ruling of the Court of Appeal based on an interpretation of the Wildlife Act 1953, because 'disturbing, molesting and pursuing' of protected species was an offence under the act.

Furthermore, the Court of Appeal argued that the DOC had no authority to grant the permits to the cage diving operators. The case is now in the

Supreme Court awaiting judgment. On 5 September 2018, the Radio New Zealand website had a big headline—'Court ruling sinks teeth into shark cage diving'. The article read as follows:

> The Court of Appeal has ruled that shark cage diving is an offence under the Wildlife Act. The ruling ends a long running legal stoush between Stewart Island Paua divers, the Department of Conservation (DOC) and two shark diving companies... PauaMAC5 said the companies were 'pursuing' and 'disturbing' the sharks, which posed a significant safety risk to nearby paua divers. In 2014, the then-minister Nick Smith announced that any shark-diving operations would need a DOC permit. PauaMAC5 supported the move but stressed that any permit must be issued with the safety of other water users in mind. DOC refused, arguing that public safety was outside its obligations under the Wildlife Act. The act specifies DOC may from time to time authorise any person to 'catch alive or kill for any purpose' any protected wildlife. However, in 2016, a high-court judge questioned whether DOC had the power to authorise permits under the Wildlife Act, because shark cage diving did not involve catching or killing sharks. The Court of Appeal has now found that shark cage diving is an offence under the act, and that the director-general of Conservation has no power to authorise the activity. Helen Cave, owner of the Seaside Hotel on Stewart Island told Morning Report she was 'really relieved' shark cage diving will be stopping. 'It's been quite a concern... I have children and grandchildren that play in the water on Stewart Island. 'She said the sharks are now conditioned to the food and people and felt it might take a while for the sharks to go away.' They've been trained—these sharks—and I'm hoping they might forget their training, but I suspect there's a bit of cunning there and a bit of residual habit. (Doyle, 2018)

Alternatively, the supporters argued for cage diving, as the Tourism Industry Aotearoa (TIA) Chief Executive, Chris Roberts, stated:

> It is disappointing to see that an activity previously approved by the Department of Conservation can no longer exist...shark cage diving is undertaken safely in many locations around the world, and there is international evidence that shark cage diving, when well-managed, does not change the behaviour of sharks. Sharks were not harmed in this activity... The activity drew people to the lower South Island and Bluff... The Wildlife Act is 65 years old and is being applied to an activity that was never envisaged at the time the Act was passed. (Tourism Industry Association, 2018)

The DOC also did not agree with the court's decision:

> On 4 September 2018, the Court of Appeal ruled that shark cage diving is an offence under the Wildlife Act, which means that shark cage diving companies will have to cease their operations. DOC's role, under the Wildlife Act, is controlling, managing, and monitoring impacts on great white sharks. We are currently investigating what the Court of Appeal's findings mean for DOC. The latest findings mean the activity is now considered an offence under the Wildlife Act and people undertaking the activity could be prosecuted. ... In 2014 we introduced a permitting system aimed at reducing the risk of harm to great white sharks. On 2 June 2017 the High Court released its judgment that DOC did not have authority to issue permits for shark cage diving... This meant shark cage diving reverted to an unregulated activity, as it was before 2014... The latest findings, as of September 2018, mean that not only can the activity not be regulated, it is now an offence under the Wildlife Act... Shark cage diving had been operating near Stewart Island/Rakiura since 2008 [2009] where there is a stable resident population of great white sharks in the waters around Stewart Island/Rakiura. (Department of Conservation, 2019)

Finally, Peter Scott (the owner of Shark Dive New Zealand) and his team decided to stop his operation altogether, and on 30 September 2018, put up a public statement on social media:

> We have decided we can't continue to fight. With the threat of prosecution for us, our skipper, our crew and passengers, hefty fines and vessel forfeiture, it just isn't worth risking continuing in the face of the illegal ruling... It's staggering to us that an argument based on illogical reasoning and fear has led to an entire industry being shut down WHEN THERE HAS BEEN NO SCIENTIFIC EVIDENCE of the accusations and in fact worldwide evidence substantiates our stance. Shark cage diving is essential for shark conservation and ensuring education about the Great White Shark population at Stewart Island has been Shark Dive New Zealand's number one priority. Sadly, the fear and hatred that the Haters have towards sharks that inhabit their waters, will continue to be treated as truth by those that are unwilling to learn otherwise. Secondary to that is as a result of the court decision, unemployment for the Bluff locals that were our very valuable crew, reduced nights for the region's accommodation providers and less visitors to the town's hospitality establishments, in a town that could do with the economic benefit these visitors provide. (Shark Dive New Zealand, 2018)

As of August 2019, Mike Haines (Shark Experience Bluff, New Zealand) still wanted to fight this ruling and completed his 2018–2019 operations. However, he altered his operation with no bait policy, with just the use of chumming to bring the sharks close to the boat. But the future remains very uncertain, with them awaiting the final verdict from the Supreme Court. My research was primarily based on his cage diving operation, while baiting was taking place as part of normal practice in white shark cage diving.

Mike mentioned in conversation (February 2017) that initially, the objection was from the paua divers, but it then expanded to the rest of the island (and now a large portion of New Zealand's general population): 'Everyone is related, and friends, and you always support your locals, as I would do.' He noted that paua divers are thought to be really adventurous people and hold high prestige in their community. Even if they don't have a high social standing, they are certainly understood to be a unique breed, considering its innate dangers (with the weather and the sharks). Also, the people who were originally from Stewart Island were either fishermen, or had something to do with the ocean, and were not really bothered with tourism.

Having said that, there is certainly an element of jealousy as to why only Mike and Peter got the opportunity to work with the sharks, and once in a meeting about cage diving, someone stood up in protest at the fact that these operators were not locals. I asked Mike about his personal relationship with the Stewart Islanders, and Mike replied, 'At the end of the day, I have known numbers of them for years; they haven't shown too much animosity against me, personally. I would be disappointed if they attack me on an individual basis. Some of the issues they have are certainly unfounded; a lot of it is about us feeding the sharks. But then again, they know that we don't, because we're not allowed to, and we don't need to. In fact, if you look at the amount of bait we have lost last season, it is pretty minuscule. And none of them ever even came to our boat to see the operation. As far as the sharks getting aggressive is concerned, if you look at the sharks before the females arrive, you will see many of them have large bite marks on them, which is why we are assuming that it's a competition between them and has nothing to do with us. Think of a pack of dogs, who are aggressive towards each other, because there is a female in heat; I think it is the same thing here. As for the safety of the kids swimming, they always knew that sharks were around; they have been doing that for 50–60 years! That's why schools used to have watchers—some kids used to watch, while others swam (I met a resident of Stewart Island who agreed with the premise). Why did they get the swimming pool? Because sharks were the issue. There was this guy, who walked from Bluff to the top of North Island, to raise the money for the pool. [It was a local publican who rode on his bicycle from Cape Reinga to Bluff in 1971 (Morris and Beaven, 2012)]. So, for them to come up and say that this is happening because of the shark cage diving is ludicrous' (February 2017).

Indeed, it seems that the sharks were one of the reasons for building the pool: 'The Halfmoon Bay Community Pool was first developed in the early 1970s as a basic, concrete, cyclone-fenced, rainwater-filled extension of the wider ocean that the Halfmoon Bay community literally has as our constantly changing vista. However, this pool of water is not so deep and doesn't contain *big biting fish*, crashing waves or seaweed' (*A Pool to be Proud Of*, 2012: emphasis added).

The owner of the supermarket in Bluff also mentioned, 'One of our workers, who is from Stewart Island always keeps on talking about it and says that as long as she remembers, when they went swimming, someone would be on a boat and when they would see a shark, they would just call out and everyone would get out. She says this goes back to her grandparents' (January 2017). Then, there is the fact that fishermen had always caught and gutted fish in Foveaux, near the island, and other places around Foveaux, which historically attracted sharks, and sharks were even known to have followed boats. Most importantly, it seems, the old fishermen knew that there were always white pointers around Stewart Island (www.whitepointer.cloud/).

Personally, there were a couple of times when I had to directly come face to face with this political turbulence of cage diving. First, when I was in the field for the first time, one day, a member of a governmental organization turned up in my doorstep. He was very polite, and we had a long discussion; we kept on communicating, and I even took his interviews. However, one day, he called me up and said, Raj, the ministry said, I cannot speak with you anymore, and you have to delete all interviews with me, and I kept my word given to him. One other time, I got invited to radio New Zealand to talk about my research and my findings. As Soosan and I reached and we sat down for the interview, we were informed that there were already calls from Stuart Island to the station, condemning cage diving and hence, pre-emptively, condemning my interview.

The mother, mix media painting, Raj Sekhar Aich

Multispecies Conflict

To understand the genesis of the conflict between the human groups, we need to situate ourselves in the human white shark conflict in New Zealand, in the Southland and Otago region. One of the important factors that seems to affect human–animal conflict is the perception of risk from a species. This perception can be often disproportionate to the actual risk involved, because it is culturally shaped. Specifically, 'Large, highly visible and potentially dangerous species are particularly likely to generate disproportionate antagonism' (Dickman, 2010, p. 461). There has been five confirmed fatal, and about seven non-fatal white shark bites on humans in New Zealand in recorded history (Shark Attack Data, n.d.). These bites are far and wide in between, but when they do happen, they are violent and often life altering, such as in the case of Vaughn Hill near Pitt Island, about 20 km from the south-east coast off Chatham Island, New Zealand, in 1996. By another freak stroke of luck, Soosan used to teach in the school where Vaughn's young girl was a student in Timaru. Soosan tells the story that the first time she saw him, there was a man without his right arm, wearing a *Jaws* t-shirt and teaching kids to swim. She contacted him for me once when I was in Timaru, and he was gracious enough to tell me his story (February 2016).

In 1996, Vaughn was diving for paua off the coast of Chatham Islands, New Zealand. Having a good dive, he was surfacing, when he suddenly realized there was no fish around. Not paying much attention to it, he continued. Suddenly, he felt a strong bump on his shoulder, and when

Vaughn and Raj, Timaru, Soosan Lucas, 2016
Global Shark Attack File, case number - GSAF Case # 1996.09.06.b

turned to his right, he saw a big black gape, a couple of feet away coming, at him... The shark looked at him right in the eye and took a bite of his arm; he hit its eye with his paua hook, but lost consciousness. When he regained his senses, he realized that he was floating on the surface quite a distance from his boat. He knew it was a shark, and it was around, and he signalled his boat to come and pick him up. As his friends came to get him out of the water, his suit got entangled, and without them knowing that he was not on the boat yet, they started off. So, he was being dragged by the side of the

The piercing blue eyes, 2017

boat and was drowning, until someone actually saw that and stopped. His friends finally got him out of the water, still not realizing the intensity of the attack because of his wetsuit, but that suit was saving his life by keeping pressure on the blood flow. They set off for the shore, and on the boat, he was in a state of acceptance of his imminent death, lost consciousness, and even had strange visions.

After they reached the port, he was put on the back of a truck and driven to the only airstrip on Chatham through hard-cobbled roads, and then flown off to the Wellington hospital. It was reported that he had lost 50 per cent of his blood and had just a bone hanging off his right arm. When they got him off the plane, the ground was coloured red like that from a freshly slaughtered sheep, and he had to be given 24 units of blood. He survived with an amputated right arm. However, to Vaughn, the shark attack itself was a very short incident; what was more significant was the process of recuperation after that: overcoming his fear of the water, and to provide for his young girl with his disability as a paua diver. His fighting spirit persisted, and he also became a champion for allowing New Zealand paua to use scuba gear while diving (which was not permitted before to control over the harvesting of paua).

There has never been a white shark bite related fatality in Southland; though there was always an awareness that white sharks are around New Zealand, particularly, in the Otago and Southland region, the general public did not properly acknowledge their lethal potential until a series of fatal white shark bites in the 1960s, off the coast of Dunedin (Jøn and Aich, 2015). A series by *Stuff,* New Zealand (Mcneilly, 2018), recounts the stories associated with these five shark attacks in Dunedin, during the 1960s, three of which were fatal. In one particular fatal attack that happened on William Blacks in 1967, his body was never recovered (an unusual phenomenon by itself), which had a devastating long-term effect on his family, as his grieving sister recalled 'the fact his body was never found... It took me actually those 40 years to get over it' (Mcneilly, 2019). These incidents scarred the community and, in some way, even brought them closer, but most importantly, it made them painfully aware of the imminent danger out there in their playground, and Dunedin became the first city in New Zealand to have shark nets in 1969. One could imagine the release of *Jaws* around 1975 on this fresh 'wound' could not have helped the fear of the white sharks in the region. And the hyperreal image of the white shark may have been perpetuated ever since. At least, in four of my interviews, I was told by residents of Bluff that they got so affected by *Jaws* that some of them stopped swimming from the wharf, and one long time diver even stopped diving in Foveaux, I would imagine a similar effect of the film on other ocean-going communities; however, I have not come across any academic publications elaborating on it, perhaps it is an avenue for further detailed research.

One unfortunate incident took place around Chatham Island, near Pitt Island, on 14 July 2019, when a paua diver was bitten from the side, and was dragged for 20–25 feet. The paua industry council chief executive, Jeremy Cooper, noted that he was saved only because he had his harness and BCD on (scuba diving equipment which has been allowed to be used by paua divers in the last five years according to the report). Now, even though obviously, this incident had nothing to do whatsoever with Bluff or white shark cage diving in New Zealand, the graphic nature of the incident, and especially since it was on a paua diver, this cannot be good for the image of white shark cage diving in New Zealand. Only time will tell the eventual outcome, and if white shark cage diving in New Zealand will ever exist again in New Zealand.

Disproportionate aggressive human response towards the species in conflict may be attributed to the feeling of frustration and lack of control that people feel in their position (Madden, 2010). Even though white sharks are a protected species in New Zealand, due to the conflict between the cage divers and the Stewart Islanders, sharks are allegedly losing their lives. In the beginning of 2016, it was alleged by the cage dive operators that a white shark was killed by the Stewart Islanders to take revenge on the cage diving operations. There was even a petition started to stop the killing of the sharks (Kirk, n.d.). As Mike mentioned to me, 'We got to stop this, but the issue is Department of Conservation is doing nothing with the information we are giving them, because they are not motivated by their hierarchy. They've had an opportunity two years ago when Zane killed a shark, shot it dead. It was on national TV! What did they do about it? Nothing! When we talked to them a little later about it, they said we can't do anything about it, because it's already been too late; statute of limitations' (Interview, Mike Haines, May 2016).

One of the reasons these killings supposedly happened was that there was a belief among the cage divers, if a shark is killed and left in the area,

Mike's interview, 2016

other sharks seem to leave the region. Whether or not there was strong enough evidence to support this claim is irrelevant here, as it has yet to be empirically tested, what is more important to consider is the effect of the idea on the lives of the sharks. If the fishermen really did believe in this, it meant sharks were being killed illegally as the result of two conflicting human groups. Furthermore, if this was true in the knowledge, then the annual migration ritual of white sharks, which might have been taking place in the region for hundreds if not thousands of years, may have been adversely affected by the actions of a handful of humans.

Human–animal conflict research indicates that public anger to human–animal conflict is not directly related to the property or material value lost. Rather, it is related more to the perception of potential risk and lack of control in avoiding it (Madden, 2010). It is the fear of attacks of these animals who are perceived as specifically targeting humans, and the intangible costs or psychological cost of danger seem more important than the tangible ones, like the tourism revenue created from cage diving (Pooley et al., 2017). Hence, whatever revenue Stewart Island's residents are/were receiving from cage diving may seem irrelevant in comparison to the loss of their perceived state of peace, and lack of physical safety.

There are also other social factors, for example, in a few African societies, there are beliefs that predatory animals like crocodiles are often bewitched by humans and directed to harm their opposing groups (Pooley, 2016). Even though literally, there was no witchcraft involved in the Stewart Island white shark case, one could imagine a similar feeling among others training and causing big biting fish to come to one's shore who were not there before. Dickman (2010) notes that people are more likely to take risks if the activity is voluntary, and not caused by others, such as, in the case of the people swimming around Stewart Island with watchers looking for sharks, as opposed to cage diving making the sharks come their way. In short, the islanders felt that they had lost their agency; they felt that dangerous situations were being created by the decisions of others.

Furthermore, as Dickman points out, our reaction to a certain threat from a certain animal is also related to the level of intrinsic dread involved, if something can be resolved calmly and in time, or has to be dealt with immediately (Dickman, 2010). And what is more urgent than an imminent attack on our children from a hyperreal monster when they are swimming in what used to be the safe backyard? Finally, Dickman also talks about how a conflict may be indicative as reaction towards certain social changes, which can be seen happening in Bluff and Stewart Island, where the residents claim everyone was happy and the humans and animals were peacefully coexisting before the outsiders (the cage diving tourists) came. In this case, I might also add, the 'others' here were not only the tourists coming to the island, who the islanders did not necessarily want, and can do without; it was the cage divers using their space and resources, and indeed the sharks themselves invading the 'safe' beaches.

Maybe, this fear and dislike of the others can be perceived as xenophobia, which is 'fear and hatred of strangers or foreigners or of anything that is strange or foreign' (Merriam-webster, n.d.).

> Xenophobia cannot be discussed without taking into account normative concepts that focus on attempts to mark 'one's own' off from 'the other.'... In social and political reality, xenophobia manifests itself in accordance with the division of the world into one's own race, nation, ethnic group, and culture, and other races, nations, ethnic groups, and cultures [Species]... Since fear and hatred, the two emotional states contained in xenophobia, are qualities that rely on subjective experience, they require cognitive signposts and social values to allow them to focus on their object. Focus on an object is less relevant when one fears something, than when one hates. Hatred of what is strange or foreign. (Wicker, 2001)

I will not get into the details of this argument in this book, but maybe we need to increase the purview of xenophobia (as in phobia between human groups) and discuss about *interspecies xenophobia*.

Hence, from a cultural perspective, a national human–animal and human–human conflict was created from fear and frustration of a group who are surrounded by large predatory fish who may pose immediate danger; who do not need cage diving tourism as a primary industry; who give more reverence to the non-tangible value to the tangible ones; want to maintain the perceived isolation and safety of their land, and are led by leaders who hold power and admiration in their closely knit community. It is not that Stewart Island does not want tourists, but they do not want the cage diving tourism to affect their 'original' state of life.

The Shark with the Pink Sunglasses

When species meet, passion proliferates. There is love that 'love in the age of extinction' that Rose (2011) has called for, a passion born of sensitivity to the mutuality that links humans to a multitude of other beings, ranging from primates to plankton to salmon to sheep to crickets to the microbes used in making cheese' (Rutherford, 2016) and might I add, big biting fish.

One day, when I was working on the boat, there was a large shout, and I realized there was a big shark on the portside. Rushing, as I looked over the boat, my pink sunglasses fell in the ocean, just as the shark swam over it. The shark never showed itself again, and everyone started laughing, that somewhere in the ocean there is a shark who stole Raj's pink sunglasses and made a run with it. It is this laughter and love that connected us together on that boat. The love for sharks, the environment, and family and friends who wanted to witness lifelong dreams coming true of their loved ones. It was a melting pot of stories about dreams, nightmares, personal quests, self-challenges, social capital, and yes adrenaline—all related to an image of an unknown fish.

One particular intriguing incident was an elderly lady on a wheelchair on Peter's boat setting off to see the sharks. I was informed that she had been out with Mike before, but had not seen sharks then; unfortunately, I had no avenue of interviewing her, yet I would only imagine it was a personal quest for her in some way. For some, it was a lifelong quest—people who waited for 20 years to do it; for others, they trained for years to get to this point. Alternatively, for certain groups of individuals, it was an occasional 'by chance' activity, like a young couple on their honeymoon in search of adventure; for some, it was merely one of the many things in their bucket list. And yet for others, like New Zealander Marvin, it was a wish to conserve the sharks: 'They need to be better represented, but how could I represent them, if I have not seen them; that's why I'm here.'

For some tourists, sharks have been a lifelong obsession—created from the appreciation of their biological and symbolic value, and even through recognition of their long lineage. Tom, a young British lawyer, was always obsessed by sharks. 'When I was young, I was always fascinated by sharks, especially maybe because I could never actually see them in UK waters. While visiting Australia, the closest I came to a white shark, was by renting a car, driving all the way down to Philips Island, and standing on the side of the rocks in hopes of actually seeing one... I didn't.' An American engineer said, 'They are my favourite animals, and white sharks for me are best of all sharks. I've seen documentaries of people diving without cages; one day, I would like to do that even if I get bitten. They are more powerful than humans in the water, and that attracts me a lot to them, and we realize we're not the boss.' As noted by a young Scot woman, 'I always wanted to see them in a cage. I'm really interested about predators in the ocean, animals who have not changed for millions of years, and are still so mysterious. When I am diving, I do think about them, but it's not because I'm scared of them, but because I respect them.' These obsessions for sharks were often related to their own child-like aspiration of adventure (heaven forbid, I blame them for that, suffering from the same syndrome myself). A Swedish student explained with excitement in his eyes, 'Since I was a kid, I always dreamt to be like Indiana Jones, and finding treasures. I quickly started liking sharks, and when I started scuba diving, I always wanted to see sharks and shipwrecks. When I was young, I subscribed to a magazine on the most dangerous animals in the world, and the white shark was one of them.'

A lot of the times, it is about the personal journey and sense of accomplishment for the tourist. A Kiwi male of 35 years said, 'This is the

The shark and the cage, 2016

culmination of a three years' dream; I started diving specifically to do this. I just want to be in the same space. I hope to feel a sense of completion! Just feel that sense of achievement.' Indeed, the same person told me after seeing the shark, 'I feel at peace, I feel relaxed. It makes me really smile to think that the shark was right by me. More than anything else, at times, it's a relief, it is something I had in my bucket list for a long time, will be glad when it's over.' And for others, it is about the social capital of diving with white sharks, in other words, the 'bragging rights'; it feels to them that it is indicative of their individuality, and represents a certain risk-taker lifestyle.

This is not only the case for divers and young travellers, but indeed for the identity of their families. A family from Los Angeles said, 'How many people can say they have actually done this! This is quite unique...you have to have a certain kind of personality to do this, if you are not a risk-taker, you probably wouldn't do this. You have to be on the other side of the spectrum to do this, because you are putting yourself and your family in harm's way.' And for some others, it is love of the nature and shared experience with other humans, 'You come out, on a rainy day in choppy seas, to see the beauty of the nature here around the islands, the birds, and how they all interact. You hear about the different experiences and share your own. You are here, and you are touching other people's lives, and they are touching yours. People from other cultures, other places, and here is a commonality they all have, the love of nature, a love for the sharks, a love for the experience.'

But there were also the aspects of coming to the boat, to be part of an important milestone of your loved one, when their dreams become our dreams. The parents came from all over the planet, but shared similar emotions about cherishing the aspirations of their children, no matter how far-fetched they may be on the face. A Japanese lady mentioned, 'My daughter loves sharks; so, when I'm with her, I spend a lot of time watching documentaries on sharks, and that's why, have come here today to be with her.' A German father: 'I came for my son, because it was in his wish list. So, if he wanted to enjoy it, I wanted to enjoy it with him. I do feel closer to my son because of this; I'm glad I was in the cage with him too.' Another American mother came to see her children achieve a dream they have always had, in an area they were always interested in. 'For my daughter, it was a dream she had had, as long as she had sense; before sleeping, I had to read her shark books. When she was 24, she said this was a part of her bucket list, which is why I thought it would be an amazing thing to do together.' Sometimes, it is memories of family members who inspired us to take long journeys. I remember my grandad, who had a huge library, where I found my first shark books, and fell in love with them. Other times, it is a guide who challenges us to think alternatively. I had this teacher, who always mentioned that sharks are badly portrayed and they're not man-eaters and they deserve our respect too, and that made me want to know more about them. There was a British couple who came back a year later, because the first time they came, there weren't any sharks. They came one more time a month before, but still did not see the sharks. One of them said, 'It's a big one! It was in my bucket list for a long time, I'm not scared, but anxious.' His partner said, 'It was not mine (the dream), but since I met him, he really wanted to do it, so it has become mine too.' Even on the day they were with me, for the first part of the day, there wasn't any shark; only because we stayed till late, were we able to see the sharks. While returning I could see his partner resting on his shoulder, and him resting with eyes closed, but having a big contented smile on his face, as our boat jumped from wave to wave.

Yes, it is love after all, and often times, one person has the ability to change our entire idea about the species, and lives that we live and perceive. It is often inspiration from one person, that leads us this far; that is why I take the act of inspiration very seriously. Ultimately, sometimes I feel, the most important reason for creative and unusual endeavours is to inspire the ones coming after us to go their own way, giving them a justification, however personal it might be to one's journey. We never know which bit of inspiration can change lives, and sometimes, the world.

Previously, I shared my thoughts on the potential reasons on fear of sharks, but it was interesting to note what the tourists felt was the reason for this fear, tourists who hailed from varied countries. People recognized that media and yes, in particular, *Jaws* had played an important part in creating the irrational fear of them. As John from Australia stated, 'It's *Jaws* and media; more people die from cows and bees, but they don't have big teeth. Whenever a shark attack happens, it gets sensationalized, but we don't hear about the hundred million sharks killed each year due to the fin trade. It is also marketing—if we repeated one thing frequently, you would tend to believe it, whether it is true or not.' And a lot of tourists seemed to know that 'we are luring them into aggressive behaviour, like jumping breaching, and it's not a fair representation of the sharks.' And yes, *Jaws* did seem to have a significant effect on people's perception and indeed fear of sharks, and white sharks in particular. 'It's quite amazing that even 40 years later, people get such a scare from the movie. I always wonder that the movie made the shark famous or was it already famous before that?' to which one other tourist noted that *Jaws* was just the catalyst; as humans, we always like our villains. All innate fears are being played out by villains in books and movies. 'When I was young, we had comic books where the heroes would be attacked by a shark or a giant squid. The sharks are always the villain; whenever there is a shipwreck, you would see sharks circling underneath.' Indeed, this image is now engrained in our common psyche. If you think even of *Finding Nemo*, the sharks are nice till they get the taste of blood; then, straight away they turn.

For some, *Jaws* had a lifelong effect, 'My mom saw *Jaws*, when she was young, she was terrified, and she never goes into the water.' Another tourist, referring to *Jaws* director Stephen Spielberg, said, 'I almost hate the man, he obviously put the fear in many people by portraying that poor creature as a killing machine, which they are not. I'm not going to go out swimming out there all on my own, just because my imagination would go wild. I'd be surprised if I didn't start playing the *Jaws* theme tune in my head when I was in the cage!' And interestingly enough, a lot of the tourists felt that the shark in *Jaws* was a male, and it was a man against the rogue male shark.

At the same time, *Jaws,* other media, and the fear of the shark transformed into people's fascination of the animal. Shows like 'Shark Week' (Robers, 2019) of the Discovery Channel and 'Blue planet' of BBC were among the few that created interest in people. The same American doctor, who had 'daymares' about sharks, admitted that *Jaws* evoked his interest in the sharks: 'When I was too young to watch it, my grandma let me watch *Jaws*. I fell in love with that movie, even if it was terrifying. From then onwards, as long as I can remember, I always used to collect shark pictures, get news about shark attacks, and so on.' Australian Philip mentioned, 'The shark programme put a little bit of fear in people, but it was piquing interest where there wouldn't otherwise be interest.' On the other hand, others noted, that even positive, and non-sensationalized documentary media representation of sharks at this point is only good to the people who were interested, which is the same argument about cage diving changing attitude towards sharks, i.e., only people who are interested in sharks come to it.

However, beyond the media, tourists feel that indeed there are some intrinsic features about the animal, and the environment it is part of, which taps into some deep-seated fear we hold as humans. Humans respond to the physical features of sharks, like their eyes and teeth: 'They look scary, they have more teeth than the entire Jackson five.' Tourists also state that they feel that there is something unearthly about the sharks: 'They look like aliens—it is the idea of the "other"; on one level, you recognize it enough with their eyes and mouth, but on the other hand, all the features are distorted a lot.'

The environment also plays an important factor in our fear of them, particularly the lack of visibility, and isolation in the ocean. It always comes down to the fear of what may be: 'Maybe it's like overcoming some elemental fear of drowning; the water is almost like a cauldron of what could be in it.' Ricky from Germany felt, 'When you look at the water and you think of the shark, it makes you feel that the water can be dangerous; it's what you don't see. In horror movies, dark shadows are scary, dark corners are scary. The sea essentially is a shadow, and you don't know what's in it'—a sensation not lost on me. A few times, I woke up in the middle of the night and there in the darkness, I would see the shapes of large shadows swimming, I would eventually rationalize with myself, the darkness of the room in my semiconscious state evoked sensations of the dark ends of the ocean, hence materializing the symbol of my fear. There was though one interesting evening, when, quite drunk, I was walking down an alley way in some country, talking to myself; when I stopped in my steps, I could see a large shark waiting for me in the dark end of the alley, I chuckled to myself, 'My friend has come to see me.'

It seemed that indeed these fears materialize from the realization that we as a species are helpless in water: 'We are out of our element in the sea. You can run through a forest if a tiger comes after you, but you cannot do anything in the water.' And most people I interviewed agreed that the murkiness or lack of visibility is what extenuates this fear, a fear which has effect on our interaction with our environment: the shark is an anthropomorphic, aquatic projection of the 'boogeyman', in this case, hunting us from both the dark recesses of the ocean and our psyche. 'I am not comfortable when I can't see the bottom, you are no longer in your environment, you have no idea what's lurking only 20 feet away from you, that can see you so much better..."I am gonna get you"—like the Grim Reaper. It's as if it knows your deepest, darkest fears, it's almost like the sea is mocking you. It is the fear of the unknown that is evoked in the water, and what is scarier than the fear of the shark? Every now and then you get that really eerie feeling, don't know what it is but something just tells you it's time to get out.' This particular 'feeling', of 'time to get out', or 'something is not right', has time and time again been recounted to me by divers. That included one man, who came out of the water in Kaikoura feeling uncomfortable, and later discovered that there was a giant white shark spotted there that week.

There was an interesting British man who told me, 'My cousin who saw the shark in Malta (one of the largest great whites ever caught of about 23 ft in the 80s) hates dark waters; he would never go in the water where he can't see the bottom.' One tourist even talked about the darkness of a cave, comparing it with the darkness of a shark's jaws: 'If you look at a shark's jaw, it is like looking into the dark abyss, and in the context that you are in a foreign environment, and they come out of the dark, and they come quick. Our unconscious minds are basically shadows, and you pull stuff out of it... and everyone is afraid of the dark.'

One of the reasons people fear sharks is the suddenness of their appearance 'out of nowhere', and the realization that they are above us in the food chain. If you're on land and there is a big predator, you can feel more secure with a sense or hope that you can run away from it, but with a shark, you are in a completely alien environment. As Shane noted, 'I have grown-up as a swimmer, but no matter how good a swimmer you are, it doesn't really matter. It is a bit of that instinctual fear, that there's something in the water which is bigger than you. You are in its environment, and it is

Shadows of the ripple, 2016

the apex predator of that environment. As soon as my head is above the water, I always have that feeling what is underneath me? I've swum near the beach before, and when I had a feeling that something is not right I came out. When you are underwater, at least you feel you are part of it, but when I am on the surface, I kind of feel I am bobbing around.'

It would appear that people were less scared when they actually saw the shark than when they didn't. A long-term dive instructor mentioned in a conversation that when we are diving, certain divers don't like diving in isolation, and middle of the ocean, because it is an alien environment, and we have no control over it. And it's that fear of what lurks below; I think it's linked to our fear of the ocean. As humans, we are in control of our environment.' Isolation does seem to be a fear. A tourist stated, 'I was always scared of them from movies and other stuff. I still won't go swimming out in the ocean by myself. If I'm with somebody, that is fine, but I will not do it myself; it's that subconscious fear. I think it's a sensible fear to have; it's that there's safety in number.' Then there was the nature of the attack, which Roseanne, an American lady swimmer, admitted, 'The fear of being eaten alive is pretty gruesome. Even if I understand all the statistics and the chances of being attacked by a shark are astronomically low, it doesn't help my fear.'

But there was acknowledgement of their perfection as a species, and their appreciation. A lot of it is love for the fish, often time using the term 'perfect'. A British tourist said, 'They have been around for so long, perfect killing machine. And there aren't many predators that size in the world. Sharks are one of the more primal ancient animals in the ocean.' There was anthropomorphic appreciation of the characteristics of the animal, often explaining how the image of the behavioural characteristics are so appealing as to become a personal totem animal: 'Well, they are menacing; they have those sharp teeth that could slice through anything like butter; and they are the kings of the sea, which makes them so respectable. Also, they are loners, and travel by themselves; I think that many people relate with it. Because sharks know what they are, that they are the apex predators out there, and they're so good at what they do. I like making my own path in the world; whales are always in a pod and reliant on each other, but sharks can go out on their own and be independent.' One tourist said, 'Animals like leopards and pumas have a magnetic presence. In Tibet, it is called 'Wa' which means magnetism, or power—sharks have the similar attribute. They are very powerful and magnetic, they come and go very fast, and you can't keep your eyes off them, you don't want to confront them, but at the same time, you cannot look away. They're quite prehistoric! Somehow, quite perfect as a species; I love form over function.' 'Form over function' comes from the architectural quote 'form follows function' by Louis Sullivan and is part of the modernism aesthetic.

The great white, specifically, has both: it's an apex predator, and it's a beautiful creature. One way to consider this is via a term I would like to propose: 'Interspecies—good gene indicator'. In the conventional understanding, a good gene indicator, or fitness indicator (Carter, Campbell, and Muncer, 2014; Gangestad, Garver-Apgar, Simpson, and Cousins, 2007; Haselton and Miller, 2006; Jonason, Valentine, Li, and Harbeson, 2011;

Miller, 2001; G. F. Miller, 2001; Roney, 2003; Simpson and Gangestad, 1992; Wilson and Daly, 2004) indicates certain physical, behavioural, cognitive, or social traits, like strength, physical beauty, musculature, intelligence, leadership, height, voice, economical affluence and provisional abilities, and creativity displaying good genes to possible mates. Of course, I do not mean it in the same sense with sharks. What I want to suggest is that interspecies good gene indicators are 'certain behavioural or physical traits in other species, which humans aspire to imbibe among themselves for their ideal self or aspire in their imagined mate.' The appreciation of the 'perfectness and strength of sharks' may be so attractive because of this kind of indicators. While this is not the place to as of yet engage properly with the details of the concept, I propose this as a seed of thought to think with in future conceptualizations of human and animal interaction, and animal symbolism and totems in human society.

Then there was the expectation of feelings arising from actually seeing the sharks that was expressed as both the love for the animal and the fear. Some wanted to see their power, some their teeth, some gain more respect of them, and some did not have any expectation. 'Absolute terror! Last night I googled videos of shark attacks and jaws for an hour. I am already mentally prepared to save my wife today.' There are a lot of unknowns, 'I think, I would be really excited, and at the same time, surprised to see them. Because I really don't know...I waited have for such a long time, and there would be so much emotion.' There was respect for the fish and their environment, and the tourist felt seeing great whites will deepen that respect. Furthermore, there seemed to be a recognition of their individuality among the tourists, and an inherent wish to be recognized by them as individuals themselves. An American tourist said, 'I hope it calms me down, having scuba dived for a long time, and always wanting to see white sharks, so I just want to get it done now.' Yet the individuality of the shark and how they interact with the cage shapes the feeling about the experience was also evident: 'I think I would be fascinated, not so much scared as long as the shark goes around and does not start shaking the cage.' Whereas, others say they don't have any expectations, 'I'm not planning my emotions for it, I just want to see how it goes. For me, ignorance is bliss; I would rather go into the cage, not knowing.'

There was hope for the adrenaline rush, 'I'm expecting that my heart would be racing, it would be a full-on rush, and it would be awe and amazement. Hopefully, it will not be hyperventilation (she chuckled).' But they did wonder how that experience would affect their later underwater experiences and how comfortable they would feel in the ocean after the encounter. But still, there was a wish of the tourists to be noticed by the sharks, 'Well, you can see the teeth in a museum; you want to look into their eyes; I want the shark to look back at me. I want the shark to be impressed by me, I don't want to disappoint the shark, because this is his space, I don't want him to think that I'm boring and leave.' With all these ideas, and images, and personal stories about an animal unseen, they awaited the meet.

The Abyssal Tourist of Foveaux

These fish—who all these humans come to see, how much do we actually know about them? Are they solitary mindless killers? Where do they stay? Where do they breed? The fact lies that we really know very little about the lives of white sharks in the region, and indeed, the world. In the recent article, 'Future Research Directions on the "Elusive" White Shark', Mark Meekan talks about our growing yet limited knowledge about global white shark ecology and biology, based on the contribution of 43 white shark scientists. They noted that the 'inherently low abundance and frequent use of pelagic habitats has historically made the white shark (*Carcharodon carcharias*) difficult to study due to the logistical and financial constraints of regularly accessing individuals' (Huveneers et al., 2018, p. 2). The review observed that from the perspective of organismal biology, we can predict that there is probably low to moderate white shark population size globally, but we do not have reliable abundant trends. We have limited knowledge of white shark reproduction strategy, litter size, gestation period, and duration of reproductive cycle.

There are also challenges of creating ecological knowledge of white sharks. Generally, due to tagging programmes, we have some baseline ideas about white shark migration of most global population, but the seasonal, ecological, and environmental drivers of the movement and population distribution are not yet properly understood. How critical areas of white shark distribution create ideal habitat for the sharks is unknown, neither are we sure where and how frequently white sharks travel beyond

The shadow just beyond our vision, 2017

the recognized hotspots. Furthermore, our understanding of white shark diet is also shaky. Then there is the lack of knowledge in socio-economic, management, and conservation concerns (Huveneers et al., 2018).

The knowledge about white sharks in New Zealand is also quite limited. The official page of the DOC website also comments about white sharks, 'Little is known of habitat use in New Zealand waters,' and, 'no data that could be used to determine population trends in New Zealand waters exists. As commercial fishers do not target this species and it is a relatively rare by-catch, non-reporting is common, making any estimates of abundance based upon reported catch or landings unreliable' (Department of Conservation, n.d.). However, the DOC completed a decade-long research project in collaboration with the National Institute of Water and Atmospheric Research (NIWA) on white shark population and migration (NIWA, 2015). In a report on 28 June 2018, the office of the Ministry of Conservation declared, 'The great white shark and basking shark have been classed as "threatened" for the first time and the future of these species is not positive [in New Zealand].' They note, 'New knowledge about great whites has confirmed an already suspected low adult population, which is either stable or in decline. A recent population estimate puts the number of adult great white sharks in New Zealand at between 590 and 750 and the total population including juveniles at 5,460 sharks' (Office of the Minister of Conservation, 2018, n.p.).'

One of the things I took from the meeting with Steve is how white shark research and the knowledge created is highly uncertain in New Zealand. The major knowledge creators are the experts from the governmental departments, which deal with the lives of the sharks including the DOC and the NIWA. Then there are others like the 'Shark Man' (Elliott, 2014), who claim to be the experts of sharks of New Zealand; international researchers like Steve Barry Bruce, a world-renowned white shark researcher from the Marine Biodiversity Hub at the Commonwealth Scientific and Research Organisation (CSIRO), and social scientists like myself.

This mixed bag of expertise often creates a complicated political dimension of white shark knowledge creation and distribution. Scientists not only disagree with each other's interpretation of white shark behaviour, but even doubt the validity of the collected data. The human conflict about cage diving in the region may allegedly be detrimental to shark knowledge creation. For example, the DOC's white shark survey in the region in 2017 did not go ahead, because of the political implications at the time (personal correspondence with researchers, names withheld). Then there are complaints, how external agents, like the 'Shark Man', burn bridges regarding the relationships that were built between the more experienced scientists and the residents of Stewart Island. (This particular individual allegedly got himself in confrontation with the Stewart Islanders, which was scarring to the relationships that other organizations were building with the them. For whatever it seems, the Stewart Islanders do not seem to be very appreciative of his 'expertise' and opinion.) Furthermore, as mentioned above, the DOC maintains that there is a lack of funding for white shark research in New Zealand, even if there was a decade-long white shark research in the region. At the same time, the cage divers allege that the DOC funds were not properly utilized for the research, rather on equipment peripheral to it (details withheld for ethical purposes).

Another group of knowledge holders are the cage dive operators who have been interacting with the sharks regularly for a decade, and I argue they hold significant ecological knowledge about the sharks, which should be considered with more credence than it often is. The problem, from this shark researcher's perspective and experience in dealing with both groups, is that the scientists do not pay much attention to the cage divers' white shark ecological knowledge gathered from their practice, while the cage operators complain that scientists' views are given value simply because they have degrees, but they invest very little time actually in the field, observing the sharks. But I would argue that cage operators are a treasure trove of knowledge about the sharks, significant for global shark knowledge.

While cage diving, Mike's spiel about the sharks helped the tourists understand the lives of the sharks in the region. Here, I situate this spiel and the knowledge emergent from it alongside the limited published academic ecological knowledge of white sharks in the region.

There are some elements where most ecologists are in agreement with the cage divers. Firstly, all agree that white sharks are long distance travellers, and are in the region by their own choice and not for cage diving. Ecologists note that white shark cage diving practices globally take place where the sharks aggregate naturally, primarily due to pinniped colonies (Gallagher and Hammerschlag, 2011). 'The industry takes advantage of the naturally higher local abundance of sharks and the reliability of more regular encounters at these sites, rather than attracting sharks to areas where they would not normally be' (Bruce, 2015, p. 2). Mike, in his spiel to the cage divers, mentions the same thing:

> Now, these islands have lots of seals in them. Around July–August, the females give birth to pups; about now, they will be eight months old, and you could see them frolicking in the waters, probably after lunch when it warms up a bit. And the pups are

> very fat, juicy, very naïve, and bite size, just the way the sharks like them. So, when we say sharks, or great whites, we're talking about the same thing, because there are no other sharks present here. The reason behind this is that the great Whites are the rulers of the ocean, and they predate other sharks. [Interestingly though, in 2016, when I was on the boat, we saw a mako and a blue shark, which has never happened before, near the cage diving operations. The only tentative explanation we could come up with was that these sharks could be there because the white sharks were not around, and there must have been an unexpected warm current which was flowing and bringing the sharks through.] So as the seal pups start getting more confident, they start swimming further and further out of the island. Of course, what happens then? (He presents a question to the tourist) ... Well they end up being food for the great whites. (Dictaphone interview, March 2016)

Alison Balance in her book, *New Zealand's Great White Sharks* (2017), worked with New Zealand shark experts—Clinton Duffy from the DOC, Malcom Francis from the NIWA, and underwater videographer and researcher Kina Skully (who was, interestingly enough, another white shark bite victim in Chatham Island. In 1995, Kina was diving for paua, and as he was surfacing, the shark bit him, much like in the case of Vaughn. Reportedly, he hit the shark with a rock and reached the surface, where his friend helped him). Alison paints a picture of the sharks as animals, who are intercontinental travellers and have come to New Zealand at a specific time of the year:

> New Zealand great white sharks are nomads that roam around the south-west pacific. They spend about four months of the year in New Zealand, from late summer to early winter; when they are in New Zealand, teenage sharks and some adult males spend lots of time at shark hotspots near New Zealand fur seal colonies, at Stewart Island and the Chatham Islands. In Spring,

> from September to December, there are hardly any sharks at Stewart Island. Once they are big enough, New Zealand white sharks make long oceanic round trips of up to 9,500 kilometres. (Ballance, 2017, p. 78)

She also points out that scientists believe that young sharks are most common around Auckland and Northland; there was a shark bite in Northland in October 2018 (*New Zealand Herald*, 2018). Indeed, a handful of academic articles published on the white sharks of New Zealand indicate such a large-scale migration trend of the sharks (Bonfil, Francis, Duffy, Manning, and O'Brien, 2010; Duffy, Francis, and Bonfil, 2012; Francis, Duffy, and Lyon, 2015). Dr Francis's team came to the conclusion from their data that there is an aggregation of the sharks around the Titi Islands in Foveaux Strait from late summer to early winter. Researchers predicted from tagging data that '...most white sharks aggregating at New Zealand pinniped colonies migrate annually to subtropical and tropical regions of the Southwest Pacific and Coral Sea and that at least some return to the tagging location. All except one shark [out of twenty-five white sharks tagged with pop-up archival transmitting tags] exhibited a period of temporary residency at or near the tag site, followed by an oceanic phase associated with movement to subtropical and tropical habitats' (Duffy et al., 2012).

The cage diving operators agree with this understanding. Time and time again, over the last 10 years, when they have gone out to Foveaux in winter, the reliability of spotting any white pointers has been drastically low in comparison to the summer. However, there are always some sharks who stay back (which may be the case for the shark that stayed near the region indicated in the last reference). 'We have two types of sharks here: the resident sharks and the tourist sharks. The resident sharks are the guys that stick around here all the year round, all through winter. They are not the sharks that we are looking for; we are looking for what we call our tourist sharks. They come here around September–October, and they come from as far as South Africa, Australia, the tropics, or even the southern oceans. They're all solitary mature males, because great whites out in the open oceans are solitary animals' (Dictaphone interview, March 2016).

An interesting anecdote can be found in Crawford's interviews with John Malcolm mentioning these resident sharks: '...you'd get the odd one all-year round. It was more of a browny and white than the black and white. You know, it was still a big White Pointer... Some of the White Pointers are more brown than black, and we used to call them the winter ones' (John Malcolm, Crawford, 2017).

This is where the agreement about shark knowledge becomes more problematic, in particular, about two topics. First, do white sharks create some sort of a loose pact and show social traits in Foveaux? Second, do white sharks procreate in Foveaux? I present here what the cage divers have learnt from observing the sharks for the last decade on these topics.

The Social Life of Sharks around Foveaux

'I've always been under the belief that great whites are a lot more intelligent than we believe. A hell of a lot more intelligent. [George Eskie] always said to me, "They're known as the 'professor of the sea'"' (Craig Hind, Crawford, 2017). Yes, it does seem that white sharks are quite intelligent, more intelligent than even we consider them now. As a matter of fact, this whole issue of 'intelligence' observed from the human perspective is another can of worms which I would rather not open now. And then, there is the matter of white sharks being 'social', which is another highly debated issue.

On a personal note, one story really struck me as I was going through Steve's transcripts:

> ANDERSON: Yeah. And we were in Paterson's Inlet and we had heard stories about these white pointers that have been coming into Oban. Swimming around Oban every day [in the early 80s].
> CRAWFORD: Halfmoon Bay?
> ANDERSON: Yeah. And people were saying the sharks come in about 11 o'clock every day. And we thought, 'No shit.' So, we parked in at Oban, and walked over the hill one morning to the

> wharf there. And we were waiting and waiting. And then just like clockwork, there was probably like 50 people on the wharf, and like clockwork, these sharks—they didn't come really close to us—but we saw enough to know that there were two small ones and one large one. And they were just swimming all together...
> CRAWFORD: They're cruising together, these three fish? They're clearly together—two smaller ones and one bigger one?
> ANDERSON: Yeah, they just did a loop, they kind of disappeared... I think they were getting round together, these animals. And it was the next day or the day after, we were back at Paterson's, and we heard the story that Joe Cave set a net... 'That's bloody terrible, why would you do that?' As far as I was concerned, they were a great attraction, you know? So, I was horrified he'd come back and get these, and pulled them up on the wharf and took photos of them, and took them to the local dump or whatever they did with them. Because they weren't aggressive, they weren't causing any problems, they were just swimming, they weren't swimming crazy. They were just like drive-by, real slow.'
>
> (Allan Anderson, Crawford, 2017)

The same incident can be again found mentioned in another interview of Crawford:

> NEWTON: Yeah, I knew about that. He [Joe Cave] got about three of them or something, didn't he?... Mother and...
> CRAWFORD: Mother and a couple of offsprings?
> NEWTON: Yeah.
>
> (Ross Newton, Crawford, 2017)

It is hard to say, if indeed these fish were mother and children, and the mother showing the children the lay of the land. What is important to consider is that the local people thought so, which means, they felt that these animals were capable of having such bonds. Indeed, it would not have been hard for us to imagine this family bond, if it was a group of dolphins doing so, because we know dolphins as mammals can create family bonds... But we are told with certainty all our lives that fish, especially, 'big predatory' fish, cannot have family bonds.

There is not enough empirical data supporting any claim that white sharks possess social structure, besides the hierarchical structure. In 2016, researchers utilizing the method of photo identification explored if white sharks in a near coastal environment of South Africa were forming intentional association among individuals, or did they simply share the same space at the same time? From their findings, they concluded that the sharks they were studying indeed were randomly sharing space and did not indicate much of a social behaviour among themselves (Findlay, Gennari, Cantor, and Tittensor, 2016).

At the same token, Findlay et. al notes in case of small coastal reef and benthic sharks, there is more and more evidence of social interaction, concluding that 'recent and growing evidence show some shark species organized into structured social networks' (Guttridge et al. 2009; Jacoby et al. 2010; Mourier et al. 2012), in which they engaged in social behaviour. Benthic species such as Port Jackson sharks (*Heterodontus portusjacksoni*) and Small-spotted Cat sharks (*Scyliorhinus canicula*) rest in groups, during which social interactions can occur (Sims et al. 2001, 2005; Powter and Gladstone 2009). Individuals of Lemon, Nurse, and Cat sharks can display preferred associations (Guttridge et al. 2009; Jacoby et al. 2010; Guttridge et al. 2011), while Blacktip Reef sharks can maintain long-term dyadic associations (Mourier et al., 2012)' (Findlay et al., 2016, p. 1763). So, even if the evidence is not present at the moment, it is not improbable that white sharks can also form association among themselves.

In the documentary 'White Shark—The Nature of the Beast' (Bhana, 2000), Bhana, a veteran film-maker, who has directed and produced 28 films exclusively on sharks, explored an experimental study when a few sharks were tagged near the Chatham Islands and South Australia to observe any potential social cooperation in moving and hunting, and their results suggested that indeed, at least in their case, the males, females, and adolescents seemed to be hunting and moving together in a group. When I inquired one of the New Zealand biologists about this documentary and the scientific validity of it, he immediately negated it as media sensationalization and nothing else (name withheld). Steve (Crawford), on the other hand, mentioned that even though it still used the same sensationalizations as other shark documentaries do, the general idea of white sharks engaging in such highly social behaviour remained quite plausible, and based on observations made by indigenous and local knowledge holders, highly probable. In a Discovery Channel documentary of 2019, two male adult sharks of similar size, considered to be siblings, were found living together across multiple years in close proximity in Western Australia, with no sign of aggression towards each other (Discovery, 2019). Finally, a very recent article, which documented the aggregation of white sharks in Neptune Island, South Australia (Schilds et al., 2019), used 'a combination of photo-identification and network analysis based on co-occurrence of individuals visiting the site on the same day to elucidate the population structure and aggregatory behaviour of Australia's largest aggregation of sub-adult and adult white sharks...and found that white sharks did not randomly co-occur with their conspecifics but formed four distinct communities' (p. 2). In other words, they might actually create some form of groups or packs, be it temporary.

Mike and the team believed that they did see some form of social interaction among these sharks on a regular basis. Mike said, 'So, the mature males measure from about 3 m [10 ft] to just over 5 m [15 ft], weighing an average of about 1.3 tonnes as they're fairly large animals. They turn up here, and as the numbers start increasing, they start congregating around this particular region here. We don't know exactly why they come here to congregate, but obviously they don't stay in one spot, they swim around about 10 to 12 km around this area [around Edwards]. It's not a big area, considering we have on an average about 120–130 white sharks in that area.' The validity of these numbers is anyone's guess; as I have mentioned before, the gap in population knowledge is quite apparent. Because a lot of estimates are made from scientific research and also from fish catch and tagging data, the observations of the operators are equally relevant in the same way.

As they congregate, they start to create something what we call loose packs and they form a hierarchy. You will see sharks with bite marks on their tail areas—that's how we can understand that they have been fighting among themselves. When you are a big white shark, the one thing you don't want to do is bite another big shark around its head, because then, you can get bitten back. And within these loose packs, they actually hunt, and have been known to hunt in packs like wolves. That means, we very rarely just see one shark at a time. Also, what we don't have here is young males or juvenile sharks. This area is predominantly an area where they come to mate. In fact, juvenile sharks here will get predated upon. We saw a juvenile shark about 2 m (6.5 ft) long with a large bite mark on his side, and he was found dead further out towards the Rupuki Islands [yet again, this was disproven in 2017, when Rocky (a 9 ft shark), and another shark (less than 7 ft), were seen near the boat for quite a few days]' (Mike Haines Dictaphone recording, March 2016).

However, Mike, Maca, and Carwyn felt something else was going on with the sharks and the bait. From their long-term observations and engagements with the sharks over a decade and about 100 trips a year on an average, they believed that the sharks strategized to get the bait. This was apparently observable when sharks had become comfortable with the boat and the bait, especially, if a particular shark repeatedly returned to the boat in successive days, which was not the case in most instances. Maca (who has been a wrangler for five-six years) thought that the shark recognized the wrangler and as much as the wrangler learnt about a particular shark, the shark learnt about the wrangler as well, and tried to 'trick' her/him. Furthermore, when there were multiple sharks at the same time (and there can be quite a few, the maximum being nine at the same time a few years ago, when Mike had to stop the operation, as it was becoming too risky), it seemed sharks did act together, with some form of loose group strategies to get the bait. Maca had experienced that a few times, when one shark distracted him by passing close to the bait near the surface, while another shark grabbed the bait with a vertical lunge. The question may be raised, if the shoaling tendency evident in this interaction is related to the sharks creating a pack, or it is multiple un-associated fish being attracted to the same area by the same olfactory stimuli? And if at all the sharks are forming packs, the question may be further raised—are they are hunting together, with pre-formed packs in Foveaux, or are they coming together near the boat for similar behavioural cues and creating the packs for the specific purpose of acquiring the bait? I can neither validate, nor disprove, these claims of group strategies; however, I have found no reason for these stories to be made up for mere amusement. And this can be empirically investigated in future research.

Mike had one more argument about social interaction among the sharks. At the beginning of the season, the male tourist sharks who came to Foveaux kept a lot of distance among themselves, but as the season progressed, they started coming really close to each other and create what the operators believed to be a loose pack. They came so close that they even ended up bumping against each other. It was important to point out, however, that according to the cage divers, this had happened even before the cage operations had begun operating regularly in the region.

Just as Jane Goodall explained the behaviour of Fifi, the chimpanzee (Klein, 2015), I would like to propose, based on the decade-long observations of cage diving operators, that the acceptability of decreasing personal space among the male white sharks in Foveaux could be attributed to an increasing social acceptance among them, had they been members of a human social group.

White Shark Copulation in Foveaux

Another debated topic of discussion is that of white sharks potentially copulating in Foveaux. The national ecologist I spoke to regarding this stated that there was no empirical data to prove or disprove this. Mike continued his spiel: 'Now, what happens next [after the males start congregating around Foveaux] is that they are all feeding themselves, getting in good condition. What happens next? The trouble makers start arriving. And who are the trouble makers?' He sends the question out to the tourists. One young man hesitantly states, 'Females?', and everyone starts laughing. 'What we don't have here is young males or juvenile sharks. This area is predominantly an area where they come to mate, but before the big females come, we need to have large males here. Because the female obviously mates with only the largest of the sharks; what they are after is obviously the best genes possible for their offsprings. The females are all mature, measuring from 4 m to just over 6 m; they can weigh up to 2.3 tonnes. So, there again, the females are eating lots and lots of seal pups. They're making themselves look pretty for whom? Obviously, the big males! So, how do two very large animals get romantic? Now, it has never been seen before; in fact, it is one of the mysteries of the shark world. What we have seen is the reef sharks mating in tropics. And we take it for granted that great whites, being sharks, would do the same thing. The male goes up to the female, bites on hard to the side of the female and gets a good grip of her. Once he gets a good grip of her, he flips her upside down. So, what happens to a shark when they're upside down? Again the question is given to the tourists. Someone replied in a query, 'Docile?' 'Yeah, they go docile or inactive or what we call tonic immobility. Once she is in that stage, he can have a way with her. Well, it is a relatively short affair. How do we know that? Great white sharks must keep on moving, as they need a lot of oxygenated water to pass through their large gills to keep them breathing. So, once they're finished, he gets out of that place fast and she goes along her own way. Of course, the female mates multiple times during the season when she is here. In fact, she could stay here for 3–4 months. Once she knows that she is pregnant, she would move to warm waters, to the top of the North Islands or even the tropics. The males would stick around here till mid-July, and then they would disappear, roundabout a one-week period. It is because, first, the females are gone, and second, the seal pups are being kicked off the island by their mothers, because in July, the females give birth again. So, the pups leave the place for about two years and then they turn up again in around September–October, and the whole process starts again' (March 2016, Mike Haines, dictaphone recording).

I asked Mike if he really believed that white sharks were copulating in this region. If so, why didn't the biologists believe that? He explained: 'First, if you think about the number of days the technical advisors and the biologists from the DOC spend here shark diving, it is not a lot. So, they are not getting to see it first-hand! Now me and Peter are spending in excess of 120 days out here each season. So, it's more than a thousand days that we've been out here interacting with the sharks. As for females mating here, we've seen it every year—females with bite marks around their gill plates. Even this year, we have seen it. You have seen it [as a matter of fact I did]. Last year, we hadn't seen any females, and there might have been a number of reasons for it, like El Nino in the Pacific. I will tell you a story about a couple of females that we had here. I was talking with one of the scientists up in Auckland, at that time two females still had active tags, one was sitting around Centre Islands, which was about 40 km from here. And the other one was sitting around Fiordland, which is about 94 km from here. They both sat there for about six weeks; we were all trying to figure out what they were up to. This was around February, when we realized that we didn't have any big males! In early March, we started getting some big males over 5 m in length; guess who turned up within two or three days? The females! The magic of it is how did they know that the males have arrived? Now these are things we would obviously love to know about, but have no idea, none whatsoever. And unfortunately, the satellite tags that were on the sharks last year [2015], have now gone flat.... So that's the end of that! When the females finally left here, one went down to the Auckland Islands about 800 km south of here. She stayed down there for two weeks, and then went back to up north Queensland. So, why she went down to Auckland Islands we have no idea; maybe, she was visiting someone down there. Now how did they communicate about this? We don't know!' (March 2016, Mike Haines, Dictaphone recording)

The question of white shark copulating in Foveaux is one we do not have answers to; however, drawing on the observations and claims of cage diving operators, I ask myself how can we definitively say that they are not? At the same time, as a social scientist, I am aware of the lack of enough documentation, although there are anecdotal experiences that can be found in Steve Crawford's documentation (*White Pointer Chronicles*). Based on the decade of observation—of the arrival of the sharks in a specific order: first, the younger males, then them getting bigger, the infighting among themselves and possibly forming loose packs, the arrival of the adult females, and the scars they have on their fins thereafter, and finally, them all leaving—I propose that the yearly white sharks' congregation in Foveaux Strait can be explained through white shark copulation rituals. Yet again, something that needs to be empirically explored.

The reality lies, that it is almost comical how little we know about them, and how much people think the 'experts' actually know. Just like in any system of knowledge creation, it is the dialectics at play, and all we can do is accept our lack of knowledge and be sceptic and not cynic.

Absence of the Sharks

I thought, all a person needs to do is hard work, honest hard work, and life finds a way. My work is to understand and create knowledge, that is all I did for the last many years. And now this house has become my catacomb again; maybe, I shall perish here. Maybe, this is the price to pay for being a scholar. Perhaps, this is the actual work...remaining awake and waiting!

'A defining feature of many angling trips is the failure to encounter! The absence of an individual on a particular occasion does not preclude the possibility of an angler's becoming fish in the hope of catching' (Bear and Eden, 2011, p.8). For the first two months of my research there were no white sharks; to be precise, there were hardly any satisfactory encounters between the sharks and the cage divers. This absence not only had a significant effect on the tourists, but indeed on the operation, let alone on myself, waiting to observe the human–shark interaction. White sharks as apex marine predators play a crucial part in the marine ecosystem. 'The loss of sharks will result in their prey species becoming abundant; this is a result of the prey being released from predation, which then enhances the prey's predation on the next trophic levels' (O'Bryhim, 2011, p. 8). Cap (Mike) also mentioned that since the Titi Islands/Muttonbird Islands (Edwards and another island in a group of islands named after the Titi birds or the chicks of the sooty shearwater) are home to a large number of seals, and the white sharks prey on the seals. The loss of the sharks may mean an increase in the seal population; not only will it have a significant effect on the population of the smaller fish, but on the muttonbirds themselves. According to Mike, an excess of seal pups may result in the seal pups disturbing the burrows of the shearwater chicks, which will not only have an ill effect on the bird population, but may be detrimental to the age-old Māori custom of muttonbirding too.

The absence of the sharks had serious implications on the shark tourism near Bluff. On my first day on the boat (24 March 2016), I saw Maca getting suited up and going in the water to look for the alleged carcass of a shark some people had dumped near Edwards. Because this had never happened before, Mike did not know what the protocol should have been for the people who did not see the sharks on a trip; so, in quite a few cases, he cancelled trips for new tourists and took the same divers out for more than six days, despite his own financial losses. The news of the absence further created ripples in the tourism circle all across Southland. Even in Queenstown, tourists got to know that there were no sharks in Bluff (which I heard from the tourists coming to the trip). On Tripadvisor, the absence of the sharks raged a storm of bad comments, contrary to all the happy ones when there were sharks. This also had an impact on the economy of Bluff, as the business that tourism brought to the locals (engaged in hospitality and food) to the visitors fell significantly.

Mike and Peter had to face a similar situation in 2014, when there were no sharks in the region for two weeks instigated by an alleged catching and killing of two sharks. However, the sharks returned after two weeks; so,

(Left) Maca getting in the water with a rope attached to a buoy, which he can use if he finds a shark, and this also helped us to keep track of him, 2016; (Right) Maca underwater, as we all hold on to our hearts, only the anchor line sticks out which was to attach to the shark, if one was found, 2016

why were they not back this season? Steve explained to me, 'The ecology of a region alters from year to year and decade to decade, and so, we cannot predict what is going on. What is important at this point for the cage diving operation is whether or not Mike will have enough sharks to keep on operating in this season' (personal correspondence, 2016). However, stopping the operation was not an option for Mike. A similar experience had been documented by Samantha Hays in 2014, when she had gone out cage diving, and there were no sharks. As reported on News Hub online in 2014:

> We wait and look, and we wait some more, and the minutes turn into hours. So, where have all the great whites gone? Mr Scott says in seven years he has never failed to find them, but in the past few days something has changed, and he is worried. Where have the sharks gone? ... The word on the street here is that a fisherman caught a great white in his net, and he's dumped it on the bottom here, which is causing us a heap of issues because the sharks that were here are gone. Is that what has happened here? Has the conflict between the tourism operators and the locals spilled over into the cynical slaughter of a protected species....' (Hays, 2014, p. 2–3)

The scientists and the cage dive operators believed from unpublished tagging data and observation that the sharks made a triangular loop around the Ruapuke Island, the Central Island, and the Edwards Island. It was assumed by shark experts of the region and the cage operators (no published data) that they took roughly three days to complete the trip. We were trying to understand if sharks are actually deterred by the smell of other dead sharks, even though I have not found any data supporting this hypothesis, at least with great white sharks. Another scientist from the DOC (name withheld) in the region also mentioned that they have heard of such incidents, but they too have found no empirical evidence on it. Allegedly, the cage divers pointed out there have been long going incidents in Australia, when some whites were killed and their bodies were dumped in consecutive intervals; others did not come to their normal region, or follow their usual migratory pattern.

Even though much evidence of that could not be found, what was interesting was the interaction of killer whales and white sharks. There have been a few reported incidents in Australia and other places where white sharks were killed by killer whales and the sharks left the region. A recent article published by Salvador J. Jorgensen and team (Jorgensen et al., 2019), investigated long term data of ecological interaction between killer whales (*Orcinus orca*) and white sharks who shared a foraging site, at the south-east Farallon Island. The data indicated that that even short visits by killer whales to the region created the absence of white shark feeding activities, and the sharks vacating the region for as long as a year. Apparently, orcas hunting sharks and change in the ecosystem had also caused the absence of sharks in Cape Town, South Africa, in 2017 and 2018. An anecdotal but fascinating account of a similar kind can also be found in Gavin Maxwell's book, *Harpoon at a Venture* (Maxwell, 1952). The author attempted at establishing a basking shark fisheries in the Hebridean Island of Soya in 1945. With trepidation, Maxwell notes, 'I do not know if one can safely draw any inference from the fact that we never saw sharks [basking sharks] when killers [killer whales] were in the immediate neighbourhood' (Maxwell, 1952).

Why this was happening, we did not have any conclusive evidence. However, there is one interesting observation to be made here, and that is, killer whales actually kill white sharks, and even consume their liver (Rigney, 2019). There was an incident observed in 1997 (Daley, 2019), and also in 2017 (Mandelbaum, 2017) in South Africa (at which point there were no sharks observed in the region for weeks). So, do the sharks leave the region because of these predators? Do they leave after one shark is killed? How do they know the whales are in the region? Or how would they know another shark is killed by the whale? All of this leads to another question: if (for the sake of argument) white sharks can understand that killer whales have killed a white shark in the region and they leave, is it not possible that they do the same when a human kills a white shark. Even if it is not related to the release of certain chemicals in the water, which is recognized by other sharks as a sign of distress. Maybe, it is some factor empirical investigators have not considered yet. Ultimately, it is just another predator hunting them! These questions remain to be answered.

Then there is the personal history of the sharks, which may also affect their interaction, presence, and absence in a cage diving situation. There was a shark that had come one day to the boat (2016), but left within five minutes. Mike thought it was one of the resident sharks; so, he had no interest in the cage, and he may have even come near the region because of the fishing boat nearby. Mike also thought that even if these were tourist sharks, coming from Australia, they would not perhaps show any interest, because they have already had experience with cage diving operations in Australia and realized that they don't get any substantial energy return (food) from the interaction with these kinds of boats. Mike noted that there was one resident shark who approached the cage one day, bit on it, and thrashed it around, and after realizing it wasn't something he was interested in, never approached the cage again. Especially because of the effects of all the stringent laws about the use of baits that are now being enforced in America, Australia, and New Zealand (Bruce, 2015), sharks who repeatedly see that there isn't much energy return for them in these settings, may not engage with them, sharks having sophisticated learning abilities (Klimley, 2013), and white sharks being global travellers. They have been known to travel from California, after feeding on elephant seals in September, to Hawaii. There is also satellite information of a shark named Nicole, who travelled from South Africa across the Indian ocean, to and from Australia—a total of 12,000 miles within nine months (Skomal and Caloyianis, 2008; Star, 2018), using their electromagnetic and other senses (Klimley, 2013). It might be that sharks are able to decide how they negotiate with cage diving operations in one part of the world, basing on their past experience with the same at other places (again something which needs deeper empirical investigation).

When the sharks were gone, different fishermen notified Mike of shark spotting. One of his friends mentioned that a big shark followed his boat off Ruapuke Island about 20 km from Edwards. Others saw them near Bishop

Island (distance not sure), and then a lot of them around Rugged Islands, about 25 km from Edwards. And after that, there are the Sub-Antarctic Islands, like the Snares Island, where the sharks go anyway, possibly because of the abundance of seals. This is only possible because white sharks are warm blooded, as Klimley puts it, 'The white shark can tolerate the cold water in temperate and subpolar latitudes because of its ability to keep warm... It is likely that the white shark evolved this capability to exist in the cold waters where its prey, seals and sea lions, are common' (Klimley, 2013, p. 101).

One more reason for the sharks not being here may be related to the El Niño. At the time of this research, there had been a drastic change in animal migration pattern at many places, like whales in Canada and North America, and other fishes, including sharks, had gone to much deeper depths. But at the same time, from what Mike said, they had sharks from December, but suddenly something happened, and the sharks disappeared and were not to be seen in the region after that. It may be that killing was a trigger incident, which pushed the sharks away towards the Sub-Antarctic, where they hunt anyway. It is important to note here that this is conjecture, but part of what an embedded researcher does, that is, to draw upon the perceptions, observations, and knowledge of the locals. There is then also the fact that global fishing might actually have decreased the number of sharks. One has to remember, one of the primary foods of large predatory sharks is smaller sharks; so, fishing and environmental degradation may not only directly kill the sharks, but also indirectly through the destruction of their food sources.

If the argument is that there was a dead shark in the region, and the carcass released enough ammonia to keep all the other sharks away, then there would be a noticeable increase in the level of ammonia in the water. Consequently, for the first two months of the fieldwork, I recorded basic chemical indicators of marine water health (i.e., nitrate, nitrite, and ammonia) to document any observable indications of such claims with rudimentary marine chemical kits. However, I did not find any significant irregularity in the levels. Furthermore, this raises the question—in such a big ocean, can minuscule secretion chemicals like this last for a long time,

The presence of other sharks like a mako shark during the absence of the white sharks, 2016

suspended in one place? Can it really cause a large scale and long term movement of a species? Furthermore, the investigative design of this study was problematic. Even if the shark was killed, how could we be sure that the chemicals I was testing were the exact chemicals which indicate the presence of the carcass to other sharks, initiating their absence. There were no data available to me of the chemical levels before the alleged killing, for any comparative analysis. Also, I was able to collect water only from the surface, because of lack of equipment, and in the training of the proper protocols of such ecological studies, this would have invariably been unable to detect if there were indeed any significant trace elements near the sea floor, where a potential carcass might have been. Hence, I do not present the readings here, but still methodologically, this effort should be noted, as to how future marine anthropological research can use ecological tools to create a holistic perspective of their research objective.

Post fieldwork, further literature survey (Klimley, 2013) revealed to me, that basing on the same understanding that rotting shark meat/carcass prohibited other sharks to feed in the region, even vacate the region. Scientists have tried to find a chemical cure to stopping shark bites—a matter of interest in the US navy, particularly after the incident of the USS Indianapolis, where 60–80 people died from being eaten by sharks during the Second World War (Oceanic Whitetip Sharks were possibly the main species involved) (Vladic, n.d.). Scientists found that the most abundant chemical from a decomposing shark was ammonium acetate [not ammonia] and it was further found that copper sulphate was also instrumental in stopping some sharks from feeding. So, a chemical shark repellent, 'the shark chaser', was made with copper acetate (as ammonium acetate was more stable in marine water), and it was mixed with black dye, so the swimmers could see the spread of the chemical. However, field experiments showed that the chemical did not work for a significant amount of time (Klimley, 2013). Even after that, more research has been done on creating shark repellents, sometimes with electro emitting bands, different visual tricks, and more composition of chemicals from shark carcass with very limited promise, especially with white sharks (Bethea, 2017).

So, if the corpse has been a trigger incident and caused all the sharks to leave, it could potentially mean the actions of one or two human individuals essentially caused the change of ecosystem of an area, which might have been a pattern going on for thousands of years. At the same time, whatever is happening could even be a chain effect to something bigger. Ecosystems do evolve from year to year and decade to decade; so, this could be a 'natural' evolution of the ecosystem in the region, which had little or no relation to human actions, at least directly. It could have also been that some orcas hunted white sharks in Foveaux, unknown to the humans, which led to the absence of the sharks. Maybe, it is the act of killing, which might be causing these absences, rather than the chemicals released from the carcass themselves, be it by killer whales or by humans. Or the most probable scenario is that there is some factor, or a combination of factors, that we have not even considered yet.

The Silence of the Sharks

When the sharks were not there during my first few months, I took it as an opportunity to connect with all the people stranded on the boat for hours and days. There was an eclectic group of tourists from March to April 2016, who came seven times to see the sharks, but remained unsuccessful. This absence of sharks further challenged the idea that the cage diving practice 'trains' the sharks of the region to come and stay where they would not come otherwise. It rather reinforces the argument (mentioned in previous chapters) that the cage operators use the opportunity of the sharks being around these pinniped-rich regions to run their business, and it demonstrates the significance of the sharks' agency in creating a successful encounter. Field notes from a few of these days depict the frustration and effect of the absence of the shark on the tourists, the operations, and the researcher, observing it in no specific order. The dates do not follow one after the other because of the weather. On many of the days, the trips were cancelled. So, for the tourists who went to the boat for more than six days, the cumulative days in Bluff was substantially higher.

There was a German couple—Stefany and Oliver, an Italian chef called Marco, and a young British woman. I got close with them in the shared sensory and social experience of waiting for the sharks in the cold and often turbulent Foveaux in a small boat with perfect strangers, and my discussions with them over those few days demonstrate a unique case study about the effect of this absence on the tourists' mental and physical beings.

The grey ocean, 2016

Journal Entry
10 March 2016
2 pm
Edwards Island

We've been here for hours today. No sharks. The sun keeps on shining. Now, it is a beautiful morning. We keep on baiting. Most people have taken off their wetsuits and put on their jackets. I can see the seabed and all the kelp below, the fish nibbling at the chum, with the sound of the seagulls feeding on the chum falling from the chum bucket. Sometimes, the tourists and I sit down and discuss different things, anything from food, to travel and politics, from Greece to Bengal, and even have heated debates about Trump and America. Divers will often boast of the things they have done, certain things will grant them more social capital than others like diving with whale sharks,

Carwyn waiting for the sharks, even if they did not come for days, 2016

hammerheads, manta rays, and deep dives, but at the top of the list was almost always the charismatic white shark.

Mike is always hopeful, at least in front of the tourists (maintaining his own social front in front of his audience) and is trying to keep everyone in a good mood. There were some dolphins yesterday near the boat, and everyone started saying that it was a dolphin watching tour and not a shark diving one. But one of the tourists mentioned, even if he did not get to see the sharks, at the end of the day, they are wild animals, and you get to come out and be near an isolated island, meet nice people, have nice conversations. So, it was definitely worth for him; it was not just about seeing the sharks. For other tourists, it is almost like a pilgrimage and they expect to sacrifice something for the experience. 'Well, if the experience of diving with sharks is really that good, then we might not get it straight away and like every other good thing...have to wait for it.'

There was a strange grin on everyone's face, when they came onto the boat with the divers having already come a few times and not seen the sharks. It is as if everyone is knowingly taking a chance even if there is no shark there today. But still, people are hopeful, and the day is quite nice, no one is getting in the water if there are no sharks, as there's really no point in getting cold. Generally, the tourists are optimistic, even after a few tries, and they recognize the unpredictable nature of it. One of the tourists said, 'Today is my fourth day on the boat. I came one month earlier; this time three days. It has been cancelled for weather two times before too. As my sixth attempt (she went on her 10th before finally seeing the sharks), I am emotionally and physically ready for the sharks.' Oliver said, 'After five days, it is amazing that we are still excited. We have seen penguins, dolphins, sea lions, blue sharks, seven gillers, albatross, barracudas, and yes, seagulls.' I said, 'Why do it again? You might fail tomorrow.' Steffi replied, 'We might not!' I asked Marko,

The tourists waiting for the sharks inside the cage, even looking at the back of the cage, 2016

'Why don't I see frustration in his eyes?' He said, 'Well, it is really hard to get me angry, and besides, I am out in the ocean every day. I like it.'

21 March 2016
11 pm

Today's trip has been especially operated to bring the people out who have not seen sharks for the last few days. But since all of them were not available, Mike has taken some extra new customers, and the ones who returned are a group of four. All of them are determined to see sharks, and Mike never had anyone coming out to see sharks for so many days without fruitful results. It became evident that seeing the sharks at that point was not about just the sharks themselves. Rather, it was a psychological challenge to many of the tourists. Maybe, it was even a sense of entitlement relating to the time and resources they have spent to see the sharks, but have not been able to.

The day again started off with an imagined experience of encountering the sharks. Stefany said, 'Oh, we will cry so much when we see the sharks.... Just with the relief.' However, as the day progressed, with no sharks around, they started losing hope and depression set in. 'I got a quote for you for your research; I thought sharks were mysterious, now I think they are assholes'— said the British girl. Each day, the same process is repeated: going to Edwards, getting the cage in the water, preparing all the divers, and waiting... but no sharks were to be seen. It is almost like people have given up the hope that they will see sharks, and somewhere, I suppose I am feeling that too. It takes a lot of calmness from skippers like Mike to come out every day for the sharks even if they have not seen one for the last seven weeks. At the same time, Mike is aware that if he closes the operation once, then his business will be done for, and he will not be able to open it ever again, considering all the political turmoil this operation is going through.

As the day turns to afternoon, the feeling of the entire boat changes, everyone starts getting more or less upset, and not much talk about

Oliver Fhaist, a marine conservationist based in Germany, and tourists on the boats waiting for the sharks, 2016

sharks actually goes on, unless I initiate it. And after day in and day out, I hardly feel the urge to ask the same question with the tentative similar answer. This PhD is certainly not going the way I expected it to, and I am starting to doubt if I will ever get it finished. To observe the impact of encountering sharks through cage diving, first there has to be sharks to have the impact.

25 March 2016
3 pm
Foveaux Strait

The depression in the boat is claustrophobic. The days generally start off enthusiastically; however, as the hours roll by, particularly after lunch, around 12 pm, it seems the vastness of the ocean and the absence of many features that we generally relate to our terrestrial existence adds to the feeling of loss. I spend most of the time either talking to the tourists or standing on top of the boat and keep a look out into the vast sea, if I can spot a big shadow or a fin. Many of the tourists tell me why they came to the 'end of the world' (I was challenged on this term by the only other Indian in Bluff, who saw it as the beginning of the world) to see the white sharks, a few of them have been waiting a long time, and a very few, all their lives. Some sleep, some keep vigil outside, and a few keep on looking at the big screen TV connected to the underwater camera attached to the bottom of the boat. The TV screen becomes a portal to the underwater world, a 'window' of possibility. It is said that one of the signs that a white shark is around is the absence of fish; birds above the ocean fly away too. So, the people watching the screen get very eager when there seems to be a lack of fish.

22 March 2016
1 pm
Foveaux Strait

It is 11:45 pm. Oliver and Steffi went to sleep (for a few days, Oliver parked his campervan right outside my house). I sat down with my audio-visual files of the day. I wanted something to happen, I wanted a 'beautiful accident'. I was rummaging through my day's video footage, looking for something, nothing but blue getting bluer, and the clacking of the cage in the water. On the last footage, I was ready to give up when suddenly, I saw him/her moving feet below the surface. I was mesmerized with all about three seconds of the creature. The shark was so close to the ground and so well camouflaged, even the people in the cage looking for them must have missed the individual.

The frustration of not seeing the sharks is particularly debilitating when they realize that the sharks where physically there in the region, but did not decide to show themselves to the cage divers. These arrivals and disappearances can be so sudden, that if you are not at the right place at the right time, and even looking at the right quadrant, you will miss the sharks. Steffi said (almost in tears), 'I don't know, I just want to see them. Today was strange, when Mike shouted sharks, I was instantly shaking, and wanted to get in the water. But in the water, after five minutes when nothing was happening, the atmosphere was starting to get down, and everyone was looking at each other—where is it? Where is it? The worst thing was, when we got out of the cage, people started saying the shark was below the bait and even jumped over it and made splashes in the water, but we didn't see it from the cage, and I was disappointed, sad, and cold... Why again?' This experience has affected her body language quite starkly. Generally, a happy Steffi physically appears to be depressed, and in a submissive body posture. 'When I was sitting on the cage I was stooping down, but I didn't feel like changing it. I was thinking that I was coming back tomorrow, but that was just making me angrier and more frustrated. I wanted to see them, but I didn't want to repeat the same experience.' There was a man who waited 20 years to see white sharks, but could not see them, because they came intermittently. 'The first time the shark came, I was inside the cage doing my diving introduction, the second time I came out to get a new pair of goggles, and the shark again came and went (he started laughing ironically). He never came back.

Sometimes, people want sharks to kill a seal just because they want to witness the wild creature. It is because we are in the wild, the rules of killed or be killed is fairer, and we can explore our own animalistic instinct that would come out as: 'I am a vegetarian, but if the shark comes for the baby seal, it's fine. Let's kill the baby seal.' When people see seals and other animals on the surface, they mention how adorable or graceful they are, but there is always an underlining wish—'I wish the shark comes and takes him.' They would treat the sharks as if calling on pets; they whistle, shout out to the ocean: 'Here sharky sharky sharky'; and quite a few times the idea of a 'shark dance' (like rain dance) to entice the sharks was gleefully proposed among the tourists. In an essence, there is a suspension of disbelief of the 'post-natural' and behaviour beyond the accepted norms of a 'civilized' Western society, if they can see the sharks.

25 March 2016
1:30 pm
Foveaux Strait

Marco sits quietly, his cheerful self-looking adrift. Mike has taken the call not to bring him out anymore, because he was out on the boat so many times for free, and Mike had to keep on bearing the expenses. Marco even suggested that I may create a small questionnaire, which checks how sad people are after not seeing the sharks. That was not actually a bad idea, but I do feel it will be quite cynical, so I won't pursue it.

A few days after leaving Bluff unsuccessfully, Steffi wrote to me about her feelings on the absence of the sharks; she titled the letter—'The ghost shark'.

THE GHOST SHARK

> *Hi Raj,*
> *After we didn't see a great white shark within the 6th try, the first feeling I had was a really deep sadness and disappointment... I*

realized how much I wanted to see them...alive and in their own element. On top came the amount of tries, the seasickness and especially that the encounter was so close but out of reach... I was and still am afraid to the bones that something could really have happened to them. Affected by El Niño, and changed currents, or the pure hate of some people who decided to see the shark as a total enemy and danger. And that we may be experiencing the extinction of a population or at least the vanishing. I was afraid and full of anger at the same time that cruelty and fear could have prevailed over fascination and appreciation.

I imagine the encounter with a great white shark kind of different (compared to experience with other sea animals). More flashed by the power, and speechless because of their perfection as the boss in the ocean... I think you feel very very small after it, but at the same time, kind of blessed and strengthened too, from them. I think the feeling of the aura will stay with you for a while, plus the adrenaline ...because you will have met this perfect hunter and survived. Maybe you will get the feeling and the consciousness back, if you don't already have it, that you are just a part of everything...

Hugs,
Steffi

One way to understand what Steffi was writing about was that 'presence and absence are in an ambiguous interrelation between what is there and what is not, absences are cultural, physical, and social phenomena that powerfully influence people's conceptualizations of themselves and the world they engage with' (Bille, Hastrup and Sorensen, 2010, p. 4). This absence of the sharks not only effected the financial and scheduling situations of the tourists, but indeed their mental and physical states, and thoughts on wildlife degradation and need for conservation. It may be argued that Steffi's longing for the experience was directly related with the absence of the sharks and effectively, the experience of that absence. In other words, the ghost shark was manifested through the longing for those 'elusive feelings' (Bille, Hastrup and Sorensen, 2010, p. 4). However, not only did this longing create a potential for further efforts to fulfil her motivation, it also created a romanticized image of the experience, and the sharks in her mind, in effect, mystifying them (be it in a hyper-positive connotation). The absence of the sharks impacted her social world, and how she interacted and created bonds with her fellow divers. Steffi's letter further raises the effect on Steffi's self-conception, a concept created not only as an animal lover, but as a marine conservationist (Steffi and her boyfriend, Oliver, had their own marine conservation group based in Germany). Indeed, I often engaged with tourists who had mentioned how they wanted to know how they 'feel' when they see the sharks under water.

Journal Entry
8 April 2016
12:20 pm
Foveaux Strait

I feel responsible personally for not being able to see the sharks, and even ended up apologizing to some tourists for it. This is a strange feeling, considering I do not represent the boat, or even part of the crew (at least officially). Mike forgot to give his spiel, even when there were no sharks and lots of time in hand. Today, I fell asleep on the boat. I never sleep on the boat, trying to be attentive to everything that is going on, but today I did. All I feel is a deep loss, a deep, deep loss...

What has my work given me? Maybe I have not given it enough? Maybe I needed to sacrifice more. Maybe I needed to be smart enough, maybe because I cannot spell, maybe it takes me weeks to read one book. But I am here now... Take it...take it all... Maybe my flesh, my soul, my nerve. There in the darkness of the corner of the house, goblins jump from shadows to shadows, and the images I have created of me, and that have been created

A blue shark (*Prionace glauca*), during the absence of the sharks in the 2016 season, which Mike had never seen before around his cage diving operation, 2016

of me, run naked. Take me...here I am with hunched up broad shoulders and veiny genitals, hairy arms, my muscles tensing up. I start punching the wall, the house trembles with each strike... Why am I left behind? I can hear the growls of the impending storm, like a wild civet tearing away flesh, bit by bit. I punch harder, the house rumbles, and my knuckles and the walls unite with red smudges...and that distinctive smell of rotten flesh oozing out as if from the scars of the wall.

The Cottage and Unexpected Company

On 26 December 2016, I left Christchurch for Bluff on the bus, hauling all my equipment along in two and a half bags, and stuffing everything else in my jacket pockets. Once in Invercargill, before setting off to Bluff, I had to visit the museum. Soosan and I had been conducting archival research in Bluff and Invercargill and contacting all the libraries and museums in hope of getting any material in regard to Bluff and white sharks. No new information was yet available.

Even though we were supposed to re-attempt the Bluff second time, Soosan could not accompany me to Bluff, because she was not feeling well. This was unfortunate, as she has been my most dependable resource in this project. Fieldwork can be extremely lonely and isolating at times, and often it is made bearable by known faces and a social support system. Theoretically, we should go in the field and attempt at cultural assimilation as much as we can; however, this can become even more complicated if the field researcher has no prior personal or professional experience of living in a certain community, although maybe these are also unique opportunities for exploration of both the self and the community they are working with.

The cottage in Bluff, 2017

Journal Entry
4 February 2017
1 pm
Foveaux Strait

I am back in Bluff and found shelter for the next few months in an old Bluffie's sleep out. The small sleepout has a wood burner, an electric cooker, a quarter size fridge, and a few unclean cooking utensils hanging on an empty cabinet.

The chair in the centre of the room is my morning equipment table, where every night, before I go to sleep, I lay out my tools and clothing for the morning. It reminds me of the cottage Charlie Chaplin used to share with the 'brute' in Gold Rush (Chaplin, 1925). I admit, I have not come down to eating shoes, but money is tight. I have always been 'muscly' and like a body builder in the off season (be it in a perpetual off season as it may). Today, I went to take one of my two weekly showers, and finally succeeded to jam the un-closable door, and get the confused water taps running in a bearable consistency of heat. I closed my eyes and as the water streamed down my body, my generally groomed body was a hairy conundrum. I always had an ambition of having a long 'explorer' beard, and was feeling pleased that I had achieved such. My hair is also getting longer, and I try to take care of it the best I can. I put an egg on it once a week, but unfortunately, since I cannot afford to add some honey in the mix, I smell like a muffin for the next two days. As my hands moved through my usually broad but now scrawny shoulders through to my sternum, I realized that I could feel my entire rib cage, which I do not know if I ever had done before. As I went into my room and looked at my chart I had been keeping, I realized I had lost 11 kg of weight in the last two months, so not only did the sharks save me from an intellectual and spiritual sense, but indeed, from a true physical sense.

I am completely running solo academically, without any support from my supervisor (so much so that six months before the submission of my thesis the university had to advise me to change my primary supervisor), and my secondary supervisor does not reply to my emails. Problems with supervisors are not uncommon in PhD scholarships, and this intellectual isolation along with the physical isolation is always important for the scholars to consider, no matter how well they might perceive their relationship with their supervisor to be, at the end, we are all flawed humans. However, this situation often makes you a stronger researcher and pushes you to innovate.

The first night before going to bed, Caroline and Pete told me not to be scared by the birds on the tin roof, and as it got dark I did start hearing them dancing just above me. When I tried to sleep, I felt a breeze touching me, not knowing where it was coming from; I did not bother and was just about to go to sleep when all of a sudden there was a loud crackle in the bed. Upon a closer inspection, I realized that the one leg of the bed had collapsed, possibly due to disuse for some time, and for modesty's sake, I would argue any implication of my body weight at the time playing any part in it. So here I was with a bed, which would certainly crash during the night, and I had to take the decision to be a little brutish myself and break off the lower two legs of the bed, rendering my bed to resemble something of a sloping hill side. In the morning when the landlord discovered my predicament, we decided to break the other two legs of the bed and create what was now equipped with a Japanese inspired sleeping arrangement on the floor. A couple of nights passed, and I still felt the breeze, and the birds still danced on the tin roof. A few more days passed, and the dancing on the top of the roof stopped. In the middle of the night, I suddenly heard a scratching next to me on the floor. It seemed the birds were walking all across the room. Considering New Zealand is famous for its birds who prefer to walk more than fly because there was historically no land predator predating on them, I kept on sleeping and let them be.

The sleepout was in the garden, and I used to walk amid rosemary bushes and apple trees. And in the gloom, usually I would see a hairy black and grey figure approaching me from the darkness. Milly was eight years old, stubborn, and almost blind, a lady of the Irish Setter breed, and considering our shared tones of salt and pepper hair, she looked much like my long-lost interspecies sibling. Every morning when I went out to the sea, she was the first individual I would see and talk to, and when I would be walking back cold and wet, she would bestow the first warmth of my temporary home. When I walked out in the garden, I would hear the shuffling of the doggy door of my landlord's house, and with her little steps, she followed me around. As much as we walked among the citrus and nectarous aroma exuberating from the vines and branches above, still we never saw the Aurora Australis, rather 'I' never did. After our walk, Milly would go back to her house, but being blind could not find her door in, and I would have to assist her to return to her bed safely.

A couple of days later, as I was about to heat up some water for my tea, I saw rat droppings on the window sill, and the realization came,

The interspecies siblings, Raj Sekhar Aich, 2017

my imagined birds, were actually rats...well that kind of explained the holes in my rice bag. But I thought maybe they will go away, and why bother them as long as they don't end up biting me in my sleep. But they did not go away, they started eating my grains, and what was worst, biting into my equipment cables—that, unfortunately, I could not live with! Even though I tried conserving everything I had in the mini fridge, including some of my camera equipment, when my landlady got to know, she was mortified and immediately got some rat poison. Milly was furiously sniffing the base of the house. But nothing seemed to work, and I started hearing, what I believed were the sound of a family of rats shuffling next to me and eating the limited belongings I had, all through the night. One day, while trying to find my glass nib fountain pen, I raised the curtains by my bed and saw a hole in the wall as big as a dinner plate. This all of a sudden explained how the rats were getting in and out of the room, and where my cold nightly breeze was coming from. I shoved some old newspaper to stop the wind, if the rats... But that was not to be.

Finally, Pete got me a rat trap… 'One of those ones that kills them'; with a little trepidation I thanked him and accepted it. That night I put the trap underneath the sink with some grains of rice. I went to bed, there was a storm out, but no breeze in my room anymore. Thinking about my sharks, and a big plate of lasagne… I drifted off to sleep. Suddenly, there was a loud shatter! As if a guillotine had fallen, fallen on my jugular, cutting my breath, the shutter was followed by an acrid screech. My eyes opened, invisible was the fight for an animal to live in the darkness in front of me, all that I could hear was the scratches of her carrying the trap, gripping on to her head, a few seconds, and everything felt silent. I never heard from them again, and Milly did not sniff the house any more like that. But I did wonder if I inadvertently killed a mother with young ones. Caroline told me that all the rats used to live up in the hill, but now because of a lot of deforestation in the area, they had nowhere else to stay, so they were coming down to the houses.

It were humans who disturbed the habitat of the rats, and now that they were in 'our' space, we had to exterminate them. As a human–animal researcher, I could not help but notice the significance of these other species who were not the focus of my research, but were having effect on the experience of my fieldwork. These animals were my companions; Milly, the dog, gave me comfort in the lonely fieldwork, and feeling connected to animals that I do, I felt a sense of belongingness in the alien environment myself, and had something to look forward to when I came back from the boat in the evening. On the other hand, the rats were the vermin and the 'other' who shared the same resources and geographical space as me. For not only were they putting my limited food ration at danger, but strangely—more important to me at the time—my field equipment, without which I could not complete my mission. In the physical 'meeting point' (Oma, 2013) created by the intersection of multispecies agents, the architecture, and the other material, we both competing for our interwoven and sometimes overlapping necessities. In this conflict and competition, I predated on her/him/them, whereas I, as the human, created the conditions of this competition in the first place by encroaching into their space, up in the hills. At the same token, in such a multispecies meeting point (i.e., the ocean), the human is at times predated by the sharks.

There beneath the shattered glass, there where the splattered water from the rain on the ground meets the exposed rusted iron skeleton of the house...in the midnight light I can almost see something enmeshed with the ground, remnants of whom is only seen by ossified grey and white fur, and I feel I can still smell his rotten flesh. A couple of weeks before I left for New Zealand in 2014, I had rescued a kitten. I had made him a nest in an old abundant fish aquarium I had from my *dadu*. Inside the shed, on my terrace, where my old paintings, some empty whisky bottles, and my punching bag lay. Only a couple of years earlier, I would have never left him like that because of the resident *bham biral* (large Indian civets, who smelt like 'gobindobhog' rice) in our area, who stayed in the coconut tree. Now, there were hardly any trees, and no *bham*. The kitten used to cry a lot and was not leaving to make his own way as cats generally do. One night, I decided that I will not feed him and maybe he will run off and start fending for himself...or I was just being mean. The next morning when I got up, he was still there, but hardly breathing, his shy little heart silently moving inside the palm of my hand. His entire body was filled with ticks. I took the dropper and tried to feed him, but that night he dozed off. But by the next morning, the ticks permeating from his emaciated carcass filled the entire area where he lay. Within the next two days they had covered the entire house, as if like the locust. I left India after a week, be it with blemishes on my skin, and no one was left in the house to care, but the ticks themselves.

The Care Package with a Bourbon Cigar

One early morning, I woke up to check my emails, and to my delight, from the Southland Museum and Art Gallery Niho o te Taniwha, curator of history—David Dudfield—had a surprise for me. They had recently acquired a photograph from the early 20th century, which raised more questions than answers. It was of a big male white shark (with visible claspers), allegedly caught in Bluff in the early 1900s, displayed on top of a sled, and everyone standing proud of themselves behind it. A body bruised with harpoon wounds, and as in many of the cases with a white shark corpse, the mouth was open with a bit of wood to display his jaws and teeth. The museum was not sure where it came from, when it was exactly taken, or any other detail. I was extremely inquisitive: Was it caught in Bluff? And why are people all around it so happy displaying it? Does this mean Bluffies had a closer connection with white sharks than most people knew? I was not holding my breath in anticipation of coming across any more details about it.

Journal Entry
29 December 2016
12:30 am
Bluff

Tonight, there is 55 knot winds in the strait, no boat trip in the morning. Mary, (Mary Leask or as she is known as, Bluff's calm voice) from her house on the marine parade, ever vigilant for the last 40 years, has been making sure that

White shark caught in Bluff wharf, early 20th century (source: Southland Museum and Art Gallery)

everyone who has gone out to the sea has come back safely, tonight is no exception. So, in the morning I can stay in bed a bit longer, but on a general day, I get up at 5:30 am, heat up some hot water to have a cup of black coffee, and carry a couple of apricots and a box of suji *(semolina) for lunch; upon return, I get one proper meal of* khichuri *(a Bengali dish with rice and lentil) and hunger has become part of the norm of existence now. But hunger, my true hunger, is for new experience, one which I have never felt before, coming from an Indian middle class family.*

Before the second fieldwork, I had about NZD 400 in my bank, another NZD 600 from tutoring at the university. This is where my white piano saved me. I sold it for NZD 600; a lady drove up the driveway, got out of the car, came in, and lifted the piano back to her car and drove away. She did not even smile, and here I was, thinking I would share some stories as to how this piano travelled with me to meet the sharks. So, I had now, the total of NZD 1,900. From which, I bought NZD 400 worth of camera equipment for the documentary, NZD 1,000 was for my rent in the field, and NZD 200 for food and transportation for four months. So, I bought 6 kg of rice, 6 kg of lentils, some salt and spices, and some semolina (from an Indian store in Christchurch) and made khichuri, *and the four months, I had one meal of* khichuri *a day after working on the boat (I cooked it at one time enough for three days). I used NZD 3 to buy a pack of frozen vegetables from the local store. I had also bought a pack of frozen paneer, which I kept a month in the freezer, for one of my university lecturers who was supposed to come and pay me a visit, but he never did. For breakfast, I used to have a couple of dried apricots I had, for lunch I used to make some semolina, which I took to the boat, and some coffee from the ship. But Soosan being Soosan sends me a few dollars now and then to get a pack of noodles from the newly Chinese owned noodle shop in Bluff and an ice cream from the shop owned by an Indian family, and sometimes a care package with fun bits and pieces.*

One interesting encounter with a Bluffie took place in the noodle shop. As I was getting my pack of garlic noodles, there was this gentleman sitting there possibly drunk. He started asking where I was from, and what I was doing in Bluff, and me being naïve as I usually am, started telling my whole life story. I was mentioning how beautiful the Titi islands are, how lucky everyone is to be here, and so on. Angrily he asked, 'How do you go there?' I tried to explain that I am a researcher and I work with Mike, and I've seen them from afar. And he said, 'I am the guardian of those islands! And Mike doesn't have the right to go there!' I tried to change the topic, but he still stuck to it. He then asked me where I lived and he would walk me back. For whatever reason, some sense kicked back to my head, and I said I have some shopping to do and some other work; then I'll get back. So, he let me be. When I talked to Mike the next morning, his advice was, 'Just stay away from drunk people who have nothing to do in life.' Another interesting incident happened while returning from an early morning jog to the Stirling Point. While I was nearing home, an individual approached me, and quite blatantly asked, 'Hey bro, do you know where I can get some drugs?' Not knowing how to react, I burst out in laughter, and apologized to him for not being able to help him out. If this later incident was indicative of the rampant pervasiveness of drugs in these regions is a matter of debate, but it was certainly curious.

The people of the local community knew that there was an anthropologist in Bluff, studying the pointers and cage diving, and they were quite amicable to me. I had random people stop their cars on the road to come up and invite me for a cup of coffee and a chat. After introducing myself, I had the owner of the only Foursquare in Bluff, Andy, invite me to his house for a dinner. The last time I was here, Mike and Carwyn took Soosan and me to one of their fire-station parties, that had a lot of beer. Even though I personally cannot stand kina (Māori for sea urchin), but I cooked it for them. And then the owners of this sleep-out, Caroline and Pete, were also extremely kind to me, and gave me fresh fish, and access to their vegetable and herb garden, and I reciprocated with some cooked dal. But did I really integrate with the society? I am not sure, as Soosan points out, initially when I came here, for a few days I did not go out, because I felt like I was the only brown man in town. One of my biggest drawbacks was I am not a big drinker (anymore), so hanging out in the town bars was not really the best option for me, which might have proved to be a source of immense pool of knowledge about the sharks in the region, but it was a conscious choice, and a right one at that.

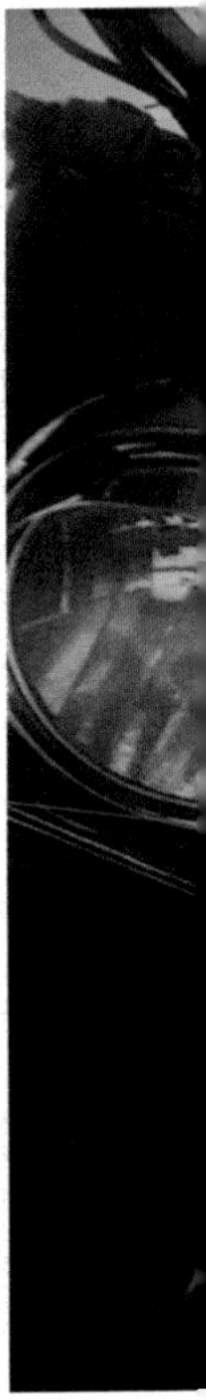

After Christmas, I got another care package from Soosan, which contained a bourbon cigar, some Homer Simpson cookies, and some sandalwood incense. Since I get scared of the darkness, Soosan had also sent a small Christmas tree light strip, which ran on batteries; so, in the darkness, white stars glowed. The same sandalwood incense had burned in the darkness on the last night of the year 2016, as Soosan had found herself on the ground,

The souls, Soosan Lucas, 2015

with intense pain. Unbeknownst to her, her appendix had burst poisoning her entire body, and she was inflicted with peritonitis. Alone and lying on the floor, she had called 111 (emergency hotline) repeatedly, and on the final time, she told them, 'If you do not come now, I am going to die.' On 1 January, Soosan was going into surgery and I received one message from her in the morning. I could not contact her anymore. Frantically I searched the hospital she might be in and contacted her other friends in the town. A day later, I was able to finally know that she survived the operation, but the doctors were not sure of the eventual outcome. I was now caught in a predicament; how could I leave my fieldwork at this time to see her? This was not helped by the fact I did not have any money to pay for the bus ride up to Timaru to go and see Soosan.

I look back and there is a giant white Mola Mola (sun fish) suspended in front of me in the abyss of the dark room. She is silent, her massive weight supported by the flickers of my hurricane, her white skin nibbled upon by fireflies. She looks at me, and swims towards me, as the storm now howls through this barren fossilized building. The distorted lights from the roof start falling one by one, and shatter at touch, trembles the sky in subdued rabid lightning. Her tiny flippers flutter, she comes close and turns her entire body sideways, and her eyes penetrate my core, I close my eyes and look away... I can't stare at her anymore, and when I look back...she is not here, and I am left in the abyss....

I can't breathe...I can't breathe...gasping for air, I run out of the room, and down to my garage. I slip and my hands fall on the ground on something that could only be dried up blood. I realize I have kneeled down on a tarpaulin, smudged in engine oil. This is the last clothing of my beast and he is not here anymore. All that remains is this tarpaulin that covered him, on the floor, engulfed in dust and splashes of Mobil, what looks like dried-up blood stains after a tiger had attacked and dragged him away into a Siberian forest. It comes to me that I had to sell him to pay for my plane fare to New Zealand. He was my Royal Enfield Thunderbird motorcycle, my most prized possession. I had painted him matt black, given him four headlamps, and meticulously crafted him bit by bit. I can still feel his skin, cold to the touch. I sit on his curved leather seat and he squeezes soft to firm; I kick him, but he stays still. The storm is picking up, the wind sweeps across the courtyard and bangs on the garage door, and I can hear '*Horibol, Horibol*'[1] chanted from somewhere close by, possibly a night-long kirtan session, mourning a passing. As my eyes adjust to the darkness of the garage, old logs, dust, and door hinges become apparent, piled like carpets of memories, one above the other. Cockroaches playing Vivaldi (an arcade game from Vivaldi browser) inside the old kettle stops me for a gaze. '*Horibol, Horibol*,' they sing, but my beast stays still. They sing stronger as the wind bangs louder, but he stays still. '*Horibol, Horibol*'—they open their breasts and cry out into the gashed

1 The calling of the name of Hari (pronounced as 'Hori' in Bengali) is a spiritual chanting.

sky, and the beast growls awake and shakes to crumble away his big chains, in a deluge of tears and spit and tyres scarring the concreate, but the door and the shackles hold him still. The garage is swept with halogen beams and a deep rumbling growl, much like the Siberian tiger, who has taken him away. The light burns at its own rhythm, intoxicated on *hariai* (a locally brewed Bengali alcohol) and stumbling with each flicker. The spiders and rats scurry to the corners; the nuts and bolts and wrench in this old garage tremble and fall, and the vibration fills up my body in an orgasmic jolt. His growling shatters the windows, and I can see the darkness blooming and deep skies ignited in the horizon. A few drops fall, and like a *nupur* (anklet) free of its chain, the bells dance in liberation, and my garage door opens finally...as I ride out into the abyss that is my city at night.

MILAN
The Meet

I cant see him, but
I see the blue
the blue where he is

A Meeting after a Hundred Million Years of Waiting

There afar, beneath the full moon, surrounded by that big neem, hogplum, elephant, and java apple trees, is a small pond, where all the ash of the cremations are put to rest. By that green sweet pond and underneath the open sky is a patch of grass, where tonight, a tarpaulin has been placed, on which old bed sheets protect some wet buttocks in dirty cotton sarees and dhotis. A little distance away, there is a big pond, in which, some of the bhaktas mix semolina, milk, coconuts, and fruits to make *sinni* (a sweet Bengali porridge, an offering for God in puja), which will be served in bowls of sal leaves. Tonight, the villagers have gathered here to sing kirtan, with harmoniums, *khartal*s (Indian finger cymbals), and dhols. Kirtan to the English is a village in the shire county of Suffolk; to Hindus, particularly to Vaishnavas and particularly Bengalis among them, it is a spiritual musical practice.

A young woman covered with her *ghomta* (head covering, usually by the end of her drape), sits in the corner, only with her lips and milky feet showing. Day in and day out taking care of her six children in sickening poverty, and an abusive husband, she had to escape to the only place that gives her solace. She starts to sing, all the other noises stop, and it is only her and her Lord and the vibrations travelling from her belly, through her breast, throat, and her lips. As she sings, warm, iridescent drops effortlessly fall from her eyes, but even she is not aware for what deep melancholy. The circle of souls raises their arms dancing with the rhythm, with their tear-embracing eyes half-closed in ecstasy. When they stop, an old man, with his

The silent stare, 2017

shaking hand, takes off his gold chain from his neck, and tearfully, places it around her neck.

Even a few months before she died, I went to my *dida* (maternal grandmother) in India. She could not lift her head, my *bouthan* (sister-in-law) and I sat beside her, she looked at me, and started to sing in her broken millennium old voice, with the tunes still intact falling like bells on marble, and all our eyes unified in tears. There she lay in bed...waiting to meet her Krishna, her eternal lover, somewhere beyond the aches of her fragile little body.

'Bon porajai sobai dakhake, amer moner anol kau na dakhe go.
Ami bondhu premanole pore'

('When the fire burns the forest, everybody watches, but no one does, when my soul burns, /Oh my friend, I am burnt in the fire of love'—Bengali folk song, anonymous).

Journal Entry
One day
4:50 pm

I am just here! Suspended in the liminality of dream and awakening, abyss and light. I bob up and down in frigid waters; cocooned in a cage, 1,600 km from Antarctica. Scuba means 'self-contained underwater breathing apparatus'. Indeed, it cages you in your own sensations and illusions, especially if you are truly trapped in a 12x5 ft cage, alone. A cage, attached to the back of our boat, the Southern Isles, and submerged in the waters where the Tasman Sea and Southern Ocean meets, with just the top hatch reaching for the blushing sun.

Your thoughts and imagination start to give rise to images and shadows of malevolent monsters looking for you amid the crimson darkness, waiting to pounce at any moment. Even if just 5 ft away from the surface and the other humans on the boat, you are isolated from the rest of the familiar universe. All you can perceive are the ethereal images created from your own demons and fantasies intertwined with what is there 30 ft in front and below of you, and things that lay unseen beyond. These images play hide-and-seek with the throbbing streaks of light penetrating the turquoise blue. As depth increases, light decreases, so does colour and contour. And only grey, green, black, and silver forms seem to be created and dispersed at the ocean floor from here. As my sinuses begin to hurt, I remember the friends whose kindness got me this far. Steve snickered—'I know why you are doing this! You want to look into her eye and know yourself!' The promise I have made to Soosan, waiting in a hospital today fighting for her life, to go to a Springsteen concert the day my fieldwork ends.

Right now, I look at my wrist dive computer—I have been under water for 40 minutes, 11.3 degree Celsius water just standing and waiting, my 5 mm neoprene suit is no longer keeping me warm. Like shards of glass, the drops capture minute newfound air pockets every once in a while. I clear the water out of my mask with forced air from my nose, and bite hard on my regulator. I feel the sharp edge of the aluminium cage through my suit; crouch down to get my balance and control my buoyancy with my breathing. We have heard of her presence a few weeks ago, a mature female great white shark. She has been spotted in the region by fishermen, but we have not seen her. I look up to the ocean surface, a decomposing tuna head tied to a rope, shivering with the waves, hoping to entice her. I see some Trumpeter and Wrasse fish nibbling on it, and the flippers of an albatross, who is having a go at it too.

Suddenly, the Wrasse disappears; there is no thing living now in front of me. I am left with the sound of my own intermittent bubbles, my weight belt banging on the back of the cage in a rhythmic drum, 'Bumb, bumb,' and the crackling and twisting of the nuts and bolts fighting with the currents trying to keep the cage attached to the boat. Her absence frightens me, as my heart rate begins to rise; I look for her through all the three windows of the cage. Behind me, I can see the large propeller, and the moss dancing on the hull of the boat. I look below the cage; suddenly, the ocean floor seems darker for a moment—then again, there is nothing. I wait. I wait...as my weight belt keeps on bumping against the back of the cage, I look to my left...and yet, nothing. I know my time is almost up, as today there is a storm forecast. Then suddenly, through my peripheral vision, I feel someone is watching me. I turn and see a pair of deep blue eyes, larger than tennis balls, looking back at me. For a moment, everything is quiet, no bubbles, no rattling of the cage. As if my heart has stopped, as has the waves, two sentient beings look at each other, as an 18-feet great white shark, Carcharodon carcharias *passes*

8 inches in front of my cage. We are connected with the movement of the waves, her eyes focused at me, but I don't know what she is thinking. I can see the small pores in her snout, the 'Ampullae of Lorenzini'. Her fine teeth don't show as much now, she moves ever so gently, and effortlessly, owning and releasing the shadows which hide her at her own will.

I bang on the cage to attract her attention, she turns back and comes towards me and now is suspended in front of me; our eyes lock and the sensations come flooding back, my entire being starts to give in. She slowly touches the cage, the cage trembles. I have felt smaller sharks rattle our entire boat with one bump, so her approach is gentle indeed. She is not a monster, she is energy incarnate, she has nothing to prove, and this is her way of interacting with and understanding her environment. And invariably she communicates with me through her body, not sure though if it is intentional or not; what is certain is that she is as inquisitive about me as I about her, be it for different reasons. She knows I am here, her sense tells her there is a beating heart, but maybe in a metal exoskeleton. Unlike most sharks, she does not bite the cage. Maybe she has had experience with human cage diving before, and realizes that she cannot get to this heart, or she realizes, the cage, or me, or both are not something she would like to taste. Maybe it is the fact that unlike what most people think, white sharks are quite picky eaters, and they are not mindless eating machines. Or she knows that I am watching her, and she does not have the element of surprise, which white sharks are known to rely on while hunting, or maybe she is just not hungry. Or it is none of the above, and we have no idea what or what culmination of factors they are.

As she grazes against the cage, I can see an intense claustrophobic, dark, never-ending abyss inside her mouth, and her mythical white triangular serrated teeth jewels the entrance. Transfixed, I look at each inch of her body pass through before me, and then below the cage. This close, I realize truly how big 18 ft is, as the blank blue canvas does not have any point of reference. Her girth is about 5 ft, more than half the size of the cage; as I see her dark bronze skin pass beneath me, it seems like waiting for a freight train to pass. I look at her iridescent skin, made of a million small teeth, as she glimmers in the sparkle of the light reflected from my metal cage. Her body curved out of bold and broad brush strokes, from her cone-shaped snout to her 4-feet-tall dorsal fin and never-ending tail. She appears to be moulded out of porcelain, but forged out of steel, each fibre of muscle contracting and relaxing visibly through her skin.

There is no lack of perfection, a design crafted over at least a 100 million years—only intense reality. A reality, which does not even allow suspension of disbelief for this poor Indian anthropologist, as he reminds himself: 'This is not a dream, this is not a dream. Take it all in, don't think...just take it all in.' If I am cursed, to perceive the universe from only the reasoning of my perspective, I would like to imagine, the meeting with her is predestined from all these million years ago. All of a sudden, at what whim, she looks towards the open ocean, flicks her immense tail, the force of which jolts my cage and I fall back... When I look up, she is gone! There is no promise of aggression, neither forgiveness, only at this moment...indifference...as I keep on bobbing, and my weight belt keeps banging on the cage. I have been under water for 56 minutes now; I know hyperthermia will set in soon, and so will the storm. I have to go now.

The Lovers

When the news came of my *dida* passing away, I was out in Foveaux Strait, near the Ruapke Island, which from my point of view looked like a desolate piece of rock in the middle of the ocean. I stepped on the deck, but with deep pain, I looked up, and there was a cow looking down at me, happy in the Antarctic winds chewing her green grass. For a moment I was transfixed as if standing in front of a Dali surrealistic painting. Discombobulated, I went inside the cabin, and my hands shaking in the cold, and head shaking in disbelief. My *dida* loved when I cooked her instant noodles...which took some time to prepare. But out in the ocean on the boat, I boil some water, soften the noodles, put the ready mix of spices, some tomato sauce from a sachet, and a pack of oregano, left over from a pizza box... I guess, this will have to do. Out in the wind, the birds feasted on them, and the rest floated gently to the bottom of the ocean with the fish nibbling on the remnants, and the cows looking on. I chuckled; maybe, now finally, she will meet and sing with her eternal love...

> *Dear Raj,*
> *I can't describe how nervous and excited I felt on that day, still feeling the disappointment from Bluff in my bones. The ship took three hours to reach the spot, and my mood was changing from 'WOOOOOOOW, we're going to see great white sharks' to 'F*** what if not...', and I started to get seasick again. Then the call... the sharks are there...and we jumped into the cage and...nothing.*

On the way to meet the lover

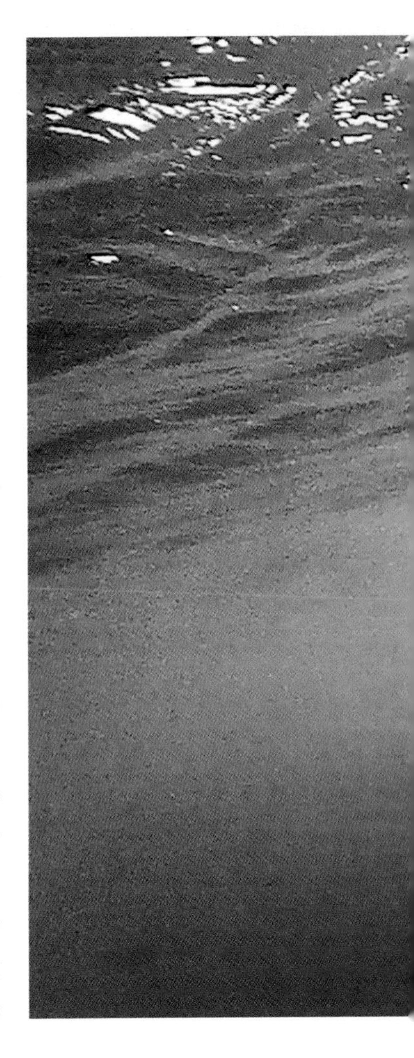

No shark...just a wall of fish and chum, it was so bloody cold and rocky. I was vomiting under water, couldn't feel my toes, and was shivering—just miserable. And then a shadow, the ghost, he was there, they exist. Just for three seconds or so and gone again. I was panically looking around, desperate to get a closer look, afraid that this may had been all, my heart was beating so fast, no sign of seasickness. A shadow...Was he back? My seasickness got me again, I was fighting against it, but needed to get up for a minute or so, then I forced myself to go back into the washing machine again.

And out of silver fish and blood, he came around cruising, slow and massive, just perfect from the left side, just so close to Olli and me. He was staring inside the cage and for an instant, it felt like he was making eye contact with me. This unbelievable deep black eye was

Transcending the separation of the cage, with the camera handle, 2017

The shark and human meeting separated by the cage, 2017

sucking me in. Holy shit! I stopped breathing, didn't dare to move. I felt so small and insignificant; he was the ruler of this kingdom, and we, just tiny visitors. Pure perfection, pure power, pure nature... his presence and energy, wow! And not a single bit aggressive, the atmosphere was very calm but overwhelming. Even now, when I think about it, I shed a tear and have a big smile on my face when I remember his presence; it was definitely a life-changing moment.

Love Steffi

Steffi had finally met her shark in Australia. She talks about 'pure nature', which is possibly indicative of her perception of the 'realness' of the experience, but I have to invariably ask myself, what is a 'real' experience of encounter with a white pointer underwater? Will the experience be necessarily the same if the circumstances and the individual shark was different? If I met her (the shark in the previous chapter), or any other white shark underwater without the cage, and unexpectedly, would my reaction to them and their reaction to myself be the same? Mike used to take his family to the Marlborough Sounds; one year, he was fishing underwater, when suddenly he turned and saw a big white shark next to him, watching him; being cool as he is underwater, he was able to just stay calm and take his regulator out and blow bubbles, and the shark left. But for weeks after that, late at night in bed, the thought kept on coming back to him 'What if...'.

Driven by an unquenchable ambition of encountering white sharks in their 'natural' environment, certain experts choose to dive with them without the protection of cages, with and without assisted breathing apparatus. Considering the predatory abilities of white sharks, such interactions come with potential for injury and even fatal outcome for both the participating species. The experts are known to be meticulous while planning and mediating through this active human–shark engagement. Effectively, they have created their own

The tourist taking a video of the shark, standing on top of the cage, Raj Sekhar Aich, 2017

rituals, norms, and understandings of the shark behaviours, which they considered to be significant for a successful dive. These include considering the individuality of the sharks, the dynamic environmental conditions of the ocean, and their own state of mind on the particular day while planning and performing such dives and the particular dive. This multispecies encounter, where the human participants have to adhere to the agency of the animal other, as well as the environment, for a successful encounter has not yet been anthropologically explored. It is an encounter that is facilitated through a practice situated in a marine contact zone, where two sentient species engage with each other in a symbiotic encounter.

Though I don't explore it here in detail, I raise it as a significant pool of knowledge in human–white shark encounter and emphasize on three crucial points emergent from it that are pertinent to our discussion. First, that the individuality of the sharks plays a significant role in shaping any human and white shark encounter. Second, the individuality (behaviour, demeanour, skill set, and confidence) of the humans interacting with the sharks plays the primary role in shaping the encounter. Third, we have to ask, what is a 'real encounter'? Is it the one when the encounter is not expected—whether on a boat or in the water? Aren't encounters like these produced and orchestrated with very careful consideration of the particular sharks, which Mike Rudson, one of the pioneers of the practice, calls 'players' (The Sharkman, 2011), that is, individual sharks who are inquisitive, but at the same time, not overtly

A white shark spy hopping and looking at the people on the boat, Raj Sekhar Aich, 2017

aggressive. Will an experience with the same shark be different, if s/he is hungry, as opposed to when they are not? So, are these 'real' experiences? The same goes for cage diving. Is it a real representation of human–shark encounter underwater? Or is a simulacrum of such an encounter? Or is it a hyperreal practice by itself, devoid of any point of reference? We might argue that it is a mediated or constructed performance of human–shark encounter, enabled by particular forms of technology and legislation (or as has happened in Bluff, stopped by legislation). It is the attempt to create or construct an encounter, whereby the sharks are invited and encouraged to perform for the cage divers, so it is more than a simulacrum, it is an artificial encounter that relies on the hyperreal construction of the sharks in the terrestrial animal's mind and culture.

Fear shapes the relationship and interactions of the humans in the cage diving practice, but the boundary between fear and fascination is often hazy. In our case, the divers, the operators, and the researcher negotiated through their fear of the sharks to engage with them, and fear (be it controlled fear) became part of the commodity the operators sold to the customers. Much like in Parreñas's (Parreñas, 2018; Parreñas, 2012) exploration of embodied experiences of humans working in orangutan rehabilitation centres in Sarawak, fear became a part of their existence, and behaviour of a well-trained volunteer towards orangutans was shaped by their fear of them, as much as the behaviour of the orangutans towards the humans shaped by their fear of the humans (Rutherford, 2016). In a similar sense, the behaviour of the humans in the cage was mediated through this fear, their body posture; the composure inside the cage was a dynamic play from the animals' behaviour and the human admiration for them, at the same time realizing how overpowering their presence really was. While it is a debatable matter if it can be said that the behaviour of the sharks was shaped by the fear of the humans or the other activities involved in the cage diving practice; however, their behaviour in particular avoidance of interacting with the cage diving practice even if they came to the immediate area was certainly an indication of their aversion towards the practice, although instead of fear, it could just have been a lack of interest.

Fear dictated the interaction between humans and alligators in their relationship in Louisiana swamp trips (Keul, 2013). 'In general, fear operated as a regulator of encounters that made experiences safe for both people and gators... Guides wanted both people and gators to have some fear, but not too little (this could be dangerous) and not too much (else no ticket purchase)' (p. 946). Guides wanted tourists to know that alligators could be dangerous, but that, 'they are not the monsters that they see on TV' (Interview C, 2010)' (p. 946). Much like his observations that humans who intentionally choose to encounter with gators choose to do so for various reasons, and often in a wish to 'confront the embodiment of their fears, but they know that such encounters can only be experienced through encounters' (p. 930–31).

However, the fear was not necessarily the same in case of the gators of Louisiana as with the sharks of Foveaux, because according to the guides, the gators alter their behaviour due to human encounter. Because the gators were regularly fed by the tourists and the guides, the guides developed personal bonds with them. They have learnt to perform for the tourist at the appropriate time and in the right manner. This was different in the case of cage diving with white sharks, because of the lack of dependability of individual sharks to return to the same spot, and the inability of the operators to 'train' them, even if the wish is the same from the perspective of the cage operators too.

Human–alligator encounter may be categorized into two sections: enticing encounters, when the reptiles were coaxed to make an appearance, strike, or jump out of the water with a food 'payout', and encounters of being, when alligators allowed people to gaze upon and approach them without any apparent motivation. Both types of encounter were enticing (as in alluring and tantalizing) because they involved animated human and alligator performances where both sets of actors were anticipating each other (p. 934). The enticing encounter seems more dramatic and better value for money of the tour, but they only do it under specific geographical considerations. Whereas in encounters of being, there were no constraints, and in this kind of encounter, the gators have more agency; that is, as if they want to be seen. In cage diving, even if the operators would ideally like to encourage an enticing encounter, where the shark could play the 'shark', often it would end up being an encounter of being, where the tourist watched them from distance (this is particularly true for the season of 2018–2019, because the cage divers were not using bait, and the sharks often stayed further away from the cage and the boat, as opposed when they were enticed to come close to the cage for the divers to see), and this physical distance

Holding the shark wrangling line, Raj Sekhar Aich, 2016

of observation can have different effect on the tourists' experience as opposed to when the sharks are close, and even playing the 'shark'.

In cage diving, fear played a similar pertinent role, where the divers, the operators, and the researcher negotiated through their fear of sharks, to engage with them, and fear (be it controlled fear) became part of the commodity the operators sold to the customers. For the sharks too, fear and apprehension about humans may even affect their willingness to engage with the operation. But one physical element that separated the alligator experience from cage diving was the 'cage'. The cage created a physical and psychological barrier between the two species; for the cage, along with the ocean, its waves, and the currents, connects two apex species in a personal, sensual experience, defined by curiosity from both the species, generally peaceful under a controlled environment. Controlled by the operators, by at least the negation of any negative interaction between the two species, when either of the species is physically harmed, but certainly affected by the shark, because they may choose not to take part in it. The bodies of the individuals (human and shark) in the encounter were being affected by the immediate, and remote movement of one another, divided by a permeable metal cage, where neither were allowed to put their sensory apparatus in or out. But the potential of danger always loomed, and palpable underlying fear is inherent in at least one of the species.

It seemed that, much like me, many tourists were more afraid until they had actually seen the shark, because yet again, the physical absence of the sharks gave the opportunity to create the monster of their nightmares just outside their gaze. The fear of being unprepared mentally when the sharks appeared, and indeed, the suddenness of their appearance and disappearance seemed to be disconcerting to people, and often the anticipation of the shark was more tense than actually seeing the sharks, 'It is kind of spooky, how they come and disappear. I was looking for it and suddenly it came from below, every time the cage banged, I looked back and I thought the shark must be there, but it was just the cage banging against the boat. It is just the way they come out of nowhere! He was 6 ft away, and I could see perfectly, but he was 11 ft away, and disappeared entirely.'

For the humans, the three windows of the cage created a frame from where to watch the sharks, so much so that for some, the entire experience seemed like watching it on TV, especially if the sharks are far away, and do not give much attention to the cage or the divers. 'When I was in the cage, I felt safe as part of the cage. Looking through the windows I felt I was looking at the shark on a TV screen—really good TV, 3D, high definition. I felt a bit isolated from the entire event, at the same time being so up close and personal, we are part of the environment, but also separate from it.' The air bubbles that often connected the sharks and the humans transcended the wall of the cage, are also an interesting part of the equation. As Mike mentioned, white sharks are attracted by the bubbles (and the commotion of the cage), and on the other hand, as noted in his own encounter with a shark when diving, he blew bubbles in the water, which deterred the sharks, and there are now experiments with bubble walls to keep big sharks away. In any case, these bubbles, which apparently attracted the sharks, created a situation of disorientation for the divers who wanted to see them as clearly as possible.

Dobson (2007) noted, when exploring shark tourism generating peak experience for the tourists, proximity to the animal seemed an important factor in shaping the experience. I also found that tourists felt that when the sharks were closer, their experience inside the cage was sometimes much scarier than if they were further away. 'If he's just out there, it is fascinating. But if he turns around comes in and starts biting everything and thrashing everything, I think the adrenaline would start running a little bit. I knew I was in a cage, and I was safe. If I was out of the cage, then I would have freaked out.'

When the sharks were actually close and paid personal attention, it elicited varied reactions in the divers. Once, when Maca was inside the cage and a shark was passing by, to get his attention, Maca banged on the cage. The shark stopped and looked at him, almost suspended in the water directly in front of him. Maca thought if he did any sudden jumping movement, the shark would be scared and swim away, so Maca got aggressive and banged on the cage again. But the shark did not flinch, didn't move at all, just kept on looking at him. Maca got intimidated, and just climbed out of the cage. That's why he said, 'Every time you think you know, you realize you don't'.

Some tourists showed a desire for attention and recognition from the centre of their attention—the sharks. When that happened, that could be perceived from the shark's gaze and bodily movement, and then the question sometimes popped in the tourist's mind—what are the sharks thinking? This personal recognition often provided a strong reminder in the tourist's mind that indeed the shark was at the top of the food chain underwater. Tourists commonly reported the unexpected thrill of having the shark seemingly stare straight at them, and felt an unexpected connection to the shark. Unlike in an uncontrolled environment, when we would generally want to escape attention from a top predator, from inside the cage, tourists often felt like they wanted to be as important to the sharks as it was to them—'It seemed that the shark actually looked at me! And made eye contact with me! I wondered what it was thinking. I was not scared, but I had chills. The shark came, looked at us, sized us all up, then disappeared in the darkness.' Dan said, 'The shark is at the top of the food chain, the king of the ocean. I was focused on it and positioning my body so that I could look at the shark, and it seemed to me that the shark was positioning itself so that it could look at me.' Mana said, 'When I was alone in the cage, he came up turned his body sideways and it was as if like he was inspecting me. So, he's on the outside looking in, and you are on the inside looking out, you both know you are there. Was I really alone? It's like a connection, connection between the two of you. But the question is: what is it thinking? I was trying to make myself more impressionable to him, I was waving my arms, and I think he was just seeing something with the big mask and bubbles.'

The tourists felt that the individual recognition by the shark of the human could also have an adverse effect on the interaction. 'When there are less people in the cage, the sharks could focus on us too. And I think there is also less bubble because bubbles can be distracting.' At least two or three times, when I was standing on top of the cage, the shark was coming to the bait, but suddenly noticed me, and with a swift turn, immediately veered away. A similar reaction of a shark was observed by a Scottish diver underwater, 'When I was in the cage, and the bait was in front of me on the surface. The shark was coming towards it, he opened his mouth, and at the last moment, he spotted me; he must have said to himself—oh no, I'm not having that—and he turned around and swam off.'

Even the same incident and the same shark could create somewhat of a different effect on different individuals—'At one point, it came right by me, and it just floated in front of the cage, and he started looking at me! That was the only time I was like: Woah! Hang on a minute (scared)!' The same experience evoked a different response from a different tourist: 'When the shark looked at me, I realized it was a dream come true and I could really feel the power of the animal. I don't think I even thought about the cold, because I was so focused.' So, the proximity of the animal and the attention from it had different effects on the humans, and possibly their attitude towards sharks; however, it was dependent on each individual—the individual shark, and the individual human.

Visual documentation was an important part of the experience; however, it was not necessarily the most important factor for the tourist. Other researchers came to similar understandings from their studies. For example, McClellan (2015) argued that photographs play an important role in how varied human groups reflect and produce their relationships with other species. McIntosh and Wright (2017) found that though photographic documentation of the wild life experiences was important as evidence, it did not contribute to a meaningful experience for the participants.

Somewhat similar were the feelings of a lot of the tourists about white shark cage diving in New Zealand. As I mentioned, most people ended up taking videos instead of photographs, because of the dynamic condition inside the cage. In one instance, I saw a diver who got in the cage and came out within a few minutes, even when the sharks were around mentioning, 'Well, I got the money shot, that's all I need.' But for others, the experience of using the camera underwater was more complicated. 'I was mucking around with my camera, when I saw the shark coming towards me, and I forgot everything about the camera.' To some, the photographs were secondary to personally having the experience, 'To me, the photo comes later; first I want to see it with my eyes, and feel it with my skin.'

Display of the images in social media did play a part, but often the images were more significant for personal memory than anything else: 'You have the memory for yourself, but you can relate to it with the photo. Either you are looking at the sharks, or looking at the back of the camera, and your reality is that screen.' Some wanted to create a balance: 'Usually, I do a lot of dives with it, then a few without it.' For a painter, the reason of taking the photographs were a bit different, 'I do like to take photos, that I could add them to my library for reference. And when I'm painting, it also helps to be able to call back on that memory.' For a father who could not have his kids accompany him on the trip, the photograph was also important, but it was only for his children.

Demystifying the Monster

One of the reasons for the lack of interest in shark conservation might be their negative image in public imagination (Thompson and Mintzes, 2002; Friedrich, Jefferson, and Glegg, 2014; Neff, 2014; 2015; Friedrich et al., 2014). And it was argued that one way of demystifying them and altering the negative attitude towards them might be direct experience with them in their environment (Bögeholz, 2006; Dobson, 2007; Miller, 2005; O'Bryhim, 2011; Friedrich, Jefferson, and Glegg, 2014). So, now the question arises, what was the effect of this encounter on the tourists and their attitude towards the sharks, because cage diving can play a significant part in the demystification of the sharks, by potentially positively effecting tourist attitude towards the sharks, through providing an opportunity for them to see sharks alive in their natural environment (even if the paradigms of the encounter are created). To address this question, I used interviews and surveys.

First, statistically I evaluated if there was any significant difference in tourist attitude towards white sharks before and after the encounter (among 300 participants: 150 tourists and 150 from general population). Some researchers found, direct encounters gave the participants a new perspective towards the animal, was significant in altering the negative attitude, and increased their show of interest in the conservation efforts. And even if participants already had a somewhat strong positive attitude towards marine environment and it's inhabitants before the shark encounter, they seemed more interested in sharks and their conservation after cage diving tours, which increased the chance of a future call for action for shark-

conservation (Apps, Dimmock, Lloyd, and Huveneers, 2016; Apps, Dimmock, and Huveneers, 2018; Cerullo and Rotman, 2000; Dobson, 2007; O'Bryhim, 2011). My findings are in similar lines with these previous findings.

I found that there were more of male cage diving participants as compared to females. There was a significant correlation between dive certification and a positive attitude towards white sharks; however, it was not the same with neutral and negative attitude. Interestingly, in this research group, there was no significant correlation found between one's level of education and their attitude towards white sharks, which may be indicative that education by itself does not help demystify any animal like sharks, or build stronger positive attitude towards their natural environment; rather, it might be more related to experience and the individual. Hence, maybe it is important to rethink the mode of education and focus on more practical experience for students to inculcate a more positive attitude towards the animals, instead of basing merely on given information to try to bring about positive change in the conservation of nature.

The general population had a less positive, more neutral or negative attitude towards the sharks than the tourists before the shark encounter. However, even then, there was a significant increase in the tourists' positive attitude and a decrease in neutral or negative attitude towards white sharks after encountering them in the cage diving practice. Furthermore, there were friends and family members who were coming with the divers who would not have otherwise come, and this experience not only helped create a stronger bond among the tourists themselves, but helped in creating appreciation in them for this animal, by providing an opportunity to come face to face with them. Hence, the shark encounter was not only effective in demystifying the sharks, and potentially beneficial for their conservation, but indeed beneficial to the tourists coming to visit them.

The interviews provided further avenues to corroborate the statistical findings. From a dramaturgical perspective, it is obvious that the sharks were not attempting to play any act, or put up any front. However, the humans were transposing the pre-perceived image of them onto them. The sharks were supposed to perform certain tricks, come close to the cage, engage with the bait, show their teeth, and perform the image of 'shark'; if nothing else, at least show up for the tourists. However, when there is inconsistency between the appearance and the manner, then it might be confusing to the audience. In other words, when the tourists looked at the sharks, they very much looked like the image they had in their mind of them from media, but the behaviour (manner) they associated with the appearance did not match (as in the monstrous fish who wants to bite everything and thrash around aggressively), and this caused cognitive dissonance in their minds—between the perceived image of the animal and the real creature in front of them. Underwater, this effect of the negative image was often manifested as intense fear and the bodily reaction of heavy breathing and disorientation; this was not helped by the fact that for many, it was their first time breathing underwater.

Often the resolution of this conflict was found in the aesthetic beauty of the animal; their apparent slow and calm movement and non-threatening demeanour affected the humans psycho-physically and calmed them down. Jake, the kiwi male, who wanted to feel a sense of accomplishment after seeing them told me, 'I feel at peace, I feel relaxed. It makes me really smile to think the shark was right by me.' More than anything else, it was, at times, a sense of relief: 'It is something I had in my bucket list for a long time. Will be glad when it's over.' An Irish woman, Louise, explained: 'I was panicking, so that my mask fits properly, the cage, the oxygen, and there was shark in the water. But then the shark came, and my breathing slowed down. I didn't think I was going to be as relaxed as I was. I used to think that sharks are ugly, big, mouthful of teeth and ferocious. Had no idea why my son wanted to do this, but after doing this...I understand.' People often were spellbound by the beauty of the animals: 'It went by like a ballerina, huge, bigger than expected, and it was so colourful, it's almost like a rainbow colour [the iridescent skin].' Bert, a British man, said, 'I thought it was going to be very intense, lot of adrenaline, but it was lot different, actually quite peaceful. A lot of waiting, then excitement when I first saw the shark's tail on the surface; nervous and shaking. When the animal came in, he was cruising around, did not have his teeth on show, didn't seem angry, just inquisitive, it had nothing to do with *Jaws*.'

The shark is as inquisitive about the humans in the cage as the humans about the sharks outside, Raj Sekhar Aich, 2017

There was the contrasting idea of how the shark will behave, often as opposed to the media presentation of them as hyper aggressive. People found them to be more curious than aggressive, 'I thought they would be biting at everything, aggressive, because you see that in all the discovery shows. But I actually found them to be somewhat timid, and curious.'Often the conflict resolution brought self-realization,' said Emma, an American woman, 'I felt free and calm. It was completely opposite to what I would have imagined. I think, I had built up this irrational, illogical understanding, or fear of what they are, and what they will do to you, and this was a great end to it.'

For some though, the experience had somewhat of a mixed effect on their attitude towards white sharks, 'I'm still intimidated by them. They've had so many scratches on them, so many battle scars. I don't think I'll feel any better in the water after seeing them, and their size. They were not aggressive, but you won't catch me jumping in these waters now.' And for a couple of people I interviewed, they seemed to be more scared about the sharks after the experience than before. 'If I wasn't scared before, I am now, because I know how big they are...shit no!' There is one more notable factor, I was able to talk to 2–3 tourists who came to the boat as observers from the surface, and they did not have an altered perspective of the animal; could it be that it is imperative to actually get in the water with the sharks to appreciate them, and just watching them from the surface is not enough? It is something that can be explored in future research.

While for some tourists, seeing the shark was an adrenaline-filled experience, for others, it was more of a spiritual experience. Sonny, the only Indian cage diver I met on the boat, said, 'When it moves in front of you, it's almost like a spirit moving, and not an animal. It was like an aura, and energy moving in front of you, and you don't see the shark as the shark. That's pure strength, it gave me goose bumps, when you think of people meditating—not to think, you're not feeling anything, and trying to grasp that energy in front of you.' There was a Canadian ex-fighter pilot who said, 'I wanted to see if there was any connection, and yes, there certainly was. I did not feel any aggression, it was a recognition, that we are all equals, I

A Canadian pilot explains his spiritual experience with the sharks, 2017

was in human form, and he was in marine form, but we are all equals.' On a personal level, I think I fell in love with this man; after his career as a fighter pilot, he spent the rest of his career as a commercial pilot. I asked him, 'Sir, you flew all your life, you were in New York in the morning and Tokyo in the evening—what is the thing, if any, you found similar in all the places?' 'The smiles... When I came out of the cockpit, the smiles were same everywhere.'

The gender and perceived gender and age of the individual shark may also affect the encounter: 'I don't know it was male or female, but I felt it was a male. That's why I kept calling him "he".' Generally, when female sharks come to the region, they are bigger than the males, and there were a handful of times, when I was able to talk to the tourists about their experience of encountering the sharks and the gender of the sharks seemed to make a difference. One British woman first saw some males, then there was a big female who came our way. She reported, 'This time it was completely different. I think I was a bit more wary this time...this is big, it's a different story! She was more interested in us! But she knew that she was the boss around. She came closer to us, swam below the cage, and constantly looked at us. There was more respect, and the other little one [little being 12 ft] almost got forgotten, and all you could think of

A French cage diver talks about his experience with the shark, 2017

was her.' Tiamat, a young Assyrian-American scholar who saw the female shark, also said with a big smile on her face and amazement in her eyes, 'She had BDE (big dick energy)'.

So, it does seem that cage diving is beneficial to the divers, often providing an opportunity of self-realization, and also taking photographic images for themselves and to share with friends and family. Notably, for most of the people, the experience of cage diving and the white shark encounter facilitated through it create a positive attitude towards the sharks (significant for their conservation efforts), and there is a substantial difference in attitude towards them before and after the encounter. Most found it to be a 'peaceful encounter', and the shark to be 'calm' and non-aggressive, unlike as promoted in media. However, we still need to ask about the perceived 'naturalness' of the encounter. When people are indeed talking of a peaceful encounter with a white shark underwater, I am not personally convinced that the paradigms of how these encounters are produced are truly 'natural'.

Yes, meeting white sharks underwater is extremely rare, and some people do have encounters with them when the sharks are 'peaceful', and the reality may stand that most of the time, pointers may pass by us undetected, because they are not interested in us or not willing to engage with us. But if a white shark is willing and interested to engage with us underwater without the incentive of any other olfactory or visual stimulus, will they not come up to have a close look at us? Even if the shark is calm, and goes for let's say an inquiry bite at us, can an untrained human handle such enquiry, and reciprocate with equal calmness? I am not convinced!

What is significant to note is that these findings are highly indicative that the diving experiences have a significant effect on positive alteration of human attitude towards the white sharks, which is essential in supporting white shark conservation efforts.

The same shark would behave differently around the two different boats (when both the operators were anchored close to each other), depending on the wrangling technique used. Did that change in behaviour of the shark affect the experience of the divers in the cage, which in effect affected their attitude towards the shark? That is, does tourist attitude depend on what the shark divers encounter and how it acts or performs around the cage? The personality of the tourist and his/her experience in the marine environment, or with large predators have effect on their attitude towards the shark?

I think there are three factors and the combination of them that shapes the experience of encountering the sharks through cage diving for the humans, and it's effect on tourists' attitude towards the sharks.

1) The personality of the tourist taking part in the practice.
2) The personality of the sharks involved in the practice.
3) The specifics of the shark wrangling strategies involved in it.

Regarding the tourist attitude towards the cage diving practice, some research has shown that tourists weighed the pros and cons of such an activity before participation, but supported the activity, and there was a cognitive dissonance from participating in the practice they know was unethical (Ziegler et al., 2018). But the scholars argued, the benefits of this practice were human enjoyment, scientific research opportunities, economic incentives to protect the sharks and their environment, education, and economic benefits to the local community. In context of white shark cage diving in New Zealand, the tourists had mixed feelings about the practice too. Often they would ask my opinion about cage diving altering the behaviour of the sharks and were inquisitive about Mike's thought on it. (It cannot be assessed definitively if the responses of the operators or myself had any significant effect on their feelings regarding the operation, as I did not assess their feelings about sharks, in particular, as opposed to that of others, who did not hear the operators or me. However, one would argue that there would be some effect of our perception and statement on them.)

Some of the tourists felt that cage diving was bad for the sharks, because in their opinion, it was training the sharks to associate boats and divers with food and stay in the same place to get food, but nevertheless they wanted to see the sharks anyway. 'By observing them and having the boat here every day, I think it is changing their behaviour and we should not do it. I think we should leave them at peace.' Although from their own experience, they could see, 'The chance of seeing a white shark is based on luck, and shark distribution is quite limited.' Some felt, it did not bring any change in the attitude, or had any effect on the conservation of sharks, and was only good for entertainment purposes, because: 'People come here for the sensation, and not to learn about them. Besides, how much can cage diving change people's minds? It is up to the individual.'

On the other hand, there were people who thought that it can be good towards challenging the negative attitude about the sharks, and even encourage people to take active steps in conserving them. 'I think, this operation can be good for the sharks, because most of us won't ever see a white shark—only in news and other videos. When they go away and hear any news about sharks being killed or finned, they may feel personally about it and want to do something.' Tourists mentioned that cage diving can bring awareness for sharks, even if they were not sure what is the effect on the individual sharks who are coming to the operation. 'Even if misguided, it is in the right direction. I don't think we're doing anything wrong with them, they are not being harmed. I can imagine leaving with a huge sense of awe, appreciation, and respect for them, and anything along those lines are good for them.' And they recognized it is important to see the sharks, and that 'seeing is learning'. 'I am quite curious about that (effect of the practice on the sharks), but after coming out here and hearing about Mike's knowledge, hearing all the pressures they are against, having you onboard, I think it's a different experience of the greater balance. I think, if these things are managed properly, there is more chance of showing that these animals are worth more alive than dead. Unfortunately, we live in a time when everything boils down to monetary gain, so if you can show that there is a portion of the population who relies on this for income, at the same time you're able to incorporate any education and science...that's the future, and tourism can play a really important role. I personally think, there is also a majority of them who would like to go their own way—to become ambassadors for the animals.' Many tourists felt that even if they don't personally do anything after seeing the sharks, they are communicating the experience to family and friends, and that gets it ingrained in other people's head too. For many others, the new knowledge did seem to bring self-realization, and many comments were along the lines of: 'I think it's a good thing. I think it's really good for the image of the sharks, once you have seen them in their natural habitat, there is no doubt at the back of your mind anymore because you know the truth, there is nothing mindless about them.'

The tourists commonly expressed the view that cage diving increased their interest in the sharks and the knowledge of their endangered status. Some explained, like the example of the New Zealand Moa (possibly used because of it being a symbolic species), that if there are no sharks for their kids' generation, the present generation will have to take the blame, and if sharks are less in number than lions in the wild, why does killing of a lion gets so much global attention, and not them? But beyond our unjustified fear of them and the image in the media, tourists do believe 'change can be brought'. Some even made reference to the situation in their home country, such as, 'In Cornwall, the waters were really bad for a long time; in the last 15–20 years, the waters have been cleaned quite a lot and the basking sharks are now coming close to the beach and the public can actually see them, which they couldn't all those years back. So, if you change the public opinion, it could change a lot of things for the better.'

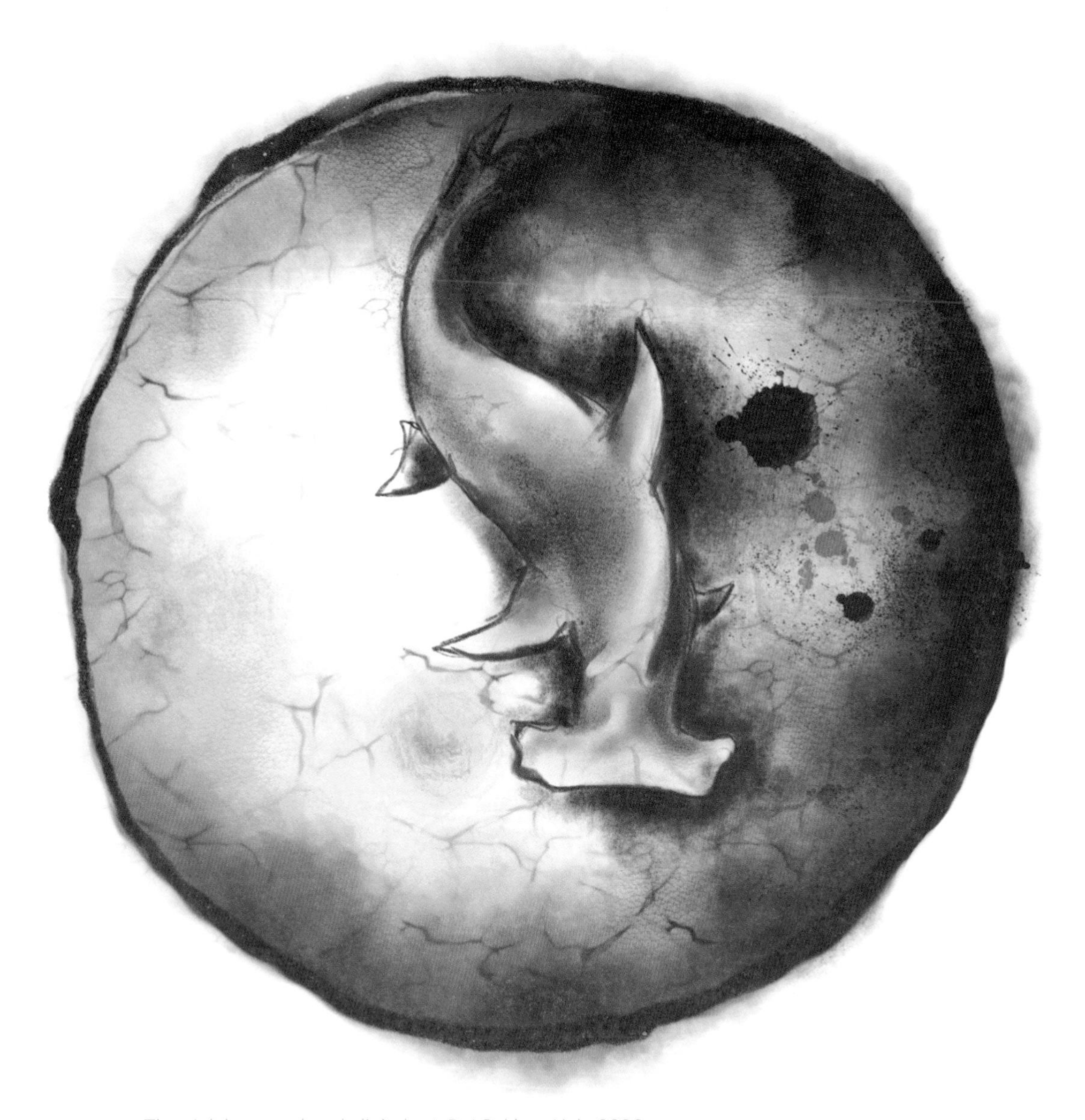

The pink hammerhead, digital art, Raj Sekhar Aich, 2020

How Else Will You See Swedish Girls in Bluff

'How else will you see Swedish girls in Bluff?'—is what one young fisherman had to say to me one day. There were four of them sitting and having a drink, all in their 20s, two of them were married, and all worked in fishing. As residents of Bluff, it seemed they were not scared of the sharks; if anything, they wanted to see the sharks themselves, 'Well, they are the big boys.' They, like most of others I have talked, felt that it is 'us' who are getting in 'their' realm, and agree with Mike, 'If a shark can drive you out of the water, then you have no business being in the waters anyway.' As we were half way through our drink, a man came in and shouted to us, 'Bloody shoot them all', but when he started talking, it was soon clear that he knew the sharks have always been there, and there is no reason to believe that there are more sharks in the water because of the cage diving, and they have always seen the sharks in the water while fishing.

As the days went on in my fieldwork, I used to get up in the morning, and go out to the sea. Soosan was still sick, in the hospital, and not seeming to get better; I was afraid of the worst. Even the doctors I talked to kept on mentioning that the situation was dire. Every morning, as the boat travelled from the new to the old wharf, in the three–four minutes, I used to write silly nonsensical poems for Soosan, something she would see when she woke up. There were poems about my squeaky rubber boots, about the sharks having a dinner party with me tonight, and ones about how I would get her marshmallows from the sea upon my return in the evening. It seemed that she looked forward to them every day:

Image, the quite ocean and the gliding birds.2018

'In this stormy morning... Hi, Soosan, it is me. The sharks are waiting, but they won't bother me, and will let me be. My Soosan is getting better, and hope she is having her cup of tea...be as it may be... When I return in the evening, I will get her Marshmallows from the sea...'

One morning, when we did not go out to the sea, I went for a morning jog to the Sterling Point, and on my way, I bumped into my old landlord, Joe (of the cottage where I stayed the first time). He mentioned that he had spent two years on Stewart Island. He couldn't understand why there could not be a compromise between the islanders and the cage divers. His friend, Trekker, was sitting next to him, an old fisherman of the region. He had seen sharks all his life in the water, many times when he was fishing, all across Foveaux, and near Stewart Island. When he saw the pointers come up as they were fishing, he felt that the sharks kept looking at them and were as much interested in them as they were about the sharks. To him too, it was unfortunate that people could not come to some form of compromise, as long as the cage divers did it in one spot, he saw no reason why they should not be allowed to, because the sharks have always been there anyway. Trekker talked about an experience years ago, when they had a couple of people in the water, and suddenly, one of the young men of 18–20 years swam up. He started shouting and asked to be picked up from the water. They had a small cage which they used to put the people in the water with.

They lowered the cage and lifted him up. When they asked him what was going on, he said that he saw the biggest shark of his life go swimming over him. He was delirious and said that he never wanted to get in the water again. So, clearly, as he mentioned, the sharks were always there and the fishermen knew of their presence.

The old fisherman Marley was the father of the owner of the Oyster Cove Café and Wine Bar and had fished all around. When Soosan was here, she had asked him why did he think that people in this region thought sharks were monsters? His reply had somewhat of a similar sentiment as Mary's: 'Probably because they bite people's arms off! (Everyone laughed.) I suppose the bloody film *Jaws* didn't help it. Kids see that, and you grow up to be an adult with that in your head, but the fishermen did not think it was very good.' However, he differs on the thought about cage diving and Bluff; he says, 'If you look at what we know now, and yet there are heaps of people (tourists) coming! They're always there.' His daughter, the owner of Anchorage, explained to us: 'Tourism is getting quite big here now, everyone wants to come to the bottom of the world. And there are lots of overseas people, more than Kiwis.' When we asked what they thought about cage diving, Marley mentioned, 'I've been talking to a few people who stayed in the hotel, and they said it's the best day of their lives! (Quite elated.)

There were some who did not even see the sharks, but it was wonderful, in fact, I have this young girl who comes to the house and helps out, and she is out fishing and everything. One day she told me that she had gone shark diving, and she said it was marvellous.' Murray totally supported the diving, he felt that not only did it bring money to the region, but it also brings more attention to the town. And this emotion seemed to hold true in some other long-term fishermen, who mentioned that it is the only way to see the sharks; as a matter of fact, my discussion with them was instrumental in them thinking that they would like to go and see the sharks themselves.

Yes, it does seem that *Jaws* had a very powerful impact on the community, even with people who have been around water all their lives. I asked Stoney about his opinion of cage diving, and the sharks: 'They're not evil, just that they put every hair of your body on end, even if you see a decent one... There was a kayak which was mouthed on a few days back, I know this because the kayak operator came to me and wanted to buy a paddleboard.' Do you think cage diving can increase people's risk in this area? 'I don't personally believe so, but I don't walk underwater; when I saw the movie *Jaws*, I didn't even take a shower for a month. I don't go diving, because I know what's out there, what my son does, and when I was his age, I did it too, but no thank you.' Indeed, it did seem his son did go underwater, and has had encounters with white pointers, which I found out at another instance.

One evening (Jan 2017), Andy and his wife, Alicia, invited me to their house for dinner. He and his wife craved stimulating conversations which they didn't often get in Bluff, and his son also wanted to see the Megalodon tooth I had. They were a young couple who had moved down to Bluff to make a peaceful life, and they liked staying here, it was a beautiful house on top of the hill overlooking Foveaux, and as they mentioned, if you know how to integrate, people are kind. In the evening, I learnt the perspective of the largest shop owners in town, and how the contemporary divers in Foveaux negotiated with the very real presence of sharks.

Over dinner, I also met a young fisherman, who turned out to be Stoney's son. He was a kelp netter and put his net from 6 ft to about 50 ft of water near the kelp. Stoney's son acknowledged, 'I am in a unique category, I am a businessman who would benefit from tourists, and at the same time, I'm a diver who dives for kina, crayfish; so, I am also in that high-risk category, but I don't see the cage diving bringing in more issues for me as a diver. At the end of the day, when I'm in the water, they are just another fin.' They often saw white sharks come, have a look and then just leave. It is not sensationalized, but when the crew members did see them, they were excited. By the same token, no one liked it if someone mentioned 'shark' in the early morning on the boat. There was a running joke, if someone felt scared in the water, others would say, 'Don't worry mate, the shark is just a figment of your imagination, it has been made up by Hollywood.' Both he and Andy were free divers, and they had similar habits of surfacing. Both made circles while going up, so that they could observe their environment, instead of being an easy target of any big predator underwater (because as I mentioned, it is generally considered that white sharks like to attack their prey in vertical strikes, and with the element of surprise).

The sensations they expressed of fear of the unknown were similar to what many of the tourists had expressed to me before, on the boat. When they got in the water, often they would feel like they were going in a domain which was not their own. When we talked about the feeling—'something is not right'—underwater, Andy mentioned, 'There is something to be said about the gut feeling, because I think you're processing a lot of information that you don't really understand, so I pay heed to it a lot', which in psychological terms can be explained as environmental cues—cues or signals present in an environment which may even be processed at a subconscious level and affect our cognition and behaviour (Lowe, Stevenson, and Bothwell, 1996; Norouzitallab, Baruah, Vanrompay, and Bossier, 2019; School, 2012; Zhu et al., 2013). Just like the tourists, they especially mentioned if the water was murky, they didn't feel safe, and even Andy had come out of the water once, because he had a feeling of being watched. Stoney's son said, 'Generally a rule on our boat is, when anyone is on the watch, they can't watch shark movies. We have sky TV, and once, we lost the remote control and I sat through two hours of shark TV. The very next day when I dived, it was absolutely clear waters and I didn't feel uncomfortable even for a moment.

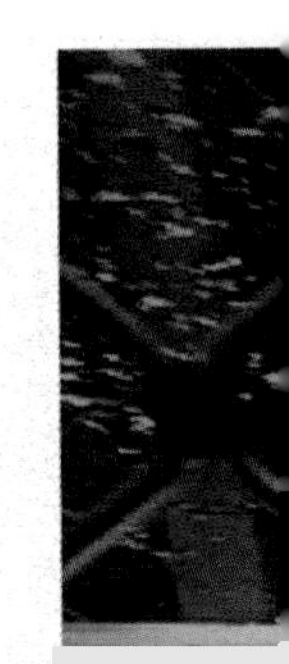

But a few months later, when I went for it again, and the water got murky, I felt uncomfortable and actually got out of the water.'

However, neither of them was sure if they wanted to come and see the sharks in the cage. Because once they have seen them, they're not sure how they might feel when they're diving without the cage. To that, Alicia pointed out, there were couple of divers in the region who had seen them out in the water, but they'd still gone in. Andy responded by expressing that they had seen the shark, in their natural environment without being attacked; so, they had a positive experience with them, as compared to seeing them from the protection of a cage. This is an important point, because it brings up the argument of a 'real' encounter, because it happens without pre-meditation, and it seems it might give individuals who have to go in the water more regularly, a deeper understanding of the sharks, and probably more confidence in the ability of their own skill, as in Mike's case. However, if this real encounter led to physical harm, it might have a different effect, as well as attitude towards one's own ability, and perception of the environment and the fish, as was the case with Vaughn, at least initially.

Alicia mentioned that she had also gone out in the trawlers and it seemed that people, including herself, dropped tons and tons of fish carcasses, to which both the men agreed that even they had done it at times. They clearly recognized that the sharks would be more interested in that than a couple of tuna heads. They also had the engine on when they were getting the fish and cleaning them in the water; so, if anything, sharks could get trained that way. [These statements again raise the argument against only cage diving causing sharks to associate boats and humans with food, while it seems that boats and fishermen dump a lot of fish carcass in Foveaux.] Here, I would again refer to Crawford's white pointer chronicle, where one fisher after the other documents how pointers used to be always around during fishing and that they were just a part of life.

Alicia then explained to me about her thoughts on cage diving and its potential benefits to the economy of Bluff better than I could ever have, and even how I may have a significant part to play in this new line of thought in the community. Since she had been in Bluff for years, she

Here I come, 2017

talked to a lot of shopkeepers, trailer park owners, and initially, many were quite sceptical about the cage diving operations, but in the last couple of years, more and more people felt that it was a good idea and supported it. That is why the debate about cage diving was strange to them. Why was it such a big issue in Stewart Island, considering the kind of money it can bring to the community? I asked her what she thought of cage diving as an industry in Bluff, and what kind of business it brought to the region. She replied, 'I don't think is negligible; it's noticeable, but not substantial yet. In other places of New Zealand, they had small beginnings too; I can't see any reason why it can't happen here as well. But I don't think great white shark diving can be as scalable as other places like Kaikoura, because all the limitations put by the DOC and other agencies. If you are only allowing, shark cage diving around Edwards and only two boats, then how many people can actually take part in that every day? Maybe, they can get a bigger boat with more cages! Maybe, it's a boutique experience. It's beautiful and it's visceral. So many jobs have been lost in this region and anything that brings in more jobs would keep three more families going, and the school going. Just depending on fishing is putting all the eggs in the same basket. You need variety, and tourism is an element of variety. Maybe, we have a bad cod year, or bad crayfish year, or something happens to the oyster bed, or if an oil tanker leaks close by, then what happens? That's why we have to have variations!'

I asked about her observation of local people's attitudes towards cage diving and that of the tourists who come to see them, to which she said, 'I think it's good for Bluff, it brings people here who otherwise wouldn't have come, spend some time in Southland; they might even go to Stewart Island. And I think Bluff is moving around; as people start seeing the benefits, that is, the money, I think there will be a gradual change. And the type of tourists that come here aren't drunken hooligans; they are quiet, they're polite, and they spend a little bit money. I talk to my staff all the time, but I've never heard any single person being actively against it. They mostly say, "Well, the sharks have always been there." I suppose it brings some money in, though people are looking.' Andy also mentioned, 'The other side of it is that having someone like you here within the community gives us an advantage. I saw you were talking to Adrian, one of our staff; now it does not matter if she's pro or against shark diving; what is interesting is that she knew that you're a shark researcher and she was having a good yarn with you—this kind of interaction helps break down barriers.'

Marshmallows from the Sea

On 8 January 2017, I came home to the cottage, dead on my feet. I sat on the bed, did not eat or look at any data collected on the boat. I felt something was not right; a distinctive vacuum glared at me from below my heart to my throat. I sat there on the bed, with all my wet clothes on, frozen face and fingers, looking out into the silence of the cabin. I can't tell you what I was thinking... It was as if someone had pulled a switch and I just had stopped working, gone dormant. The late evening turned into night, then late night; the stars were not glowing in my atramentous room, there were no rats, no storm tonight; it was all silence.

Somewhere in that silence in my mind, two little kids were playing together. One is a raven haired bachcha (kid) under her *shiuli* tree by her warm *pukur* (pool/pond), amid the soft, muddy soil of Dumdum, Kolkata. Next to her is a little blond bairn, playing among the hard-cobbled rocks of the seaside town, on the Southbank street of Blackpool. She picks the flowers and weaves garlands for the neighbourhood puja, as it is Dasami, the day when Goddess Durga goes back from her maternal home to her drunken husband, Lord Shiva, in the mountains. Women, in golden and white and red pleated sarees wrapped around their glossy and wrinkled skins, play with vermilion. In the meantime, this little girl steals a few flowers from the neighbour's garden, mashes them in water to make fragrance, her angelic fingers lift the small aromatic bowl, and she tries to sell it to the neighbours.

Gouri, the beautiful daughter of my *dida*, sees fairies in the *pukur*, dancing, with open hair, and glowing with light; they call her in the middle of

A full rainbow in Bluff, 2017

the night to come and play with them. The family members got so scared of these visions she had, they started to lock the doors. Daughter to the single mother, Jene Marie, this little goldy (Soosie) was lost in her dreams, dancing on the road, dreaming that some Hollywood producer will see her in the street, and make her the next super star. Little Soosie has magic surrounding her...her steps, her eyes, and her mom understood this side of her and taught her to accept her abilities, rather being scared of them.

When she was 16, the pearly skinned Gouri fell in desperate love with a passionate painter, with a body and skin of obsidian, wild creative instincts and destructive impulses of Shiva. They got married and she came and settled down in this home, a home filled with people, my uncle and his wife, *dadu*, mumma, my *pishi* (aunt), and their kids. And this house, and the people in it were her life and her universe, even if there was not a lot of money, it never suffered from the poverty of love. At that same time, women's liberation was striding forward on the other side of the globe, and Soosan, a dashing tall Viking, strode down the abbey road in her red high heels with the beats of this liberation and her own. Earning hundreds of thousands of pounds, butting heads with the dunderheads of the male dominated work force, an industry leader, and a successful business owner, her light fell in every room and heart she entered, and every lips and forehead she kissed.

In this house, she got tormented by the love of her life. And her tears painted the floor in cryptic dialects, visible only to weary eyes. And as the years rolled by, one by one, everybody left, only remained me and him, then him and the ticks, and now—him and this lurid chasm of an empty skull, as if inviting some tantric to pour *mahua* (alcohol made from mahua berry) in him and consume ecstasy. But then there was a hero waiting just by her, he held her hand, and together, they started the journey of bringing up her two nutty (one in particular) young boys, one day after another. She slipped once on her path momentarily, giving in to the allure of mediocracy, and ended up being in the New Zealand wilderness, away from friends and family, and anything that was known. All was lost but her spirit, and a beautiful little girl, whom she protected against a monster in the true sense in sheep clothing. A monster who only left destruction and putrid hate in his path. And life came down on her after that, wave after wave after wave, to teach her what lesson, I don't know. I suppose the only one that came out of her challenges is the lesson for others like me, to learn from the strength of her indomitable spirit. And this loss and hate, never could take the smell of the blooming flowers of these blossoms, at either side of the globe.

I must have fallen asleep sitting upright, I don't know when, but sometime in the night, I could hear someone crying, somewhere far away. It seemed like a known voice, but I could not figure out who it was. I tried to go back to sleep. But I felt the intermittent crying of the voice all through the night, and it got louder and louder, and suddenly I heard it coming from the chair next to my bed. I opened my eyes and saw someone was sitting on it. He had long blond hair, with the head stooped over covered by his hands. He was naked, with big strong arms filled with tattoos, and lactating breast, skin wrinkled like he was 100 years old. He sobbed slowly, as if he has lost something very precious, his drips started to fill the floor, and the

The flower of the west, 2015

water in my room started rising, with small waves created around the old fireplace. The shadows and the images in the shadows started smudging in the water, like ink on handmade. I touched me, my cold hand fell on my heart; shivering, I woke up, and the only thing I heard myself saying was, 'Fuck the research!'

Ms Soosan Lucas is, unequivocally, the strongest human being I have ever had the privilege of knowing, if my readers have been with me this far, I can tell you with all my honesty, the challenges she has overcome every day is nothing short of miracles. I would literally never have even survived in her shoes, let alone prosper, which she does every day too, but that is her story to tell. She is not my friend, not my research assistant, she is my Guru-bon (sister connected through the teacher). She gave me food, when she did not have any, shelter, when she hardly had one herself, and most importantly, she would bear with my manic-depressive tendencies, and here I was thinking of the fucking research... when she might fucking die... I knew I had to go and see Soosan, what if anything happened to her and I never went to see her, what would I tell myself for the rest of my life? What if she survived, and I never went to see her, what would I tell myself for the rest of my life? And just as such when in life the questions are clear, there is clarity in one's response... I have to go and see her, although I had no idea how.

The incident that ensued the following day was bizarre to say the least, and even though I am not a 'believer', all I could think of was the term Darrel my friend uses a lot, 'synchronicity'. I went to the boat as usual, as I did on any normal day. The back of my head, I had Soosan in the hospital and I *had* to go see her. I started my interviews and interaction with the tourists; everyone was excited to see the sharks. About this time, I met a bearded angel who had come to meet with the white sharks. Patrick, a lawyer from Christchurch, was about to go back to Christchurch at the end of the trip and was incidentally about to stay in Timaru for the night in a friend's home (incidentally Patrick was also the name of one of our dear professors who passed away three months ago at that time). I asked Patrick if he could give me a lift, and Patrick said yes to the proposition whole heartedly.

Gouri, the flower of the East, Samir Aich, 1985

So, when we got off the boat, we walked to the car park behind the Bluff lodge, I put a denim shirt on, took one in my back pack, an extra pair of underwear, my wallet, and got back in the car. We drove through the winding roads towards Timaru, and I treated Patrick with a coffee, which is basically all I could afford at the time, with $23 left in my account. We talked about how Patrick as a young lawyer, aims to help his community, and the challenges he faced; there were discussions about how he felt about a shark who had a lot of scars on his face, and looked like a fighter that he admired.

I reached the hospital at 2 am, they knew I was coming, and had kept the house keys for me. As I walked through the empty corridor of the hospital, the receptionist had a kind smile on her face, as if she knew how I had reached here. She led me to Soosan's room. The door opened, and in the soft light of the room, I went to her; she was in the bed, frail and sleeping, with drips in her hand. I had never seen her like that, my eyes swelled up. I sat next to her, kissed her cheeks, and went away. When she woke up, she felt that she had had a dream that Santa Claus had come to see her, but only in the morning, when she looked at her desk by the bed, she realized there was a pack of marshmallows.

Soosan seemed better every time I went to see her, and I got her a small light which glowed in the night, and she used to share the marshmallows in the purple light with the other roommates, with the toy of Maui, from the animation film Moana, which had just come out (and apparently who I had the stark resemblance with at that point). When they were giving Soosan two intravenous antibiotics every hour, she used to sit and sing the *Mission Impossible* music in her mind, like miniature Tom Cruise was getting inside her veins and was healing her. She used to tell her body, 'I love you', and 'Thank you for protecting me all these years'. The hospital food was horrible; so, the sandwiches I smuggled for her helped too. The day I had to go back to the field, I looked at her, she was in her hospital clothes, looking at me with her teary eyes, but she and I both knew I had to go back, and that she was going to be okay, and on I went to board the bus to Bluff.

Rocky, the Excited Shark

One reason for conflict about cage diving is the effect of cage diving on the sharks, and effect of that on the humans who share the same water space. Not only is it significant to consider because the politically heated debate it raises nationally and internationally while considering cage diving, but because it deals with real effect on both the species sharing the water space, a debate which might potentially cause in ending the operations all together. Even though I am not an ecologist, as a social scientist involved in the entire debate, it is important for me to present here my understanding about the matter.

To contextualize the discussion, I start with some observations by Steve Crawford, who was kind enough to explain his thoughts on the topic. He thought that there were credible arguments on both sides of this debate (cage diving is increasing chances of the others who share the water space of Foveaux Strait being attacked by white sharks, because the sharks are being attracted here, where they were not before). Evidence from his primary data suggested that white sharks have always been seasonally abundant in New Zealand waters. However, Steve agreed with many of the indigenous and local knowledge holders who stated in the interviews that white pointers are 'not silly', that is, they are much more intelligent and curious creatures than has been previously recognized by many researchers in the science knowledge system. Steve thought that there was abundant evidence supporting the idea that white pointers exhibit strong behavioural conditioning to circumstances where humans accidentally

A shark passing in front of the bait, 2017

or intentionally provide fish (both sensory cues like blood/burley and consumables like flesh/frames). Like all vertebrates, white pointers have the behavioural capacity to learn, and to change their behaviours in response to changes in the spatial-temporal distribution and intensity of such cues. Steve thought that the 'key ecological uncertainty' in this case was the degree to which experience and behavioural conditioning with specific human-proximity food cues (line-fishing, trap-fishing, net-fishing, spear-fishing, fish-cleaning, fish-processing, freezer works, and any such eco-tourism activities including but not limited to cage tour dive operations) has *significantly increased and a lasting effect* on a white pointer's tendency to investigate and potentially test human circumstances encountered at a later time and place. He thought that the science knowledge system (and the government management programmes that claim to be based on such science) needed to immediately address this key ecological uncertainty with rigorous experimentation. If and only if the evidence from such research shows a significant effect in this regard, then science-based government agencies should be required to explicitly evaluate strict regulations for any circumstances that introduce such human-proximity food cues.

In discussions with Steve, I created a table to identify differences in possible behavioural cues present for white sharks during white shark cage diving and fishing activities in New Zealand.

Behavioural cues	**Fishing**	**White Shark Cage Diving**
Accumulated stains of previous blood/guts on Vessel	Yes	No
Burley/chum	Yes	Yes (While on station)
Fish frames/carcasses	Yes (While cleaning)	No (According to regulations)
Active wrangling of tow-bait	No	Yes (While sharks in proximity)
Humans on deck	Yes	Yes
Humans in water/cage	No	Yes

It is probable that there are two behavioural cues, which are present during fishing and which are not during white shark cage diving in New Zealand, and two the other way around, while two are similar in both the cases. First, the dynamic play of shark wrangling and all the different electro, olfactory, auditory, and visual stimuli that are created from humans in the cage and in the water are only present during cage diving, and not during fishing. Furthermore, humans on the cage diving boat intentionally interact non-aggressively with the sharks (which may not always happen in fishing boats), particularly because white sharks spyhop, and they can see the humans interacting with them, even if during both the operations, there are humans on the deck, and there is chum created in both the cases, intentionally or unintentionally. On the other hand, there are accumulated stains of previous blood/gut and a significant amount of fish carcass disposed of, in case of fishing boats, which does not happen during cage diving, where they receive energy return (feeding on the fish carcass) when they investigate the vessel which is actual energy return for their investigation of the vessel, as opposed to the other cues. All these factors have to be carefully considered and experimented with to understand how a shark may behave differently to a cage diving boat, as opposed to a normal fishing boat.

Taking the above into consideration, I explore the current literature on the topic, and my studies on shark behaviour and ecology around white shark cage diving in New Zealand. In 2015, Barry Bruce, from CISRO Australia, was entrusted by the Department of Conservation, New Zealand, to create a summary of the white shark cage diving industry globally and the potential effect of it on other human water users in New Zealand (Bruce, 2015). Bruce, as a global pioneer of white shark research, has been working with them for decades, and even been the shark advisor in the Cousteau's white shark project. Arguably, this research becomes one of the corner stones of any investigation into the effect of cage diving on the white sharks in New Zealand. The summary of the findings was: 'Overall, it would seem unlikely that white sharks exposed to cage diving activities are any more or less likely to present a risk to divers, swimmers or surfers in areas away from cage diving sites than any other shark' (Bruce, 2015, p. 4). There was also a study on South Africa's white shark cage diving industry and for concern to other water users in 2006. It was noted, 'Whilst the authors are confident that cage diving activities are not contributing to the recent rise

Fishermen gutting fish in Foveaux around the cage diving operation, as the animals feast on the entrails thrown overboard, 2016

in attack rate, we recognize that even a perceived link may have dramatic consequences on the white sharks national conservation status, and the stability of an economically viable non-consumptive industry' (Johnson and Kock, 2006, p. 57). Even if the chance of risk on the other water user did not seem to increase, Bruce proposed three strategies to control any potential impact of the operation on the sharks, and I contextualize these strategies, with the current mode of practice in New Zealand.

Limiting Effort (e.g. the number of operators)

When my research started, there were only two operators who were allowed to operate cage diving boats around Edwards Island within about 1,000 ft from the shore. By the end of writing my book though, there was only one left. There were stories of a plan to use a glass bottom boat to bring tourists to see the sharks, but that never became a concrete idea because a 5-m shark could have easily broken the glass.

Spatial/Temporal Closures (restricting the activity to certain sites or zones and time periods)

As mentioned, the only island around which the operators were allowed to work is Edwards. This decision was reached in a mutual consent between the operators and the DOC. Not only was this island at least 10 km away from Stewart Island, but also historically, there was always consistent observation of sharks around this island. Also, the area around the island is quite shallow, about 30–40 ft; so, even if the sharks are near the bottom. There is a possibility of observing them, once the tourists are in the cage. There was, anyway, restriction on time periods, because the sharks only seemed to be reliably present around Foveaux Strait from December to July every year, and due to the turbulent nature of the strait, on many days trips had to be cancelled.

Controls on Equipment Used or Other Operational Restrictions

It is significant to note, Bruce maintained that the New Zealand Code of Practice regarding the use of bait allowance in white shark cage diving was the most stringent in the world (Bruce, 2015). The operators are only allowed to use one bait a day, and if the bait is lost because of the shark, then they cannot use any more for the rest of the day, though there have been debates for allowing at least three in a day. The operators are also not allowed any artificial method of attracting the sharks, for example, audio stimuli. Among the three operations allowed to practice cage diving in Australia, one can only use audio stimuli, like music, to attract sharks. And hence, there are claims that white sharks like hard rock music. A more precise interpretation might be that the loud vibration in water is able to intrigue the inquisitive white pointer.

Ecological Effects of Cage Diving on Sharks

Bruce rightfully pointed out that although a lot of studies on the effect of cage diving on sharks refers to them as shark behaviour studies, most of the empirical studies were essentially 'ecological' in nature and investigated 'aspects relating to the swimming behaviour of sharks, including swim speed, swimming depth, localized movements around cage diving sites, patterns of residency, habitat use and influences on broad scale movements/migrations' (Bruce, 2015, p. 12). He mentioned this had led to misinterpretation in research objectives, 'It has been assumed that where changes in behaviour have been studied and documented, they relate to sharks being conditioned to associate vessel and humans with food or become permanently more "aggressive"' (p. 12). Consequently, any consideration on the effect of cage diving on the white shark ecology and behaviour needs to consider four factors.

1. The large-scale movement
2. Localized movement (around the boat, and in the local area)
3. Immediate ethological effects on the sharks
4. Long-term ethological effects on the sharks

The Large-scale Movement of White Sharks

It is pertinent to point out, most scientists note that white shark cage diving practices globally take place where the sharks aggregate naturally, primarily due to pinniped colonies (Gallagher and Hammerschlag, 2011). 'The industry takes advantage of the naturally higher local abundance of sharks and the reliability of more regular encounters at these sites, rather than attracting sharks to areas where they would not normally be' (Bruce, 2015, p. 2). It was evident that they were not just around those islands, but travelling far and in between. From the limited empirical data of the white shark movement pattern, however, we think that white sharks are long distance travellers, and come to the region particularly during summer, from data gathered even after the long establishment of cage diving around Edwards (Ballance, 2017; Bonfil, Francis, Duffy, Manning, and O'Brien, 2010; Duffy, Francis, and Bonfil, 2012; Francis, Duffy, and Lyon, 2015). It is safe to say the operations have not at least affected the large-scale movement of the sharks in New Zealand, at least to the extent they reside in the region, specifically due to the cage diving operations.

Influence on Localized Movement

Temporarily altering the behaviour of sharks is one of the key elements of a successful and economically viable shark cage-diving operation and an essential element for client satisfaction. Sharks are attracted to the vessel and contact time is encouraged so as to enable clients to view sharks that would otherwise not be reliably seen. However, wildlife tourism that involves provisioning (feeding), attraction or some form of reward for the animals involved can often result in changes to behaviour in target species that last over varied time-scales and may give rise to unintentional effects on those species and the ecosystem within which they reside...' (Bruce and Bradford 2013, p. 902).

It does seem that there is a chance that the localized and small-scale movement of white sharks may be effected by cage diving activities. Gallagher et al. (2015) in a paper on 'Biological effects, conservation potential, and research priorities of shark diving tourism', presented a meta-analysis of relevant information in these areas and evaluated the potential effects of dive tourism on shark behaviour, ecology, and subsequent human dimensions. They pointed out, 'Shark diving tourism creates scenarios with novel stimuli for the animals to react to; as such, the ability to learn is central to this issue. Broadly, repeated exposure to stimuli creates the opportunity to learn via: (i) associative learning, which is the learning of an association or relationship between two events (e.g., operant and classical conditioning, observational learning); and (ii) non-associative learning, which is when learning occurs as the result of the presentation of a single stimulus' (e.g., habituation and sensitization; Lieberman, 1990).

Furthermore, they mention that sharks certainly have the cognitive ability to associate with provisioning activities, and more so, since it is done at a specific place at a specific time. They concluded that the effects of cage diving operations on the movement of the fish are dependent on the species, and the scale of the operation, and operations which provide baits can actually alter small-scale and short-term movement of the animal, but not long-term movements. Huveneers et al. (2013), using radio-acoustic positioning system, monitored the fine-scale movements of 21 white sharks, to explore the effects of shark cage-diving activities on their swimming behaviour and space use. They found that the operations indeed influenced the fine-scale three-dimensional spatial distribution and movement rate of white pointers around Neptune Islands. However, individual variations of the sharks played a significant part in this equation; so, they noted that the relationship of the sharks with these kinds of operation are dynamic and complicated in nature.

Immediate Ethological Effects on the Sharks

Few global researchers document the effect of cage diving on the ethology of the white sharks and do so primarily through the analysis of tagged sharks, and a few times, on visual observations from boats. Shark cage diving may not be the only reason why sharks congregate around a specific site; instead, it might be related to factors like interaction to other species in the region, including other sharks and animals they feed on, individual motivation of the sharks, stimuli habituation, searching pattern, and seasonal movement. As one report noted: 'Sharks are probably present at Seal Island for the sole purpose of hunting seals. Consequently, if chumming had a direct impact on the sharks, this would likely be most evident through a change in some element of their feeding behaviour. Data collected over the field season suggest that predation rate remained relatively consistent throughout the study period (A.A. Kock, unpubl. data) and that no decrease occurred during chumming periods [in South Africa]' (Laroche, Kock, Dill, and Oosthuizen, 2007, p. 207). A recent study (Huveneers, Watanabe, Payne, and Semmens, 2018) suggests that tourism activities as cage diving can alter instantaneous activity levels of white sharks, including vertical lunges and instant accelerations; however, the overall impact might be insignificant in the interaction with individuals that are infrequent.

Long-term Ethological/Ecological Effects on the Sharks

Bruce and Bradford (2013), conducted a comparative analysis before and after the sustained cage diving efforts to investigate the long-term impacts of cage diving provisioning on the behaviour and movement patterns of sharks in South Australia. They discovered that the number of sharks reported by cage dive operators significantly increased after 2007, and indeed, shark cage diving operations influence localized movement of sharks.

In lack of any more empirical data on the subject, particularly in New Zealand, it becomes important to note the observations and understandings of the cage operators, even beyond recognizing the political position of Mike Haines on the matter. I enquired about his perception of the effect of white shark cage diving in New Zealand on increased risk of other water users in Foveaux. First, he pointed out, 'Eight years this has been taking place. Has there been any attack? No! Has there been any close calls? No!'—which is a matter of fact. He then points out what many of the fishermen had indicated both in the literature and my interviews: 'What issues there has been? Some of the fishermen from Stewart Island say that big boats are being attacked by white sharks. But great whites are inquisitive animals, they bite on things to check what they are, they come up to our boat and take a bite to see what it is. Three–four weeks back, a video was sent to me, from 'Fangaroie' with a short shark approaching the boat and there was a bait hanging from the side. Is that our fault too? I've seen fishermen drop a lot of fish carcass in these waters; wouldn't that train sharks to associate boats with food? It's a common practice around here, cod fishermen, for example—if they catch a ton of cod, 30 odd per cent of that is just waste. Because cod is gutted, and the rest is thrown over the side. Over the years, we have had so many boats come next to us and clean their fish. The reason they do it is: firstly, they get calm conditions where we are, and secondly, they like an opportunity to see the sharks themselves. This has also taken place just outside Half Moon Bay [at Stewart Island] for years, and the local islanders blame us for attracting sharks into the bay, which is pretty strange. Because we're not beating, and we cruise around 15 km, and sharks can't follow that speed for a long time. But sharks can follow cod fish boats, because they travel around 4–5 km. On the other side of it, people who have stayed in Stewart Island, and Bluff always had issues with sharks, going back 70 years if you asked them.'

I asked Mike about sharks potentially being trained in the operation to associate boats and humans with food. 'If we had the same shark coming here day after day, month after month the likelihood is that we would condition them; you can do that with a bird, a cat, or a dog. I'm sure it is possible with sharks, just like we see with other fish. Every day, we come here and start the burly, we see the fish are there like Trumpeter, Rass, the Blue Cod. But the thing with sharks is, we are getting an individual shark for two or three days, across to three weeks. If there were more of them, and they came regularly, I'm sure they could have been conditioned. And trying to train them? I have been trying to do that for years; that never works. I know they train dogs with clickers, I thought I'd try to do that with sharks' (we both started laughing) (Dictaphone interview, Feb 2017).

In consideration of the literature, I wanted to explore both the ecological and behavioural aspects of the effect of cage diving on white pointers in New Zealand. For the ecological aspects, I wanted to observe the effect of cage diving on short-term movement of the sharks, if the same individual came repeatedly to the operation, if the time of first observation with the animal

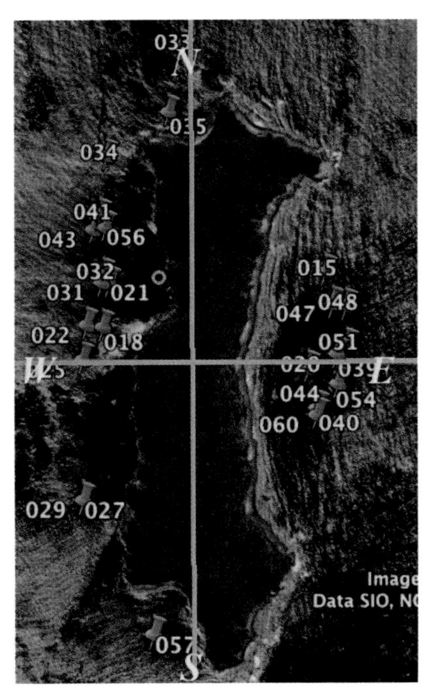

Edwards Island, dissected in to quadrans to chart the encounter with white sharks, Google maps, 2017

was similar, and if the side of the island they were encountered was also similar. From the behavioural perspective, I wanted to explore if the sharks were getting more aggressive over time in relation to the bait provided in cage diving.

I used photo identification method to create a database of the sharks. In 19 days, I was able to positively identify four specific sharks who came to the site more than once, among the total of nine sharks. The results indicated that only one shark came to the boat in a repeated manner over the days the data were collected; so, individual differences were observed to be a significant factor, instead of generalized association behaviour, when concerned with return to the cage diving boat.

I was interested in whether there was a higher chance of seeing the sharks on any particular side of the island. The plan was to divide the area surrounding Edwards into four quadrants and to pin GPS coordinates on the first point of contact with the sharks, to assess if the sharks were present in any specific side of the island in a predictable basis. However, there was a lack of foresight on my part, because depending on the environmental condition, the captain anchored the boat on one of the sides and waited, and if we did not get the shark on one side and if the weather was permitting then we moved to another site. Consequently, the results show that most of the times, the boats were on the eastern or on the western side, that was where the boats were most sheltered from prevailing winds, and hence were anchored.

To test the effect of cage diving on the behaviour, I used a method based on Mandy Bromilow's method of observing white shark behaviour around a cage diving vessel (2014) when bait is provided. I divided 'white shark behaviour' into a few phases:

Phase 1: *Investigation*: Slow Swimming: The shark swam slowly at the surface towards the prey, or just past it; Turn-About: After swimming past the prey, the shark turned back toward it slowly or moderately quickly, with little change in speed.

Phase 2: *Pursuit*: Jump: The shark leapt partially or completely out of the water, attempting to attack the prey; Horizontal Lunge: The shark accelerated quickly along the surface toward the prey with its jaws open and its back partially out of the water; Vertical Lunge: The shark accelerated quickly toward the prey from below with its jaws open.

Phase 3: *Prey Capture*: Horizontal Surface Grasp: The shark slowly approached and grasped the prey while swimming horizontally near the surface; Vertical Surface Grasp: The shark slowly approached the prey from below, grasping the prey in its mouth at the surface; Lateral Snap: The shark attempted to bite the prey, either successfully or unsuccessfully, with a sudden lateral snap of its jaws as it swam along the surface.

Phase 4: *Prey Handling*: Carrying: The shark held the prey in its mouth while swimming slowly with large amplitude tail beats; Thrashing: The shark held the prey in its mouth while shaking its head and body from side to side in an attempt to remove flesh.

Phase 5: *Feeding*: The shark repeatedly bit the prey, either at the surface or underwater, consuming pieces of flesh.

Phase 6: *Release*: The shark released the prey from its mouth, leaving the prey partially or completely intact (Bromilow, 2014).

I utilized her method to make observation of the behaviour of one particular shark I named Rocky, who came to the boat the maximum number of times in a span of 19 days. Rocky was one of the smallest sharks that Mike and his team had ever seen (about 8–9 ft). Initially, when he had come to the boat he was very aversive of the cage and the bait. However, over time, I saw the change in his comfort level with the boat and the bait. As he became more and more confident, he started approaching the boat closer and closer, and started investigating the bait more and more.

I have to mention here, the empirical validity of the data collected from these ecological and behavioural studies may not be very strong, because I am not a trained ecologist, neither was I proficient in testing fine tuned ethologial dimensions of the sharks. Rather, as a social scientist working from a marine anthropological perspective, testing out the methodological dimension of such research, and their possible use in social scientific exploration was more relevant to me. Having said that, the indicative data set that was created from my studies did seem to agree with Bromilow's findings, that the sharks showed more increase in investigative behaviour and not so much for aggressive behaviour, and the movement of sharks did seem to be more related to the individuality of the sharks, rather than being directly related to the cage diving operations and the bait provided. But to have a deeper understanding, we need more detailed ecological investigation, particularly in the context of New Zealand. Furthermore, as I made the observation earlier, the demeanour and strategies of the particular shark wrangler on a particular day, with a particular shark, also had significant effect on the behaviour of the sharks.

White sharks are extremely smart creatures, and it is very much possible that they learn to associate food with boats if they get the right behavioural cues. However, from my observations and observations of cage divers, it seemed unlikely that individual white sharks were trained by white shark cage diving in New Zealand to stay close to Edwards or getting too comfortable to the boat or more aggressive, particularly if the operations are strictly controlled and there is not repeated interaction with one individual shark for sustained time. It is unlikely that white shark cage diving in New Zealand is increasing the risk of 'attack' on other water users any more than other activities which may produce sustained behavioural cues to white sharks, like fishing activities around Foveaux. Furthermore, from the literature at hand, it also seems unlikely that white shark cage diving in New Zealand is affecting the long-term movement of white sharks around Foveaux Strait. Having said that, it is important to note that there is evidence that operations like cage diving can indeed influence short-term movements of white sharks, and there should be strong legislation to control too much interaction of individual sharks to the operations. But to be clear, my findings are indicative only and deserve more structured empirical investigation.

As for the methodological dimension, any future researcher attempting fieldwork on marine anthropology ideally should train themselves in the basics of ecological study, and perhaps, even take courses on them before commencement of the fieldwork. They should make sure that they provide themselves specific research questions to answer, and have all the tools, and other logistical supports ready. Photo identification is an excellent method of data collection about individual marine animals, and researchers should understand it in depth. Researchers should keep in mind, if they are participant observers, perhaps they would have to set aside specific times in the exercise, where they can focus on the finer details of an experiment or study. Finally, the tools of such studies should be properly studied beforehand, and the researcher should be proficient in tabulating and recording the findings.

Unfortunately, neither I, nor my supervisor at the time, had the right mind to engage ourselves with any of these pre-fieldwork strategies to make sure proficient data collection and analysis was possible. However, these studies, I would argue still hold significant value, reminding the future researcher the challenges of such work, and to encourage deep engagement with them before commencing fieldwork.

The stars and whale shark, digital art, Raj Sekhar Aich, 2020

Babu Ghat

Babu ghat, my beast stops, I smell of Mobil, and for some reason, cashew nuts. The mucky waters of Ma Ganga glisten and flow by her own grace, filled with ash, corpses, flowers, food, and shit. Straw skeletons of a deity flow by and parts of the skeletal breast remains afloat above the water surface, creating momentary islands for crows to rest on. Sweet and pungent winds from her ripples blow across my face. Some of the vendors are still sleeping on the ground under old blankets or on top of their vending carts. I buy a cigarette and a cup of Bengali *dudh cha* (milk tea) and look into its brown skin. The plump joggers come here every morning for a jog, and as a tradition, finish it with plates full of sweets and fried *kochuri* (or kachori, a kind of fried Indian bread). The show has not started yet, but still she flows on and I cannot help but look for a shark fin somewhere hidden in her curves, even now.

Ma has protected me and been my guru, and even spoilt me all my life, I have found her all through the world, on land and water. She has been there for me in the form of my mother, *dida*, sister, *bouthan*, friends, lovers, kind neighbours, and even rivers and idols. And she has been on two legs, on four, and even propelled forward by giant fins. And I have caught myself uttering her name in an early-morning somnolent state; singing in front of some isolated island while no one watched, and while calling out to Milli—both the old lady from bluff and the younging who now lives with Soosan in feline form.

I know I have wronged you, Ma, so many times, but you have always forgiven me, and as much as I have had to leave you time and time again—

Photograph (1) of white shark caught by Mary's father (date unknown)
Photograph (2) of white shark caught by Mary's father (date unknown)

as now, I have always known that I would find you again, be it in some unknown form and some unknown place. There may be no true path, there may be no grand adventure to have, if a perspective is mirrored through the grand creations of the cosmos, and the futility of the microscopic human existence. But my life—insignificant as it may be, the adventures I have had—infinitesimal as the may be, the knowledge I have created—trivial as it may be, none of this would have been possible without you. Your indulgence seeds my afflictions towards my curiosity, which then is given credence by your presence and absence.

A couple of days before leaving Bluff, Mary came to pay me a visit with an envelope in her hand. It was her father's photograph...and yes, it was also that of a great white shark. The first photograph was from 1955–56 on the Bluff wharf. Mary's father, Alfie Ryan, and uncle, Jimmy Roderique, had caught the white pointer around 'smoky' Stewart Island on the fishing boat *Eleanor*, but she had always thought it was a Grey Nurse shark. In the second photo, Jimmy was seen cleaning the shark, he was the person who had caught it. The shark was so big that it was towing the boat the other way and Uncle Jimmy was thinking of cutting the line. Mary tells me that they must have ultimately shot the shark and carried it home by tail, because she thinks sharks die when the water passes through their gills the other way.

Untitled image of a Bluff fisherman with a large white shark jaw in the early 1900s (Source: Bluff Museum)

The effort Soosan had put to convince the Bluff Museum curator had borne fruit too. There were a few photographs of Bluffies with white shark jaws, and one particular newspaper article which answered some important questions. The Bluff history group had also sent me the same report, be it in a different iteration. So, at some point, there were actually shark-catching contests in Bluff, and an alleged 18-feet white pointer was caught at the port in 1910 and was supposed to be the biggest shark ever caught in Foveaux (which is obviously an overestimation of the size as happens in most cases with white sharks). Mr Vic Metzger caught the shark, with a special hook and a leg of mutton near the wharf, and the old custom office, about where Mary's father's shark was displayed. The carcass was first

The same spot at the wharf where some 60 years ago the shark was hung, 2017

Bluff's 18ft shark

THE recent shark-catching contest at Bluff produced some big specimens but nothing comparable with this 18-footer which was caught at the port in 1910. Caught by Vic Metzger, who had an engineer's shop in Bluff, it is the biggest ever landed there. Mr Metzger made a special hook and using a leg of mutton as bait he set his line in a barrel just off the wharf where the old Customs offices are tidy today. After being caught the shark was put on display behind "Kelly" Metzger's shop before being taken to Invercargill by a travelling showman and exhibited at the Invercargill show.

Well known personalities in the picture are from left to right: Norman Anderson, Mrs G. Wheeler, O. F. "Kelly' Metzger, Sergeant Tonkinsor of the constabulary, a womai passenger off the Tarawera Charles Bradshaw senior (re puted oyster-eating champion Constable Beadle, another pa: senger off the Tarawera, (R. George, "Bonny" Parson: Rose Randall and Jim Young whose father was headmaste at the Bluff school.

Early Bluff Newspaper Issue 3
A paper of Bluff History Group
(undated article

RECORD SHARK CATCH

The big shark shown in the photograph was caught in January 1910 and was 18 feet long although its weight is unknown now. This is the biggest shark ever taken in Foveaux Strait and was caught by the late Mr Victor Metzger. The shark caused considerable excitement at the time and was exhibited in the yard at the back of Mr O F Metzger's shop in Gore Street. Mr Metzger sold out before going to the First World War and the shop was subsequently destroyed by fire. The wall of the Golden Age Hotel is seen on the left. The onlookers from left to right are: Norman Bradshaw, Mrs Whealler, Kelly Metzger, Sgt Tonkinson, Miss Annie Bradshaw, Charles Bradshaw, Const. Beadle, Mr Moffet, G R George secretary of the Harbour Board. Behind him is Bonny Parsons, next are Miss Randle (later Mrs Hankey) and Jim Young.
After the shark had been exhibited in Bluff it was taken, says Mr OF Metzger to the Invercargill show grounds where the travelling showman, Peter Farrell, put it on exhibition, charging one shilling admission fee "to defray expenses".
A bait to attract the attention of the crowd was a young man who held various parts of women's underwear – corsets, and so on – which were supposed to have been found inside the shark when caught.
Peter Farrell was known throughout New Zealand as a showman of the top rank. He was an uncle of Mr William Urwin and Miss Gladys Urwin of Bluff.

Left: Newspaper article about white shark caught in Bluff wharf, 1910 (Source: Bluff Museum)
Right: White shark caught in Bluff wharf, 1910 (source: Bluff history group, no date)

displayed in Bluff, then displayed in the Invercargill show ground with the admission fees of one shilling. Even here, the marketable commodity of 'the man-eater' was recognized, and women's underwear were displayed alongside, which were supposedly found inside the shark to intensify the idea of the beast preying on the beauty.

Furthermore, a blog of an archival researcher interested in white sharks in New Zealand (Lyes, 2013) revealed a similar incident from 1904. Lyes focused on a particular image of a white shark in Bluff (image no. 145) and brought forward a few newspaper reports of the time. *The Southland Times* reported on February 1904, that due to the warm weather at the time, a large shark was seen in Bluff wharf, and two other sharks were caught

White shark caught in Bluff, unknown source, date unknown

from the wharf. The same month, another large 'brute', 15 ft in length, that was prowling the wharf for some time, was caught. Lyes also pointed out to another report in the *Taranaki Daily News* which wrote of potentially the same incident. The report mentioned that it was usual to see sharks in Bluff around the summer, but the sharks caught in those few months were exceptionally bigger (it should be noted when the report says that there were sharks in Bluff during the season, maybe the distinction was not always made what species of sharks they were—they were just big sharks; so, there could have been more white sharks swimming there). Apparently, a large shark caused the fishermen quite a lot of anxiety, as it followed around the boats. And the fishermen even threw away food into the water in fear. Finally, it was caught as it was swimming near the wharf, and it was 12 ft and 2 in long, and 7 ft in girth.

On getting these resources, I wanted to tell Mary of my findings of white sharks in Bluff; so, I called her at her home. Her son picked up the phone, and upon hearing my name, he thought I was a telemarketer, and with some stern language, hung up the phone on me. Albeit, after knowing that it was me, he was deeply embarrassed, and we all had had a laugh. Mary revealed to me that the custom house mentioned in the photograph was the shed just next to where we anchor our boat every morning. She was shocked and said, 'But we always used to swim in there', and it just goes to show that the sharks were there, from a long time ago.

Do these bits of information prove that white sharks are always around Bluff, and cage diving by any chance cannot attract white sharks to its shores because they are already there? The answer must be—'No!' But what it indicates is that Bluff may have had a longer history of interaction, cohabitation, and conflict with white sharks, than even the community is aware of, and even if the sharks do come to Bluff now, it is not necessarily always related to cage diving. But the question it raises is—did they come till recent times? Do they come now? If not, why did they stop, considering there were more people and ships in the area before? Maybe, it was all the fishing boat and ships with whale and seal carcasses at the wharf—some indication of which can be found in Crawford's interview, where old fishermen talk of the whaling around Bluff. Truthfully though, none of the questions can be answered conclusively, but at least I could console myself that no matter how improbable it was encountering a white shark in Bluff, it was not impossible; so, my fear while cleaning the boat was justified, and at least, there were more of them there than in my bathtub, be it a 100 years ago.

Incidentally, in 1877, and other years, there were many recorded shark attacks on the swimmers at Ganga's ghats, like Babu Ghat in Kolkata. Although, in another expedition, I would go in search of them in the Sundarban delta, what I found are stories for another book. But this photograph I took of a 65-year-old lady, who was bitten in around 1984, should be enough to elaborate the seriousness of a shark bite on the human body in the Sundarbans. And much like in the case of Vaughn, more than the trauma of the bite, what was challenging was the struggle to make a living after it, particularly in the poverty ridden Sunderbans.

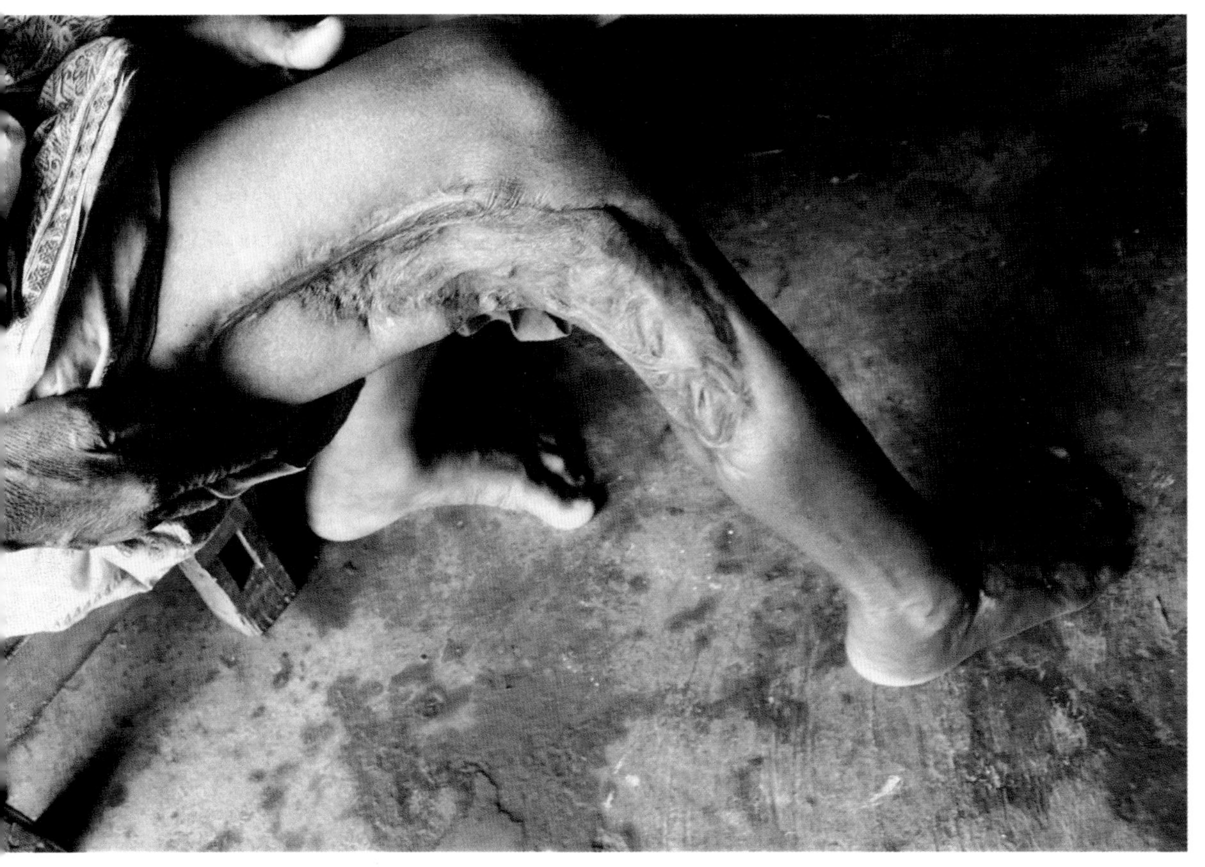

Kamot bite in the Sundarbans, 2019

So, then again, looking out at Ganga for my fin is not preposterous, late may I have been by a 100 years or so, and rain, fins, love, and fear connect my miles and fathomless temporality. It is 4 am, and the crows are calling now, I kick him alive again...

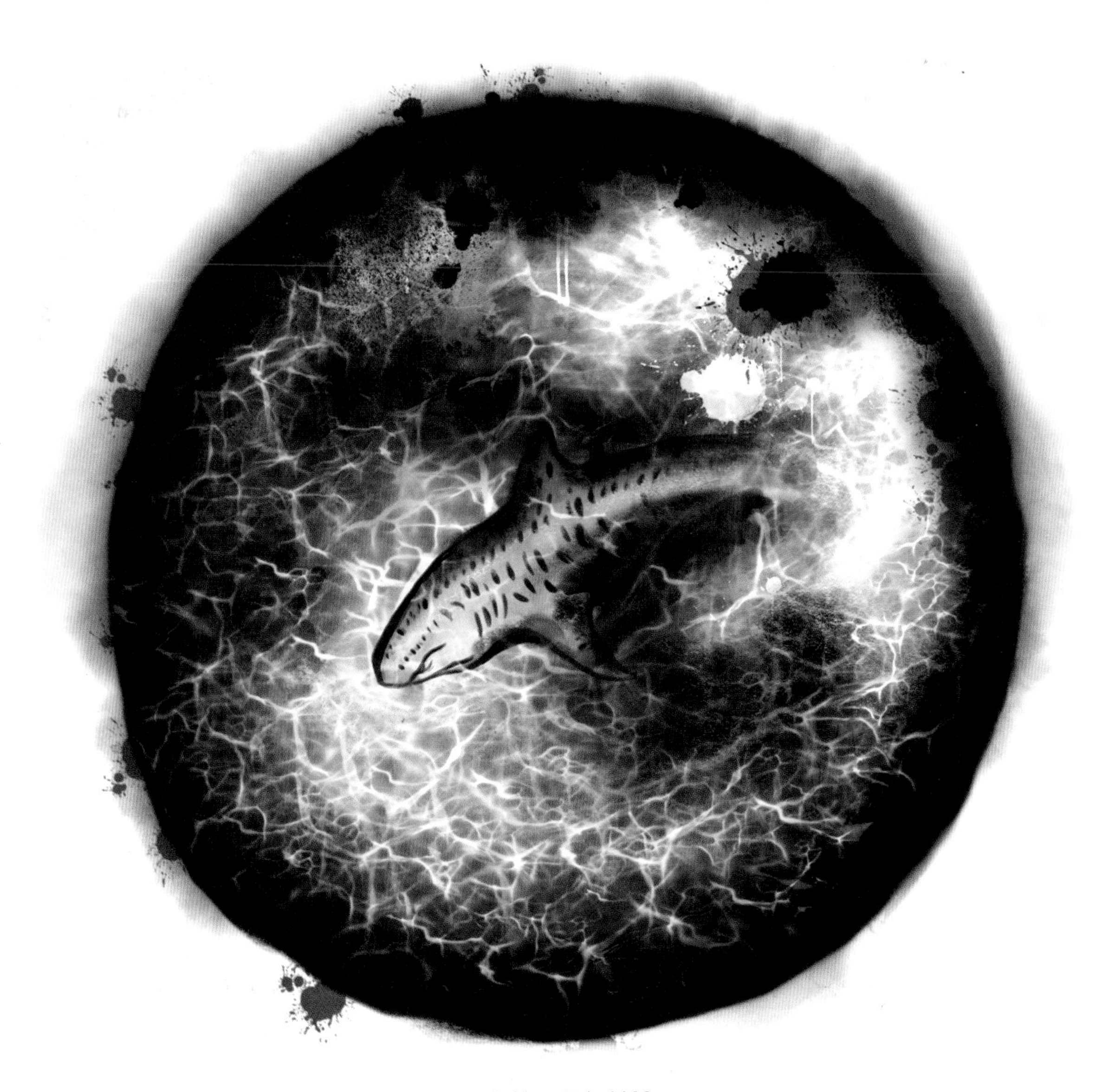

The thunder tiger, digital drawing, Raj Sekhar Aich, 2020

The Rhododendron Forest and the Deserted Meat Factory

I see myself riding into my courtyard—and here he comes, with the thumping of his beast. He is tired, his eyes are deep caves, hair unruly, and you are still standing here, and you have been with me all this time. But there are more journeys to be taken, and more walks to be had.

As I walk through the Glory Track, the heavy gust of the southern winds seeps through the old bush. The cold drizzle, sound of the crickets, the touch of the water on the leaves seem so surreal, yet it is something I had known before, seen before and felt before, maybe in a different continent. As anthropologists, driven to know the unknown, we go out to the field, the ocean, the bush, the forest, and the bustling city, to find ourselves. Maybe, in the similar vein, was my own journey to New Zealand, then to Bluff, and finally, to the sharks. I know, even if I had deep discussions and spent a lot of time with the humans, part of me was not being able to fully integrate with the Bluff community, and I could never be a Bluffie. I am not even sure if I integrated fully with the cage diving operators; if anything, I felt closer with the tourists who came to see the sharks, with all their dreams and hopes to come face to face with them. On the other hand, I felt I integrated with the soil, the trees, Milly and other dogs of the community; often stranger dogs will come up to me from some street corner; or porch, wagging their tails, as if I were their long-lost friend. I even felt a deep connection with the rats who shared my room and had their demise in a guillotine, and in some senses with the sharks. There is no information I need to gather today, there is no data that I need to collect, as I walk through this path, very possibly

Bluff woods, 2017

for the last time in my life. I can be here as long as I want, not anymore, not any less. Today for the first time, here in this unknown country, and as the unknown person who I am now, I feel unafraid. Right now, I don't have the fear of my freezer running out, having high cortisol, hunger, or the fear of not being able to finish my research; I feel sheltered among the blossoms of red rhododendrons.

This was the final time I went for a walk in the Glory Track, but I still had not completed the full Foveaux walkway around the island, somewhere in my heart I was still afraid to do it from the experience of the last time, and unwilling because Soosan could not do it with me, as she was still recovering in the hospital. A few weeks before I left, my friend and brother, Jon Jeet, a visual artist from Christchurch came down to Invercargill to visit his old stomping grounds. Jon took me out to the city for a beer and we headed up to the hustle and bustle of the night that was Invercargill, in Jon's car with one door that did not lock. We were looking for some laughter and some cheap whisky maybe, maybe even some clichéd trouble. After a few drinks in a few holes here and there, we walked up to Jon's car and he said, 'Come on brother, let me take you home.' On the way, I explained to Jon how I got scared on the path to the Bluff walk. He said, 'Lets walk together and see how we feel, ha?' We reached the Stirling Point and started walking, feet by feet into the inky darkness. We walked quietly, both drowned in our own thoughts, until we came up and sat beside a placard which sang odes for the southern right whales, whales which brought the Europeans to the southern shores, whales who had thereafter been pushed to a threatened status. Over the last decades, their numbers have been increasing in the wild, but they never came back to the shores of Bluff. We sat down and Jon asked, 'You afraid, brother?' I said, 'Not as much.'

I don't know what happened that night; maybe it was his fatherly protectiveness, maybe I felt I was walking with someone who was of the land, or perhaps for his Indian and Māori lineage, he was a living connection between my two realities. The fact remained that I had to complete the Bluff walk myself before I left, to give me the scope of the full land and an opportunity to see the almost mythical Ocean Beach freezing works on the other side of the hill. As usual, I was prepared for everything, had all my equipment charged the night before, including my cameras and GPS, because I wanted some important shots for my documentary. In the morning, I put on my all-weather suits and started walking early; I am an extremely slow walker and with the dislocated hip injury acting up again for the last few weeks, I knew I needed to give myself all the time I could.

As I started off, there was no one there, except a couple of ducks sitting by the road and enjoying the rising sun. It was a 15-minute walk to Stirling Point, and then three hours or so around the entire hill, coming to the Ocean Beach Road. Initially, for the most part, the walk was peaceful—by the hill, under the bush, with birds singing and a light breeze on the leeward side of the hill, but after the Lookout Point, you turn to the windward side. As I walked on, I really appreciated the name: 'Those sheltering hills' (Bremer,

1986); yet again, I suppose, the stronger the storms, the stronger our shelter feels. I was batted by howling winds; I knew a heavy southerly was coming, and I could see the dark clouds gathered in the distance approaching.

Thankfully, I was prepared; when the storm and heavy rain hit, it was hard walking in a straight line, but I still kept going; funnily enough, that was the most practical thing to do at the point. There were no trees on that side, and long sloping hills covered with tall yellow grass, flask, and sharp rocks. The familiar sound of birds was absent on this side, just the stormy winds lashing on the waves below the jagged cliffs. On my headphone, an old song recorded by my father came up. I felt his voice guiding me through this barren wind-slapped shores. No one knew where I was and I doubt anyone would be coming this way for the next two days because of the storm; so, I better take every step carefully across the slippery ledges, I thought. I kept jumping from rock to rock, and across the grassland, till I reached the old abandoned freezing works. Locked, and no one allowed in, I could see how all the meat and blood would have flown into the water from the factory, and I could imagine the gumboots stomping on the land, and all the little wings and big fins feasting there in its heyday. The freezing works not only brought humans to the shores of Bluff, but also the ships and sharks, but now it lay in its own frozen state. The ships are resting in the cemetery, the people have left, but I wondered if the sharks still came to the shores. I could not help but ask myself, have there never been white sharks among them in the recent times?

I would stumble upon this excerpt later in Crawford's manuscript, from Allen Anderson's recollection of some fishing expeditions in 1982...

> ANDERSON: Oh, Ocean Beach - that's Bluff. So, the freezing works discharged offal straight into the ocean.

The silent and locked Ocean Beach freezing works, 2017

CRAWFORD: This is a freezing works for lamb?
ANDERSON: Yeah, sheep and that. And it discharged like whole stomachs and that, because you could see these stomachs like with the bladders and that, just bobbing in the water. And the water was red. And every time we used to go past there,

> we used to get up the front like that and scan for them, look for them. You know look for the White Pointers, because the skipper and everyone said there's always White Pointers there, and keep a keen eye and that. And I don't remember how many sightings we had, maybe two or three. But I definitely saw sharks there you know? But big sharks. Seemed like White Pointers. Just on the surface, just cruising around. So that was obviously a real hot-spot for them [sic]. (Allan Anderson, Crawford, 2017)

As I came out of the other end of the track, I realized that the entrance of the Ocean Beach freezing works was actually at the beginning of the road, by my house, and hence the name. I did not need to go all across the island to see it. As I realized this, the two ducks who were still sitting beside the road, started laughing uncontrollably and flew away.

I hit on the rusted tap with one of my iron weights, to salvage some water. After a few hits of rancid brown water seeping out like pus, I wait a few moments, and out comes clean water. I put my face under it...the warm water drizzles over my face, my eyes, my hair, and leaned over shoulders. I am still thirsty, there is a smudge of Horlicks left in my flask, I lick it off, as much as possible. The bricks and the rusted old paint are humming with a touch of indigo sunlight and I walk up to the terrace to welcome him, naked and dripping.

Memories Flicker

Memories flicker like the light of the decaying bulb, images of the bareness in front merge with memories of light and colour; vermilion, azul, and saffron drip over monochrome frames. The Durga pujas, the parties, the kirtan sabha, and night-long debates, the gathering of my friends, and more. Before that, my brother's, my father's, my *dadu*'s. Before that, a *hogla pata* bush with foxes, snakes, and spirits—both the unfathomable and the drinkable kinds, and supposedly just a hundred years earlier, Royal Bengal tigers prowled these bushes. Perhaps, in some other dimension somewhere, the party is still going on.

It is 7 pm, the headlights light up the black and orange gates, it opens, and the cars drive into the driveway caressing the checkered cement. The light splashes on the mezzanine fish pond, where two large Rohu fish swim and a blonde cat leans over looking at them in desperate anticipation. It is a two-storied house and the courtyard is filled with plants and terracotta. If you look up, one can see the reflection of blue lights from the window above. A loud big man in shorts and a T-shirt covered in paint comes down and, with his deep voice, starts laughing, and invite them up. There is a new guest tonight, she is wearing pink salwar kameez, and has a grey streak in her hair, and long long...red heels.

As they walk up, the stairs are lit with a red hanging light and in the mezzanine, there is a light table with a full-size fibreglass Batman bust painted in gold. From here you can smell sandalwood incense and hear whispers of Offenbach 'Barcarolle'. They walk up, and stand in the doorway;

Memories of the ocean, 2017

to their left is the bathroom with the white ceramic bath, and mirrors all around, reflecting the blue light from the ceiling. To the right, a dark corridor with strange memorabilia like a crossbow, a Roman legion, a helmet, and an antique gramophone.

A step further into the left now is a kitchen, seems like a typhoon had blown through, with flour, tubes of acrylic paint, bottles of cheap wine, and a recipe for fried chocolate, all lie around like a forbidden orgy. In the front is the library, dark and scattered with books, and dumbbells. But on the right is where the party is going on now. Intermittent lights light the room, the hardwood central wall has the bar inside the cupboard, filled with a crystal decanter and glasses of many sizes, gathered together from various broken

sets. There is a deep violet mini grand piano and above it is a katana displayed in a glass shelf.

The ceiling has one giant fan; it moves ever so gently, so as to not commove the flames of the small kerosene lanterns spread across the room. The floor is of white marble and the edges of the walls are filled with large stones gathered from riverbeds. The wall to the left and back is painted black, filled with swords, knives, and a large camel whip. The reflection of the whip is seen in the huge mirror multiplying the flickers of the lanterns, and in the centre of it, an etching of a giant shark jaw. Besides the mirror, photographs of Rabindranath Tagore, Homer Simpson, and Calvin and Hobbes; below, a lightbox on which a framed photograph of Bruce Springsteen and a half-drunk glass of whisky sits—all of them seemingly share the drink. The wall to the right of the room has a large window looking into the corridor, and the rest is painted white with a large photograph of this man scuba diving, and above it, an underwater titanium knife.

There are red and black cushions laid by the wall, where everyone sits and laugh, the room and the house seems to hold more people than it can sustain. There are painters, lawyers, police officers, dancers and actors, a few scientists, and a few who blissfully do nothing. Some of them talk about the new film they are working on, some about global politics, while some of them about nice asses. The house seems to be like a permissive guardian, where we can do whatever we want, and be assured that she is looking after us.

She walks into the dark corridor and approaches the room at the end, which seems like his bedroom. A soft yellow night lamp creates blasphemous plum red shadows amid the ruffled-up bed sheet. There is another giant mirror on the wall by the bed, in which she finds her reflection drowning amid oil painted waves. The rest of the wall is filled up with photographs of him, one where he is rock climbing, other where he is paragliding, and one with a few tigers.

He comes and stands next to her. She says, 'You seem like a very self-obsessed man.' He smiles, 'No, I like it, you are unapologetic about it.' 'So, are you a painter? A scientist or a singer? You are complicated!' He smirks and fumbles, 'I am just confused, and because I hide it well, people think I am complicated.' We are all but ships lost in the ocean, looking for a light house. For those who have God, or whatever they choose to call it, they can never reach this light house, because God is always one step away, but this light house, though never reachable, is always a marker for direction. For obsessive people like myself, and I am guessing you too, our obsessions are only our light house, and the shores are reachable. But I doubt, how many of us actually jump down on the cool sands once we reach, rather, make wave towards the next obsession, and the next journey. Ultimately, it is all a practice in the innate wish to keep silent.'

Do you at least enjoy the process, the journey?' 'Naa... Not most of the time!' 'Then why do you do it?' He remains silent... There is only one painting in the room, that of a golden shark. She stands there looking at it, the waves stand still in her glass. She asks... 'There is so much terrorism, rape and violence around us, then why do you paint this golden shark? Where is the memory of our time in your work?' He remains silent, she looks at him, hesitantly he replies, 'As today's time once shall not be, withered out of existence just like this leviathan that shelters us tonight, my love, shall remain only the romance of this all-encompassing loss. Pain has an invariable path of osmosis, don't you think? Transforming from the pain of every day losses to the impotency of not being able to be one with the waves flowing in the universe. All this pain we see and feel is only taking a sip of that ever flowing melancholy stream. No matter how much blood shed, her golden water shall never turn red, that is where my golden shark swims, and shall for ever'.

They walk back to the living room, and for hours, they dance, drink, and he sits by the piano and sings, and someone brought a cake which, believe it or not, they cut with the sword. But then things quiet down, everyone sits

The house, Raj Kamal Aich, 2019

in a circle, with the flicker of the lantern, the sound of the ceiling fan, and the rustling of the leaves of the mango tree next to the window, and talks about their dreams and loves lost and distances travelled.

Kuntal says, 'The house itself is a creative entity, it is a microcosm among the hard glass and cement of Kolkata, you both complete each other. There is no noise here, there is music, whatever does not touch your soul is only noise. What I also remember is the smell of incense, when I went to Ireland for my studies, I used to work all day, and study all night, I brought along some incense with me, and I used to have them lit as I studied. Now that I think about it, that smell probably gave me some sense of security and comfort, maybe subconsciously it reminded me of this house. Sumorto said, 'It is this *alo-andhari* (a mix of light and dark) and this house has many images, I think different for different people, I think that is the point, here we could have been whoever we truly were, so how we saw it, depended on who we were. Like I always admired your martial arts practice, I think it gives each person a different perspective. Like the warrior monk. Although I don't know if I can really call you a "monk". But both you and the house gave a sense of safety.' Avik says with his loud laugh, 'It is the blue light, God I know you always wanted to make a blue film.' And Sourab da said, 'It is forgiving. No matter how I felt all through the day, when your brother and me were in the school, we came here and felt that it was all forgiven. But truly saying, I am not sure your brother forgave this house. Your mother was thrown out of here, so when he got a job after school and university, he wanted nothing to do with this house. You were too young to see the atrocities this house held...he was not.'

At the end of the night, he stood there—alone—a little bit tipsy, scruffy-haired, with the last ember of the smoke lighting his lips, watery eyes, silent. Their car moved out of the driveway. She looked back again, and he had vanished in the darkness of the house; all that was left was her smell, like up in the mountain, a *dabanol* (forest fire) tonight, in the sandalwood forest.

Making Sense of it All

The *ghughu* (spotted dove) cries sitting on the top of the betel nut tree. The tree sways from side to side with the shy touch of the sun. The 'khash khash' sound of the dry edges of the leaves and the smells of unripe mangoes and neem leaves blow with the breeze. The first rickshaw of the dawn: a thin man, with pit-stained blue-and-white striped shirt and drowsy red eyes pulls a voluptuous middle-aged woman, with the rusted chain just about holding on, and the sound of thick hips clad in a yellow cotton saree rubbing against the nylon purple flowery seat cover.

My denim shirt is crumpled from my Ma holding on and not wanting to let go. As uncouth as usual, with the top three buttons undone, a crocodile tooth pendant on striated pectoral muscles covered by grey and black chest hair and a tattoo of an orca traversing its ridges and that of the shoulders peep through. These blue shoulders are flanked by a head of long grey and black hair with a few emergent *jatas* (dreadlocks) and amidst this all sits a little Bengali boy's clean face with heavy eye brows and, from what I have been told, rose bud lips. (A clean shaven brown face is the way to go while crossing international borders apparently.) If you travel down from the shoulder, you reach the broad wrist wrapped with a copper wristlet and a GPS watch, underneath which a faded shark tattoo decorates a veiny hand filled with cuts and bruises, and calloused stumpy knuckles, and because of these cuts and bruises and the broken bones, this body of mine bares, my friends called me an abstract painting. Beyond that lies a little bit of a tummy like that on an offseason boxer, a result of gorging on Bengali food

The early morning song over the Bluff hill, Soosan Lucas, 2016

for the last few months. My jeans have some remnants of yellow and red oil paint, reflecting the morning sky, where the clouds ripple like waves in a far-off coast, as their embrace of the white sands is emulated by the rustle of the neem leaves.

The mournful kirtan of the night has stopped, and on someone's radio, Amar Pal sings, '*Rai jago go, Jago shamer mon mohini binodini rai*'—Raga Bhairavi to awaken the neighbourhood (classical songs that were sung in olden days to awaken the community). The husky voice radiates through the wet streets, from house to house; here comes the *manjira* beating with my heart beat; now comes the flute tune rising and drowning with my breath; and finally beats of the *khartal* (Indian cymbals) at rhythm with minute vibrations in my hands that we call the warriors' shakes.

I laid down all the photographs, and printed sheets of my findings on my Japanese bed, a picture started to evolve. Cage diving with white sharks in New Zealand is a multidimensional, multi-agent practice, which is generally beneficial to the lives of the sharks; the humans choosing to engage with them (including the operators, the tourist, and the researcher); global shark knowledge; and the community in which it is based on—if proper strategies are initiated and maintained. It is a constructed practice, which has the potential of demystifying the hyperreal monstrous image of the species, a species which has a long lineage, and is much smarter and possibly has social traits (even if it is not the same as understood from a human perspective) than we give them credit for. Finally, it is a very interesting example how a 'human' practice is not just the product of human actions, but is often the coming together of multiple species, material, and technologies.

Bluff is a unique place; historically, she had a geographical isolation from the rest of the country, which might have extenuated the sense of psychological isolation that is palpable in the region. In its heydays, she was a 'big little town', bustling with life and amenities, especially because of the Ocean Beach freezing works. The freezing works not only brought humans to the town, but also sharks to her shores, and it seems Bluff has had a closer and longer interaction with white pointers, more than what the locals recognize. However, after closing of the freezing works, the town lost a lot of its people, and eventually, facilities. Some people love it the way it is, with its isolation and peace, while some others believe that there is need for other industries to be developed in Bluff, because she has all her eggs in one basket (fishing). One such emergent industry had been white shark cage diving off the coast of Bluff, in Foveaux Strait.

The white shark cage diving in New Zealand is a multi-agent practice, which is only successful when all the sociopolitical actors like the varied bureaucratic groups and conservation groups; the operators and the divers; the environmental conditions, and the sharks themselves interact together in a dynamic harmonious relationship. For the operators, the daily objective of the practice is that the tourist can get in the cage and see the sharks at least once, and they have their own norms and methods to make this a reality. However, even if the humans want to go in the water to see the sharks, the sharks and the environmental conditions decide the outcome of the day. Consequently, it is best understood as a multispecies activity, which depends on the active agency of both the species and is affected by and affects both the species at the same time.

However, this practice, and the human–shark interaction facilitated through it, has been surrounded by controversy in the region, the Bluffies are divided in their opinion about cage diving and the benefits it brings for the local economy. On one hand, some of the original residents like Bluff's seclusion, don't want much tourism in the region, and feel that cage diving is only asking for trouble, with the increased potential of a shark attack on the residents. Furthermore, the hoteliers who owned the more expensive accommodations observed that most of the people who come to their establishments are not interested in cage diving, and cage divers often cancel the booking at the last moment because of trip cancellation. So, cage diving creates more problems than financial benefits for business in the town.

Alternatively, there are others who believe that cage diving is bringing social and economic exposure to the society. Particularly, the younger residents feel that the cage diving practice brings in people from other countries and cultures which is great social exposure for them, an emotion also shared with other long-term residents of the town too. Shop owners and the hoteliers with cheaper accommodations revealed that indeed there is avenue of cage diving bringing in revenue to the region. If not a lot, it seems a steady stream of people come to the region because of cage diving.

In the Bluff Lodge, which is the backpackers' den in Bluff, the caretaker, 'K', took me around and gave me a detailed look at the facilities, priced from $25–$60. A lot of people who come to do shark diving are young travellers, and don't want to spend $150, in the more expensive places. It was also a delight to see how happy she (K) feels when she hears the happiness in tourist stories after seeing the sharks, or even people who have come multiple times to see the sharks, and finally were able to.

However, one major problem was due to the competition between the two operators, for the boats used to leave Bluff very early in the morning, and people could not avail the services of the town. Cage diving can indeed be beneficial for the community (if at all, the operations continue), particularly, if the current operator and the other business owners work together in consideration of time and prices, and create a battery of services for the tourists who come to see the sharks, so they spend more time and money in the town. This is an opportunity for the community to build a new industry, an industry which not only brings people from all over the world to see the sharks, but with proper promotion can bring a lot more focus to the entire Southland region of New Zealand, from global scientists to tourists alike. Furthermore, it may even generate more work and social opportunities for the younger residents of Bluff who crave it.

Mike charged $500—$600 for one trip per passenger and, considering he could take about 1,500 tourists out in a year, it is a lot of revenue that he earned from the operation—particularly now, since Peter has decided to stop his operation, if Mike can keep on operating.

Then there is the matter of tourists availing other amenities in the town, including the few restaurants and couple of pubs and grocery outlets. One problem with this is either the restaurants are too cheap, or too expensive, and there are no places with decent ambiance, serving food at a reasonable price, which the tourists often expressed to me, and I agree with personally.

Cage diving has emotional and social benefits to the tourists who come from all over the world, and drive down to Bluff and brave Foveaux for one chance to see the sharks. Though for some tourists it is an incidental experience, for most of them it is a long-term and planned endeavour. Instead of just individuals willing to take part in a random adrenaline-filled experience, it was often a unique opportunity for people from all around the planet to have one encounter with the creature of their dreams and nightmares. It was the coming together of humans for the love of the animal, the environment, and each other, and a pilgrimage to fulfil a lifelong dream or witness the dreams of a loved one coming true. And when the encounter actually happened, it seemed to have a profound effect on the humans: highly emotional, self-reflective, and even to some, spiritual. Hence, white shark cage diving in New Zealand not only created opportunity for the social benefit of doing something 'extreme' and different for the tourists, but indeed emotional benefit by achieving an aspiration that many of them have had for a lifetime.

Besides this potential benefit of altering the negative attitude towards the sharks, and in the process conservation efforts of the species, as mentioned earlier, when my fieldwork started, during the first two months, there were no sharks. The operators felt that this was because some groups who were opposing cage diving had killed a shark and dumped the carcass in the water where they were operating, and there is a common belief that secretions from a white shark carcass drive other white sharks away from the region. So, the first day when my fieldwork started, Maca, the deck hand was sent in to dive in the water to check for the carcass. Thus, it is evident that the white shark cage diving in New Zealand creates an immediate awareness of any potential illegal killing of white sharks in the region. The operators have more interest in conserving these sharks, because a live shark had more value than a dead one, which is often true in other wild life tourism enterprises, and certainly seem to hold true for shark tourism (Cisneros-Montemayor, Barnes-Mauthe, Al-Abdulrazzak, Navarro-Holm, and Sumaila, 2013). As a matter of fact, the cage operators had requested the DOC to put cameras on the island, but their request was not responded to. There had been a lot of information that had come to me about the potential killing of sharks in the region, but none of it was proven either way, so I will not dilute the matter by providing anecdotes and hearsays, but it is clear that there is certainly more awareness and concern about the matter, because the operators' business depend on them, and they are the Kaitiaki or protector of the sharks.

Cage diving is primarily the method in which photographic and videographic images of the sharks are taken, edited, and distributed through certain orchestrated productions—focusing on the sharks lunging at the bait or biting the cage. In the same manner, more and more images of sharks can be taken, edited, and shared in the public domain, which does not focus on human–shark interaction, rather on the calmness of the interaction, which is the case most of the time. These images may be the antithesis of the ferocious man-eater image produced by the media. They may focus on their aesthetic beauty and inquisitiveness about the humans, rather than always on the open jaw. As long as there are organizations and individuals who intentionally produce and distribute images of the sharks focused on their aggression, there is also the need of images that focus on the alternative perspective, to be undertaken in an effort of demystifying the sharks, and cage diving seems like the safest and most viable avenue for doing this. However, it is about time to discuss governmental and private intervention with some form of censorship or at least strong disclaimers when videos, images, and stories of shark filled with misinformation is shared in mass media, if countries are serious about natural conservation. Much like some social media groups putting disclaimers of 'potentially misleading' information, in case of 'alternate facts' in America. Much like there is care about being sensitive while creating and disseminating content about certain religious groups, and disclaimers on use of cigarette on Indian TV. Why should there not be disclaimers about misinformation that is spread through this hyperreal image of sharks in media, which is potentially related to the devastating effect on an apex species, which in turn has effect on the global marine ecosystem. There is a need for immediate accountability!

Let me be clear, sharks did not need humans to survive, till we started to destroy them and their habitat; they were doing fine for more than a 100 million years. It is unlikely that practice of cage diving was initiated specifically thinking about benefits for the sharks. Even though white sharks it seems might have been considered a menace for a long time, the hyperreal image was constructed and promoted by recent media, which was only possible by the use of cage diving in the first place. If a conflict was not created from the practice of cage diving in Foveaux Strait, people would have not killed sharks (if at all they did), as a reaction to the practice. Furthermore, if not regulated and executed properly, it can lead to immediate physical danger for the sharks too. In early December 2019, a shark got stuck in the cage window in Guadalupe, allegedly, the shark died after 25 minutes of the incident. It seems that the operators had been warned beforehand about the design of the cage, but the warnings were not paid attention to. But since the damage is done, maybe now it can play an important part in some efforts of undoing it.

In the current state as it stands, white shark cage diving in New Zealand can demystify the sharks in public perception through direct encounters. Survey data in conjunction with interviews indicates that indeed cage diving had a positive effect on the human attitude towards white sharks. Even the people who have been obsessed with them their entire lives, or are trained divers, often held a mystified and negative image of sharks in their mind, which cage diving certainly does seem to challenge—an important attribute to support their conservation efforts, and beneficial to white sharks as a species. On the other hand, a majority of the people who come to dive, do have somewhat of a positive attitude towards them anyway, more than the general population. But there are friends and family members who are coming with the divers who would not have otherwise come, and this experience not only helps create a stronger bond among the tourists themselves, but helps in creating appreciation in them of this animal, by providing an opportunity to come face to face with them. This is yet again significant to shape public opinion effective in creation and maintenance of legislation promoting white shark conservation efforts. Furthermore, three different factors may impact the effect of the experience on tourist attitude towards the sharks, shark wrangling, the personality of an individual shark, and personality of the individual human.

White shark cage diving in New Zealand is potentially beneficial to the global shark knowledge database. Considering the operators are with the sharks for more than 100 days in a year, they themselves are a pool of knowledge about shark behaviour and their lives in the region, particularly concerning shark procreation in Foveaux Strait and white shark sociability. Basing on the observations of the operators and my own, I proposed first, that the yearly white shark congregation in Foveaux Strait can be explained through white shark copulation rituals. Also, the acceptability of decreasing personal space among white sharks in Foveaux can be explained through increasing social acceptance among individual sharks.

I also made some comments on white sharks getting trained by cage diving and any possible increased risk for the humans sharing the same water space, considering the significance of this question in the national dialogue about cage diving. Much like Laur Kiik admits the problems of his own lack of training in natural science for his ethnographic investigations (Kiik, 2018), I was not intellectually and logistically equipped to investigate the topic from an ecologist's perspective. What I was more interested in was to explore how such a topic may be addressed methodologically by a social scientist working particularly in marine environment. This was too important a question and too good an opportunity not to engage with—if nothing else—at least to provide scope for future marine anthropologists to learn from my mistakes.

First, the limited amount of literature on cage diving with white sharks, particularly in the region, does not predict any increased danger to the humans sharing the same water space. Furthermore, white sharks have been historically present in the region. As far as training white sharks to come to the boat, there seems to be no evidence to support that cage diving can alter the 'large scale' movement of the white sharks; rather, the practice itself revolves around the life cycle of the sharks in the region and their large scale annual movement, potentially for mating and feeding. However, it is true that it can certainly alter their small scale movement—supported by the mere fact that the human–shark encounter facilitated by cage diving is so unpredictable, and that in 2016, for two months, there were no sharks, and again in 2017 for some time. These may be considered as observations which indicate otherwise.

Furthermore, there is the estimate that there has been increase of pinnipeds in Foveaux in the last few decades (Bradshaw, Lalas, and Thompson, 2000), which may have increased the number of sharks too. Most importantly, it seems counterintuitive that the fishing boats that have always caught and cleaned fish in Foveaux and around Stewart Island, and it did not increase the number of sharks in the region, nor numbers of sharks approaching boats, but an operation which is not even allowed to 'feed them' has increased it.

As mentioned, being primarily interested in the methodological questions, I arranged a few studies to observe the effect of the practice on short scale movement of the sharks, and effects on their behaviour. I found no specific recognizable pattern in the time of the first contacts with sharks. Furthermore, from photo identification data, I found that there seemed to be only one shark who came to the boat in a repeated manner over the days the data were collected. Finally, I assessed if there was any increase in aggression around the cage diving operation. My data indicated that the behaviour and movement of the sharks seemed to be more related to their individuality, rather than being directly related to the cage diving operations and the bait provided, and even if the enquiry approaches towards whether the bait increased in the frequency of appearance of a particular shark, the attacks did not. But the dynamics of the shark wrangling and the relationship created between the shark wrangler and the shark also had effect on shark behaviour. All these findings are just indicative and have to be tested in controlled experiments. But at least, according to the literature on the topic, observations of the cage divers, fishermen, and myself, there did not seem to be much evidence indicating that white shark cage diving is training the sharks and creating more danger for humans using the waters.

On the other hand, the cage diving boat engages in strategies particularly to attract white sharks and keep them near the boat. It is important to understand white sharks are extremely smart predators and should be treated with utmost care to minimize any significant positive correlation with humans and food. Indeed, if we contemplate on it from a human visual perspective, the scene of the cage diving is constructed like those in classical hunting stories like that of Jim Corbett (Corbett, 1944, 1948), where cattle is left in a cage to attract tigers or leopards. Similarly, the human in the cage might look like a bait to attract the sharks, which reinforces to the observer subconsciously that we are their food and they want to eat us. Then there is the practice of wrangling which might actually alter shark behaviour at least in short term, and more comprehensive investigation of shark wrangling techniques employed by cage divers and their effect on shark behaviour is hence required.

Furthermore, if my child is in the water and I believe there is a practice few kilometres away which might be bringing 'big, biting fish' to my shores, when I believe they were not there before, I too would not want the operations to go on. And realistically, precisely because there is no ecological or behavioural investigation focused on the effect of cage diving on the sharks in Foveaux, there is certainly need for more social and ecological research on the topic to create a clearer picture of the side of the Stewart Islanders. However, the media presentation of the shark–human conflict, and promotion of the hyperreal image of the monster-shark is certainly adding fuel to the fire. The innate desire of the residents of the region to hold on to the original 'paradise' and keep the 'others' (the tourists and the sharks) away, may also be a consideration for resistance towards this operation.

There are some points I have to note from a marine anthropologist's perspective. Any future researcher attempting fieldwork on marine anthropology ideally should train themselves in the basics of ecological study, and perhaps even take courses on them before commencement of the fieldwork. They should make sure that they provide themselves with specific research questions to answer from the ecological dimension as much as social scientific dimension, and have all the tools and other logistical supports ready. Researchers should keep in mind if they are participant observers, they would have to set aside specific times in the exercise, where they can focus on the finer details of an experiment or study. However, that may be the de facto situation while participant observation; because the researcher is participating in the work at hand, it may be unprofessional or even irresponsible to take focus out for specific observation, especially at the heat of any particular activity where her/his work is significant for it to function properly. Finally, the tools of such studies should be properly studied beforehand, and the researcher should be proficient in tabulating and recording the findings.

There are a few questions that I have raised in this book that opens room for future research. First, there is scope of future detailed anthropological research focusing on the conflict raised in Stewart Island about cage diving. If my research was focused on exploring the cage diving conflict in more detail, I would have had to spend as much time with the Stewart Islanders as I did with the cage operators and Bluff, and it would have been a different research all together. But my research was not focused on this. Furthermore, because I was working with the cage divers, probably I would not have been welcomed in Stewart Island. I observed that level of excitement of white sharks around a bait in cage diving practice was directly related to the suddenness of the action of bait retrieval and the proximity of bait retrieval from the body of the shark. This can be explored in future ecological, ethological exploration through controlled experiments.

In discussions with two–three tourists who came to the boat as observers from the surface, I found the encounter did not have any effect on their attitude towards the sharks. Which raises the question that is it imperative to get in the water with the sharks to appreciate them? In future anthropological or tourism research, it will be interesting to understand that not only the encounters themselves, but what precise facets of human and non-human animal encounters are more significant to have effect on tourist attitude towards them. I proposed three factors (and the combination of these) that may have significant impact on the tourist experience during cage diving, and consequently, their attitude towards white sharks:

1) Shark wrangling (the practice of trying to control shark behaviour with bait).
2) The personality of an individual shark.
3) The personality of the individual divers.

This provides scope of future research for more detailed ecological investigation. I documented the dreams of the cage divers about white sharks. In future psychological research, it might be interesting to know what kind of dreams people who are petrified of sharks and would never consider diving with them have, if any, as opposed to the ones who choose to interact with them in activities like cage diving, and situate the discussion in psychoanalytic paradigms of phobia, fear, and subconsciousness

The Shark experience team, 2017

research. I proposed that the white sharks may be able to decide how they negotiate with cage diving operations in one part of the world, based on experience with other places, this opens up scope for future ecological research with the help of satellite tagging and observations from cage diving boats. At least in four of my interviews, I was told by residents of Bluff that they got so affected by *Jaws* that some of them stopped swimming from the wharf, and one long time diver even stopped diving in Foveaux. I would imagine a similar effect of the film on other ocean-going communities. However, I have not come across any academic publications elaborating on it; perhaps, it is an avenue for more detailed anthropological or media studies research on the topic.

The colour, mixed media painting, Raj Sekhar Aich, 2005

The Promise of Springsteen

On the day I was about to go back from the field, I got up at 5:15 am, squeezed in all my stuff in my two and a half bags, had my coffee, and was ready for the road. Pete and Caroline had graciously offered to give me a ride to the town, from where I was supposed to take the morning bus back to Timaru. As I was stepping out of the house, Milly saw me, but when I tried to pat her, she ran away. Caroline told me, Milly saw my bag and was upset because I was leaving. I don't know if that was truly the case or if it was my smelly travelling gloves, but I had to leave without saying goodbye. We passed the cottage Pete and Caroline were renovating for tourists. Caroline asked, 'What do you think we should name it?' I replied, 'Sun Rise.'

Caroline had been in Bluff for the last 24 years, but she knows even then, she is not a Bluffie; to be a Bluffie you have to be born in Bluff. 'Bluff is a strange place; if it does not like you, it spits you out. Thankfully it did not spit me out over the years. I came with my ex-husband from Auckland because he was interested in deer hunting, which they did from on top of helicopters. For $30,000, we bought 60 acres of land. When I came to Bluff, I fell in love with it—it was so peaceful and beautiful, and those sheltering hills were kind to us. It has been a great place to bring up kids; my daughter, from a very young age, knew how to work the land and had her own horse, which she used to ride around. My son was always aloof; as a kid, he used to sit on the hay stack, lost in his own dreams. But I did not have to worry about them even if they were out till tea time.'

When I reached the Invercargill bus stop, with my two and a half bags, and one more filled with cameras, I was hungry for familiar faces, Springsteen, and yes, food. Soosan and I were supposed to go on a pilgrimage to see the Boss (Christchurch Bruce Springsteen, 2017), but the doctors were not sure if Soosan would be able to. I had $5.32 in my bank account, so breakfast was out of the question, but a steamy white chocolate latte was a fair game. I just remembered that Pete had handed me a small package that had come to me the night before, and yet again, it was from Soosan. I opened it and it was a nautical key ring that Soosan had sent for Mike and a pack of spicy nuts; I confiscated the key ring for my own pocket, and disembowelled the nuts. As I was engrossed in my nuts, a young man at the bus stop enquired for a lighter, and was delighted when I said, 'keep it', and was even more delighted when I offered him some nuts from my limited ration. Scot was a Bluffy in his 20s, moving to Nelson for a job he just got in a factory; he said it was good for him to go, as it kept him out of trouble, and also, it would be good to get some money in hand. He asked what I was doing in Bluff. I said I was researching the white pointers. I asked, 'Have you ever seen one?' Scot replied, 'Ya, I used to work on fishing boats, and one time, me and my dad caught a 14-feet one in the net, about four to five years ago, somewhere around Stewart Island. We cut the jaw and the teeth; then it was sent to some research or something. We put the jaw in the ground, and let nature do its thing and clean it.' The bus approached, we shook hands and took separate seats.

As we started to drive away, leaving the antiques shop and the fancy waffle shop, a well-endowed woman stood there wearing a teal t-shirt, with printed on her bosom—Fake Love. I kept on thinking maybe, this was my last dance, my last adventure. Maybe, I was leaving this man I created bit by bit for all these years and was abandoning him somewhere in Foveaux Strait surrounded by giant white sharks, and I would never see Bluff, or even sharks ever again. Always running after the adventure out in the wild, maybe he is lost again. I still feel like peeping through the cracks of a mud house, looking at inside the premises where they make colourful idols and

The iridescent skin, 2017

wanting to be part of this play of creation. I feared, I did not end up being the scholar at the top of his game as I had dreamed of. Soosan had gifted me the Springsteen autobiography (Springsteen, 2016) which I always carried with me. It is a talisman for me, but it almost feels too precious to read. Even to this day, I carry that hard cover wherever I go, but I have not read it beyond the first few pages. I had also not properly listened to any of his songs for more than two months now, as if holding on to an orgasm for the ultimate crescendo. I could not help myself but play it in my Mp3 player, but every time I tried, I could not go on more than the first two sentences of 'Drive all night...', as my senses would not be able to cope and would bring me to tears. I am hungry; I am seeing Soosan for the first time after she has been out of the hospital, and she is cooking pork with cracklings. I was going back, back to the known, however unknown it was, and my friends and family, however distant they were.

Sharks saved my life, they gave me direction, and quite literally by losing all the weight. And maybe, this book is still in similar vein, not for the rest of the world, but to remind myself that I really did this. Am I less afraid of them? I am not really sure of that, but I know there was no monster looking back at me across that cage. Jenny in Napier told me that I am afraid of them, because I am a shark myself (something that had been told to me at least three other times over these last five years), and my fear is the fear of self-acceptance... I don't know, but for now sitting in that bus, all I could think of was their iridescent skin sparkling with the shards of light penetrating the turquoise blue ripples above, and maybe that is the way forward.

Spread of unbiased factual knowledge is pertinent, however, at the end, 'Beauty will save the world' (Dostoevsky). I do not think that mere information can alter the shark image, for mere presentation of false knowledge did not create this image in the first place. We have to share stories and images of the beauty of the sharks, challenge the mass global image by personal stories of love and admiration, and how they are related to our art and culture, for in the intersection of these expressions lies knowledge that cannot be merely expressed in verbal communication. We have to first understand and acknowledge the deep-rooted reasoning of our fear of them, and from that, create a culture of dialogue to heal our scared minds. Maybe, approach it as a social cognitive behaviour therapy, starting from small groups of individuals, then moving on to the mass society. Maybe, it is important to ask the experts not to take merely the knowledge-based approach and preach to people about the importance of the sharks and statistical probability of shark attacks. More than preaching, maybe it is about sharing our fear and our sense of beauty, more than knowledge, maybe the path is personal emotional connections, then and only then can the demystification happen, and the connection be made.

The taxi is here, a freshly bathed yellow Ambassador with blue stripes. A Bihari gentleman, ready for his first trip to the Kolkata Netaji international airport. Why am I going again? Near the end of 'Limelight' before his

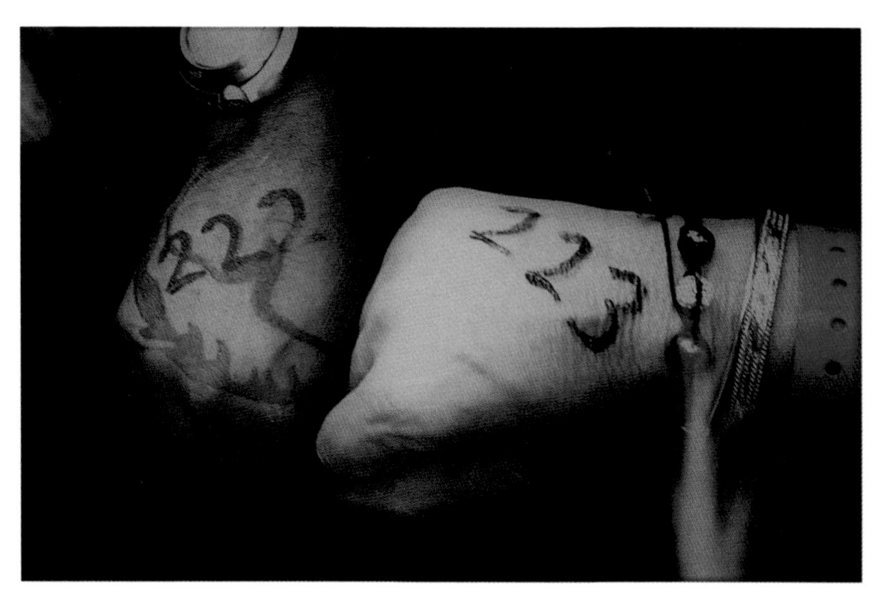

Concert entry stamps, Christchurch, Soosan Lucas, 2017

final show, when Terry asked Calvero, why did he insist on going on with the show when he said he hated the theatre, Calvero replied, 'I do, I also hate the sight of blood, but it is in my veins.' Maybe it is in 'my veins' too, but no matter how many times you do it, you still get a churning in your stomach, although you get better at distracting yourself from it. To some, 'home' is a land; to some, the people in it; for others, the buildings, and to the unfortunate lot—anywhere but here. My home is now my work and these two and a half bags, and I put them in the car. You are standing there, quiet, my love, my shelter, oh my shackles, the trammel of your fragrance still holding on to me, beyond the liberation of miles and dreams. The terrace walls have been broken and removed, as if peeling off the saree, leaving the bare skin before cremation, and maybe, my final liberation is with you set aflame and the embers being one with the dust, or maybe, somewhere beyond. I sit in the car—it smells of rubbed leather—and the man smiles and twists the ignition...

Epilogue

On 11 October 2019, it was declared by the Supreme Court, that the court of appeal had made a mistake, and that cage diving should be allowed to operate. *Newshub* had this report:

> The decision means that now, the company Shark Experience has the all-clear to forge ahead with the summer season—its busiest time. 'It's a huge relief, now that we have that decision we can now get on with business,' says Mike Haines, Shark Experience Ltd director.... Cage diving uses burley and bait to attract sharks, but that tactic alarmed paua divers and in 2013 the industry group Paua Mac5 called for new rules... Eventually, the Court of Appeal agreed it was an offence because operators were 'pursuing' the sharks to deviate off swimming patterns... But Shark Experience Limited took the case to the Supreme Court, which today agreed with the Court of Appeal's finding that hunting or killing had a wider meaning – such as pursuing sharks. But it said that the ruling had also made the mistake of declaring shark cage an offence under the Wildlife Act, concluding the court: 'erred in issuing a declaration that shark cage diving is an offence under the Wildlife Act. 'So Shark Experience is now looking ahead, after barely surviving the last year.'

(https://www.newshub.co.nz/home/new-zealand/2019/10/huge-relief-for-businesses-as-supreme-court-overturns-shark-cage-diving-decision.html)

Head mounted cameras to take point of view shots, 2017

THE DOCUMENTARY

The book comes with a sensory documentary,
that can be accessed from Youtube.com.
https://youtu.be/WS0Xx4K3_BU

I have to thank my primary editor, Mr Sandip Chowdhury, and Soosie Lucas as the art director and logistics coordinator for their crucial support, their invariable inputs which has helped me shape this documentary. The documentary follows in the footsteps of a novel approach in visual ethnography that arose around the 1980s and 1990s, when ethnographic film-makers started showing deep interest in the subjective and reflexive genre of film-making. By the 1990s, there was a growing interest in the embodied and sensory experience evidenced in reflexive and phenomenological approaches to visual ethnography emphasizing on individual experience (Grimshaw, 2001; Loizos, 1993; MacDougall, 1998; Ruby, 2000; Pink, 2006).

Currently, there is a growing movement of sensory ethnographic film making (Barbash, n.d.; Carta, 2012; Nakamura, 2013; Pavsek, 2015; Pink, 2014). As the lab manager of the Harvard sensory lab, Ernst Karel puts it:

> The practice of making nonfiction work which goes under the names media anthropology or sensory ethnography is based on the understanding that human meaning does not emerge only from language; it engages with the ways in which our sensory experience is pre- or non-linguistic, and part of our bodily being in the world. It takes advantage of the fact that our cognitive awareness—conscious as well as unconscious—consists of multiple strands of signification, woven of shifting fragments of imagery, sensation, and malleable memory. Works of sensory media are capable of echoing or reflecting or embodying these kinds of multiple simultaneous strands of signification. Experiencing them constitutes an intellectual challenge for the viewer, who must actively bring their critical faculties to bear on the experience of the work, in effect to complete the work through their experience of it. (Wright, 2013, para. 7)

Recent endeavours aimed at synergizing human—animal research with varied dimensions of art including visual art forms such as photography and sensory ethnographic film making (Galloway, 2018; Kirksey, 2014; Riley Koenig Crystal M., Koenig Bryan L., and Sanz Crickette M., 2017).

These approaches are particularly of interest to this research, as the experience of cage diving, and that as the researcher in the field cannot be comprehensively explored merely through written documents and needs to be presented in other formats, which engaged the senses more widely and intensely. Consequently, I have created a full-length documentary, which is proposed as a practice-based element of the book, and not merely a complimentary material for the written words. The documentary explores the sensory embodied experience of the research in the field, and the cage diving experiences created from the interaction of the sharks, the cage, the boat, diving equipment, the people, and the environment of Foveaux Strait.

Done from the first-person perspective, the documentary hopes to elicit sensory expressions which do not necessarily constitute of how we expect

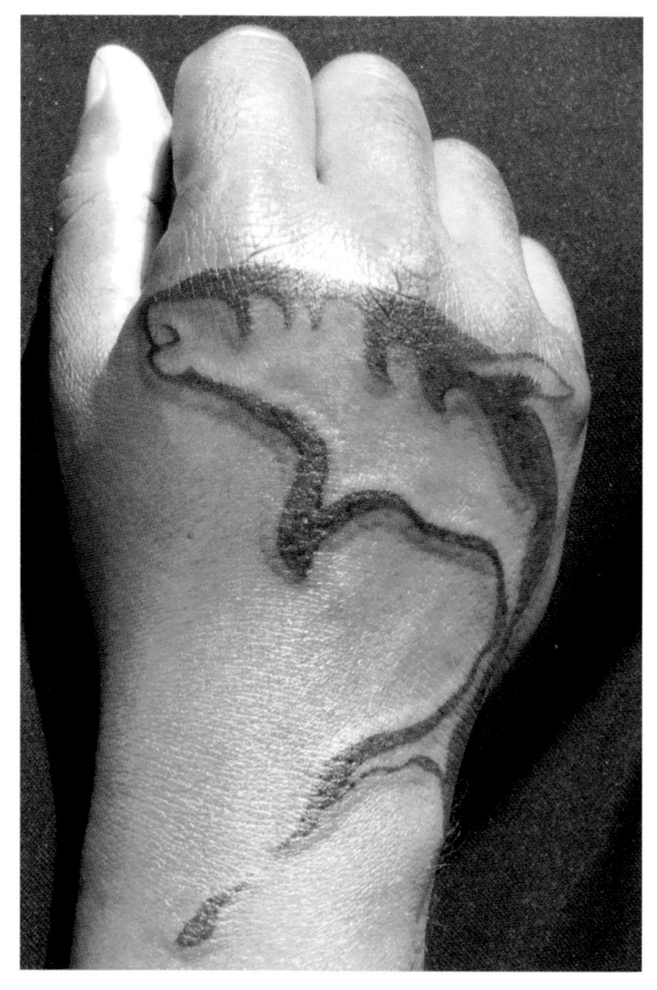

The birth of the shark tattoo, Invercargill, 2016

to feel or experience. It explores this dimension of dreams coming true, and love and fear for an animal these tourists have never seen in person, but often heard, and thought of for a long time. Indeed, the documentary brings forward the passion and admiration, and even spiritual connection with a species, who has been entrenched in the global mass consciousness as 'mindless man-eaters' for the last few decades. All of this is expressed through the eyes of the researchers as the media through the ever-evolving field site of Bluff, and Foveaux strait.

References

60 minutes. (2011). *The Sharkman*, Retrieved from https://www.youtube.com/watch?v=HpMFSL7fivI

Abram, S., & Lien, M. E. (2011). Performing nature at world's ends. *Ethnos: Journal of Anthropology*, 76(1), 3–18.

Acuña-Marrero, D., Cruz-Modino, R. de la, Smith, A. N. H., Salinas-de-León, P., Pawley, M. D. M., & Anderson, M. J. (2018). Understanding human attitudes towards sharks to promote sustainable coexistence. *Marine Policy*, *91*, 122–128. https://doi.org/10.1016/j.marpol.2018.02.018

Adams, M. (2018). Towards a critical psychology of human–animal relations. *Social and Personality Psychology Compass*, *12*(4), e12375-n/a. https://doi.org/10.1111/spc3.12375

Adorni, E. (2017). Power Feels before It Thinks: Affect Theory and Critical Animal Studies in Religious Affects by Donovan O. Schaefer. *Relations: Beyond Anthropocentrism*, *5*, 85.

Agrawal, N., Rajvanshi, S., & Asthana, A. (2017). Intraguild interactions between five congeneric species of Thaparocleidus (Monogenoidea) from the freshwater shark Wallago attu Lucknow, India. *Journal of Helminthology; Cambridge*, *91*(6), 718–725. http://dx.doi.org.ezproxy.canterbury.ac.nz/10.1017/S0022149X17000049

Ahčan, U., Radonić, V., Militarov, A., & Leban, M. (2014). Treating a patient with lower limb injury from shark attack – a case report. *Zdravniški Vestnik*, 83(1).

Aisher, A., & Damodaran, V. (2016). Introduction: Human-nature Interactions through a Multispecies Lens. *Conservation & Society*, *14*(4), 293–304. https://doi.org/10.4103/0972-4923.197612

Aiyadurai, A. (2016). 'Tigers are Our Brothers': Understanding Human-Nature Relations in the Mishmi Hills, Northeast India. *Conservation and Society*, *14*(4), 305–316. Retrieved from JSTOR.

Akerman, K. (1995). *The Use of Bone, Shell, and Teeth by Aboriginal Australians*. Lubbock, Museum of Texas Tech University: Lubbock.

Allan, A. (1981). *See Our Island: Stewart Isaland*. I.P. Lilly.

Allen, M. (2017). *The Encyclopaedia of Communication Research Methods*. https://doi.org/10.4135/9781483381411

Alvarez, P. (2010). *Cultural Anthropology*. Retrieved from http://www.culanth.org/fieldsights/280-the-multispecies-salon-ii-gleanings-from-a-para-site

Ames, K. L., & Schlereth, T. J. (Eds.). (1985). *Material Culture: A research guide*. Lawrence, Kan: University Press of Kansas.

Amin, R., Ritter, E., & Kennedy, P. (2012). A geospatial analysis of shark attack rates for the east coast of Florida: 1994-2009. *Marine and Freshwater Behaviour and Physiology*, *45*(3), 185. https://doi.org/10.1080/10236244.2012.715742

Amin, R., Ritter, E., & Wetzel, A. (2015). An Estimation of Shark-Attack Risk for the North and South Carolina Coastline. *Journal of Coastal Research*, *31*(5), 1253–1259.

Anastassakis, Z. (n.d.). Remaking everything: The clash between Bigfoot, the termites and other strange miasmic emanations in an old industrial design school. *Vibrant: Virtual Brazilian Anthropology*, 16(0). Retrieved from https://www.academia.edu/39736550/Remaking_everything_the_clash_between_Bigfoot_the_termites_and_other_strange_miasmic_emanations_in_an_old_industrial_design_school

Andersen, J. C. (2011). *Myths and Legends of the Polynesians* (Dover ed edition). New York: Dover Publications.

Anderson, B., & Harrison, P. (2010). *Taking-Place: Non-Representational Theories and Geography*. Retrieved from http://ebookcentral.proquest.com/lib/canterbury/detail.action?docID=539832

Anderson, D. J., Kobryn, H. T., Norman, B. M., Bejder, L., Tyne, J. A., & Loneragan, N. R. (2014). Spatial and temporal patterns of nature-based tourism interactions with whale sharks (Rhincodon typus) at Ningaloo Reef, Western Australia. *Estuarine, Coastal and Shelf Science*, *148*, 109–119. https://doi.org/10.1016/j.ecss.2014.05.023

Anderson, S. D., Chapple, T. K., Jorgensen, S. J., Klimley, A. P., & Block, B. A. (2011). Long-term individual identification and site fidelity of white sharks, off California using dorsal fins. *Marine Biology*, *158*(6), 1233–1237. https://doi.org/10.1007/s00227-011-1643-5

Anderson, S., Box, P. O., & Goldman, k. (1996). Photographic evidence of white shark movements California waters. *California fish and game ,82*(4):182-186 199

Andreotti, S., Rutzen, M., Wesche, P. L., O'Connell, C. P., Meÿer, M., Oosthuizen, H., & Matthee, C. A. (2014). A novel categorisation system to organise a large photo identification database for White Sharks Carcharodon carcharias. *African Journal of Marine Science*, *36*(1), 59–67. https://doi.org/10.2989/1814232X.2014.892027

Andrews, E. (2019). To save the bees or not to save the bees: Honey bee health in the Anthropocene. *Agriculture and Human Values*, *36*(4), 891–902. https://doi.org/10.1007/s10460-019-09946-x

Andrzejaczek, S., Meeuwig, J., Rowat, D., Pierce, S., Davies, T., Fisher, R., & Meekan, M. (2016). The ecological connectivity of whale shark aggregations in the Indian Ocean: A photo-identification approach. *Royal Society Open Science*, *3(*11), 160455. https://doi.org/10.1098/rsos.160455

Annual Review of Anthropology. (n.d*.). Human–Animal Communication*. Retrieved April 24, 2018, from https://www-annualreviews-org.ezproxy.canterbury.ac.nz/doi/10.1146/annurev-anthro-102116-041723

Apps, K., Dimmock, K., & Huveneers, C. (2018). Turning wildlife experiences into conservation action: Can white shark cage-dive tourism influence conservation behaviour? *Marine Policy, 88,* 108–115. https://doi.org/10.1016/j.marpol.2017.11.024

Apps, K., Dimmock, K., Lloyd, D., & Huveneers, C. (2016). In the Water with White Sharks (Carcharodon carcharias): Participants' Beliefs toward Cage-diving in Australia. *Anthrozoös, 29*(2), 231–245. https://doi.org/10.1080/08927936.2016.1152714

Apps, K., Lloyd, D., & Dimmock, K. (2015). Scuba diving with the grey nurse shark (Carcharias taurus): An application of the theory of planned behaviour to identify divers beliefs. *Aquatic Conservation: Marine and Freshwater Ecosystems, 25*(2), 201–211. https://doi.org/10.1002/aqc.2430

Aquarium, M. B. (n.d.). *Study Finds White Sharks Flee Feeding Areas When Orcas Present*. Retrieved April 18, 2019, from Monterey Bay Aquarium Newsroom website: https://newsroom.montereybayaquarium.org/press/study-finds-white-sharks-flee-feeding-areas-when-orcas-present

Archambault, J. S. (2016). Taking Love Seriously in Human-Plant Relations in Mozambique: Toward an Anthropology of Affective Encounters. *Cultural Anthropology*, *31*(2), 244–271. https://doi.org/10.14506/ca31.2.05

Archer, A. P. (2018). The men on the mountainside: An ethnography of solitude, silence and sheep bells. *Journal of Rural Studies, 64,* 103–111. https://doi.org/10.1016/j.jrurstud.2018.09.005

Armitage, H. (2015). Shark attacks: Is it safe to go back in the water? *Science*. https://doi.org/10.1126/science.aac8881

Aronoff, J., Woike, B. A., & Hyman, L. M. (1992). Which are the stimuli in facial displays of anger and happiness? Configurational bases of emotion recognition. *Journal of Personality and Social Psychology, 62*(6), 1050–1066. https://doi.org/10.1037/0022-3514.62.6.1050

Aronsson, P., Gradén, L., Linnéuniversitetet, Fakultetsnämnden för humaniora och samhällsvetenskap, & Institutionen för kulturvetenskaper, K. (2013). *Performing Nordic heritage: Everyday practices and institutional culture*. Performing Nordic Heritage (New;1;). Ashgate. https://doi.org/10.4324/9781315599991

Arva, E. L. (2008). Writing the Vanishing Real: Hyperreality and Magical Realism. *Journal of Narrative Theory, 38*(1), 60–85. https://doi.org/10.1353/jnt.0.0002

Arzoumanian Z., Holmberg J., & Norman B. (2005). An astronomical pattern-matching algorithm for computer-aided identification of whale sharks Rhincodon typus. *Journal of Applied Ecology*, *42*(6), 999–1011. https://doi.org/10.1111/j.1365-2664.2005.01117.x

Ask the Psychologist (n.d.) *Fear of Sharks in the Swimming Pool*. Retrieved June 20, 2017, from: https://askthepsych.com/atp/2009/07/14/fear-of-sharks-in-the-swimming-pool

Austin, A. (2018). *Swimming with sharks at National Aquarium in Napier.* NZ Herald. https://www.nzherald.co.nz/hawkes-bay-today/news/article.cfm?c_id=1503462&objectid=11964141

Azzarito, L., & Katzew, A. (2010). Performing Identities in Physical Education: (En)gendering Fluid Selves. *Research Quarterly for Exercise and Sport, 81*(1), 25–37. https://doi.org/10.1080/02701367.2010.10599625

Badmington, N. (2003). Theorizing posthumanism. *Cultural Critique, 53,* 10–27.

Bailey, D. E. (n.d.). *Hyperreality: The merging of the physical and digital worlds.* Retrieved July 1, 2020, from https://www.academia.edu/8597462/Hyperreality_The_merging_of_the_physical_and_digital_worlds

Baker, L. (2001). *Playing with sharks in Kaikoura.* NZ Herald. Retrieved from https://www.nzherald.co.nz/travel/news/article.cfm?c_id=7&objectid=200276

Baker, P. (2015). *"The Evolution of Underwater Photography."* Retrieved 10.10, 2015, from http://museum.wa.gov.au/explore/month-shipwrecks/patrick-baker-evolution-underwater-photography.

Baldacchino, G. (2010). Re-placing materiality: A Western Anthropology of Sand. *Annals of Tourism Research, 37*(3), 763–778. https://doi.org/10.1016/j.annals.2010.02.005

Ballance, A. (2017). *New Zealand's great White Sharks: How science is revealing their secrets.* Retrieved from https://ipac.canterbury.ac.nz/ipac20/ipac.jsp?index=BIB&term=2470362

Banerjee, B. (2013). Utopian Transformations in the Contact Zone: A Posthuman, Postcolonial Reading of Shaun Tan and John Marsden's The Rabbits. *Global Studies of Childhood, 3*(4), 418–426. https://doi.org/10.2304/gsch.2013.3.4.418

Banerjee, S., & Manna, B. (2017). Wenyonia sanyali sp.n. (Platyhelminthes: Cestoidea) from Chilloscyllium griseum (Bamboo Shark) in West Bengal, India. *Proceedings of the Zoological Society.* https://doi.org/10.1007/s12595-017-0238-7

Banks, M. & H. Morphy (1997). *Rethinking visual anthropology.* New Haven, Yale University Press.

Banks, M. (2001). *Visual methods in social research.* London, Sage.

Barbash, I. (n.d.). *Sweetgrass.* Retrieved May 16, 2018, from http://sweetgrassthemovie.com/

Barbosa, D. R. & Franco T. C. B. (1991). "Ana´lise e interpretac ̧a˜o dos dentes de sela´quios." Reunia˜o Cientı´fica da Sociedade de Arqueologia Brasileira *6*(36).

Barker, S. M., & Williamson, J. E. (2010). Collaborative photo-identification and monitoring of grey nurse sharks (Carcharias taurus) at key aggregation sites along the eastern coast of Australia. *Marine and Freshwater Research, 61*(9), 971–979. https://doi.org/10.1071/MF09215

Barnbaum, B. (2012). *The art of photography: An approach to personal expression.* Santa Barbra, CA, Rocky Cook inc.

Barrett, L. F., Mesquita, B., & Gendron, M. (2011). Context in Emotion Perception. *Current Directions in Psychological Science, 20*(5), 286–290.

Barua, M. (2014). Circulating elephants: Unpacking the geographies of a cosmopolitan animal. *Transactions of the Institute of British Geographers, 39*(4), 559–573.

Barua, M. (2015). Encounter. *Environmental Humanities, 7*(1), 265–270. https://doi.org/10.1215/22011919-3616479

Bates, T. (2013). HumanThrush entanglements: Homo sapiens as a multi-species ecology. *PAN: Philosophy Activism Nature, 10,* 36–45.

Bateson, G. & M. Mead (1952). *Trance and Dance in Bali.* USA: 22 min

Baudrillard, J. (1988). *The evil demon of images: Dedicated to the memory of Mari Kuttna, 1934-1983.* Sydney: The Power Institute of Fine Arts.

Baudrillard, J. (1993). *Symbolic exchange and death.* London; Thousand Oaks: Sage Publications.

Baudrillard, J. (1994). *Simulacra and simulation.* Ann Arbor: University of Michigan Press.

Baudrillard, J. (2001). *Selected Writings.* Stanford University Press.

Baudrillard, J., & Poster, M. (2001). *Selected writings* (2nd ed., reexplained). Cambridge, UK: Polity.

Baum, J. K., et al. (2003). Collapse and Conservation of Shark Populations in the Northwest Atlantic. *Science*: 299- 389.

Baynes-Rock, M. (2013). Life and death in the multispecies commons. *Social Science Information*, 52(2), 210–227.

Baynes-Rock, M. (2019). Precious reptiles: Social engagement and placemaking with saltwater crocodiles. *Area*, 51(3), 578–585. https://doi.org/10.1111/area.12484

Baynes-Rock, M., & Thomas, E. M. (2015). *Among the Bone Eaters: Encounters with Hyenas in Harar*. Pennsylvania State University Press.

BBC. (n.d.). *Blue Planet II* . Retrieved July 30, 2020, from https://www.bbcearth.com/blueplanet2/

Bear, C. (2011). Being Angelica? Exploring individual animal geographies. *Area, 43*(3), 297–304. Retrieved from JSTOR.

Bear, C. (2013). Assembling the sea: Materiality, movement and regulatory practices in the Cardigan Bay scallop fishery. *Cultural Geographies, 20*(1), 21–41.

Bear, C., & Eden, S. (2011). Thinking like a Fish? Engaging with Nonhuman Difference through Recreational Angling. *Environment and Planning D: Society and Space, 29*(2), 336–352. https://doi.org/10.1068/d1810

Beaumont, N. (2001). Ecotourism and the conservation ethic- recruiting the uninitiated or preaching to the converted? *Tour, 9*(4), 317–341.

Becerril-García, E. E., Hoyos-Padilla, E. M., Micarelli, P., Galván-Magaña, F., & Sperone, E. (2019). The surface behaviour of white sharks during ecotourism: A baseline for monitoring this threatened species around Guadalupe Island, Mexico. *Aquatic Conservation: Marine and Freshwater Ecosystems, 29*(5), 773–782. https://doi.org/10.1002/aqc.3057

Beierl, B. (2014). An Intercultural Understanding of How We Imagine Nonhuman Animal Others. *Society & Animals, 22*(5), 537–540. https://doi.org/10.1163/15685306-12341340

Beil, L. (n.d.). Shark *Fight: Scientists Complain about Rival Great White Tagging*. Retrieved June 23, 2017, from Scientific American website: https://www.scientificamerican.com/article/shark-fight-scientists-complain-about-rival-great-white-tagging/

Belk, R. W. (2008). Hyperreality and Globalization: *Journal of International Consumer Marketing*. https://doi.org/10.1300/J046v08n03_03

Benezra, A., DeStefano, J., & Gordon, J. I. (2012). Anthropology of microbes. *Proceedings of the National Academy of Sciences of the United States of America, 109*(17), 6378–6381.

Benford, R. D., & Hare, A. P. (2015). Dramaturgical Analysis. In J. D. Wright (Ed.), *International Encyclopedia of the Social & Behavioral Sciences (Second Edition)* (pp. 645–649). Elsevier. https://doi.org/10.1016/B978-0-08-097086-8.32046-3

Bennett, J. (2010). *Vibrant matter: A political ecology of things*. Durham, NC: Duke Univ. Press.

Benzecry, C. E. (2017). What did we say they've said? Four encounters between theory, method and the production of data. *Ethnography, 18*(1), 24–34. https://doi.org/10.1177/1466138115592423

Bernard, H. R. (2006). *Research methods in anthropology: Qualitative and quantitative approaches* (4th ed). AltaMira Press.

Bernard, H. R., Ed. (1998). *Handbook of methods in cultural anthropology*. Walnut Creek, AltaMira Press.

Berry, C. R. (2019). Under surveillance: An actor network theory ethnography of users' experiences of electronic monitoring: *European Journal of Criminology*. https://doi.org/10.1177/1477370819882890

Berta, A., Churchill, M., & Boessenecker, R. W. (2018). The Origin and Evolutionary Biology of Pinnipeds: Seals, Sea Lions, and Walruses. *Annual Review of Earth and Planetary Sciences, 46*(1), 203–228. https://doi.org/10.1146/annurev-earth-082517-010009

Bertoni, F. (2012). Charming worms: Crawling between natures. *Cambridge Anthropology, 30*(2), 65–81.

Bertoni, F., & Beisel, U. (n.d.). *Society and Space Commentary*. Retrieved from http://societyandspace.com/material/commentaries/more-than-human-intelligence-by-bertoni-and-beisel/

Bertrand, G., & Malcolm, F. (1997). *Sharks and Rays of New Zealand*. Canterbury University Press. https://ipac.canterbury.ac.nz/ipac20/ipac.jsp?index=BIB&term=625763

Bethea, C. (2017, September 6). *Did This Scientist Just Create the Ultimate Shark Repellent*? Retrieved September 2, 2017, from Outside Online website: https://www.outsideonline.com/2237371/it-just-consumed-me

Bettencourt, A., Costa, C; Bettencourt, A; Senra, M. (2019). The symbolic meaning of cattle and sheep/goat in the Bronze Age: Faunal inclusions in funerary contexts of South-Western Iberia, *International*

Journal of Osteoarchaeology, 2019, 1-11. https://doi.org/10.1002/oa.2756.
Bhana, M. (2000). *"White Shark" The Nature of the Beast*. Retrieved from https://www.nzgeo.com/video/white-shark-the-nature-of-the-beast/
Biancorosso, G. (2010). The Shark in the Music. *Music Analysis, 29*(1–3), 306–333. https://doi.org/10.1111/j.1468-2249.2011.00331.x
Biddle, J. (2007). *Breasts, Bodies, Canvas: Central Desert Art as Experience*. Sydney, UNSW Press.
Bille, M., Hastrup, F. & Sørensen, T., F. (eds) (2010). *An Anthropology of Absence: Materializations of Transcendence and Loss*. Berlin: Springer
Birke, L., & Hockenhull, J. (2012). *Crossing Boundaries: Investigating Human-Animal Relationships*. Retrieved from http://ebookcentral.proquest.com/lib/canterbury/detail.action?docID=999476
Bivins, M. (2011). *Embodied Experience Defined*. Retrieved June 14, 2018, from Maureena Bivins PhD, LAc website: http://maureenabivinsacupuncture.com/embodied-experience-defined/
Blake, H. (2010). World's 10 worst shark attacks—*Telegraph*. https://www.telegraph.co.uk/news/worldnews/africaandindianocean/southafrica/6984067/Worlds-10-worst-shark-attacks.html
Bleakley, A. (2012). The proof is in the pudding: Putting Actor-Network-Theory to work in medical education. *Medical Teacher, 34*(6), 462–467. https://doi.org/10.3109/0142159X.2012.671977
Bluff history group. (2004). *Some Notable Bluff Dates*. Early Bluff.
Bluff history group. (2008). *Home*. Retrieved May 16, 2018, from Bluff History Group website: http://bluffhistorygroup.webs.com/
Bluff history group. (n.d.). *Bluff Heritage Trail.*
Bluff New Zealand History (n.d.). *Bluff*. Retrieved May 16, 2019, from https://www.bluff.co.nz/a-look-at-bluffs-past
Bluff Oyster & Food Festival. (2019). *Next Festival!!! 25 May 2019 !!!* | Retrieved May 15, 2019, from http://www.bluffoysterfest.co.nz/
Bluff. New Zealand. (n.d*.). Giant anchor chain*. Retrieved May 16, 2019, from http://www.waymarking.com/waymarks/WMP95R_Giant_anchor_chain_Bluff_New_Zealand
Boellstorff, T. (2006). Visual Anthropology. *American Anthropologist, 108*(4), 881–882. https://doi.org/10.1525/aa.2006.108.4.881
Boessneck, J. & Von den Driesch, A. (1992). "Tell El-Dab'A VII. Tiere und historische unwelt im nordost-delta. IM 2." Jahrtausend v. chr. anhand der knochenfunde der ausgrabungen 1975–1986. Fische. Untersuchungen der Zweigstelle Kairo des O ¨ *sterreichischen Archa¨ologischen Institutes 10*: 42–48.
Bögeholz, S. (2006). Nature experience and its importance for environmental knowledge, values and action: Recent German empirical contributions. *Environmental Education Research, 12*(1), 65–84. https://doi.org/10.1080/13504620500526529
Bonanni, L. (2006). Living with Hyper-reality. In Y. Cai & J. Abascal (Eds.), *Ambient Intelligence in Everyday Life:* Foreword by Emile Aarts (pp. 130–141). https://doi.org/10.1007/11825890_6
Bonfil, R. (1994). *Overview of world elasmobranch fisheries*. Retrieved from https://books.google.co.nz/books?hl=en&lr=&id=upthrn838t0C&oi=fnd&pg=PA1&dq=Overview+of+world+elasmobranch+fisheries&ots=RdJfCaccIX&sig=rS9pwvop28HsZw41Wp0qKkL3_2I
Bonfil, R., Francis, M. P., Duffy, C., Manning, M. J., & O'Brien, S. (2010). Large-scale tropical movements and diving behavior of White Sharks Carcharodon carcharias tagged off New Zealand. *Aquatic Biology, 8*(2), 115–123. https://doi.org/10.3354/ab00217
Booij, J. (2015*). Shark Dancer.* Huawei cellphone commercial. Shot in Bahamas.
Boonman-Berson, S., Turnhout, E., & Carolan, M. (2016). Common sensing: Human-black bear cohabitation practices in Coloradoâ"ScienceDirect. *Geoforum, 74,* 192–201.
Bossolini, J. (2007). *Ethnography*. from http://www.multispecies-salon.org/ethnography/.
Bougen, P. D., & Young, J. J. (2012). Fair value accounting: Simulacra and simulation. *Critical Perspectives on Accounting, 23*(4), 390–402. https://doi.org/10.1016/j.cpa.2011.05.004
Boyd, M. (2017). Painting with Horses Towards Interspecies Response-ability: Non-human Charisma as Material Affect. *Animal Studies Journal, 6*(1), 129–154.

Bradley, D., Papastamatiou, Y., & Caselle, J. (2017). No persistent behavioural effects of SCUBA diving on reef sharks. *Marine Ecology Progress Series, 567*(Generic), 173–184. https://doi.org/10.3354/meps12053

Bradshaw, C. J. A., Lalas, C., & Thompson, C. M. (2000). Clustering of colonies in an expanding population of New Zealand fur seals (Arctocephalus forsteri). *Journal of Zoology, 250*(1), 105–112. https://doi.org/10.1111/j.1469-7998.2000.tb00581.x

Brandon, G. (n.d.). *Sharks world's largest sharks*. Retrieved from https://www.youtube.com/watch?v=VWyiQrcAbmo

Brandt, K. (2004). A Language of Their Own: An Interactionist Approach to Human-Horse Communication. *Society & Animals, 12*(4), 299–316. https://doi.org/10.1163/1568530043068010

Braun, V., & Clarke, V. (2006). Using thematic analysis in psychology. *Qualitative Research in Psychology, 3*(2), 77–101. https://doi.org/10.1191/1478088706qp063oa

Bremer, J. E. (1986). *Those sheltering hills*. A history of Bluff. Bremer, J, E.

Turner, L., Sellbach, U., & Broglio, R. (2018). *Revolution; in The Edinburgh Companion to Animal Studies*. Edinburgh University Press.

Bromilow, M. (2014). *Feeding behaviour of white sharks (Carcharodon carcharias) around a cage diving vessel and the implications for conservation*. Thesis, University of Michigan.

Brooks, K., Rowat, D., Pierce, S. J., Jouannet, D., & Vely, M. (2010). Seeing Spots: Photo-identification as a Regional Tool for Whale Shark Identification. *Western Indian Ocean Journal of Marine Science, 9*(2), 185–194.

Brown, C. (2018). The social lives of sharks ProQuest. *Australasian Science, 39*(2). Retrieved from https://search-proquest-com.ezproxy.canterbury.ac.nz/docview/2007002812?pq-origsite=summon

Bruce, B. (2015). *A review of cage diving impacts on white shark behaviour and recommendations for research and the industry's management in New Zealand*. Department of Conservation, New Zealand.

Bruce, B. D., & Bradford, R. W. (2013). The effects of shark cage-diving operations on the behaviour and movements of white sharks, Carcharodon carcharias, at the Neptune Islands, South Australia. *Marine Biology, 160*(4), 889–907. https://doi.org/10.1007/s00227-012-2142-z

Brunnschweiler, J. (2015). *Shark Attacks and Shark Diving*. Wilderness & Environmental Medicine, 26(2), 276–277. https://doi.org/10.1016/j.wem.2014.11.002

Brunnschweiler, J. M., Payne, N. L., & Barnett, A. (2018). Hand feeding can periodically fuel a major portion of bull shark energy requirements at a provisioning site in Fiji. *Animal Conservation, 21*(1), 31–35. https://doi.org/10.1111/acv.12370

Bubandt, N. (2019). Of wildmen and white men: Cryptozoology and inappropriate/d monsters at the cusp of the Anthropocene. *Journal of the Royal Anthropological Institute, 25*(2), 223–240. https://doi.org/10.1111/1467-9655.13023

Buchli, V. (Ed.). (2002). *The material culture reader*. Oxford; New York: Berg.

Buckley, R. (2012). Rush as a key motivation in skilled adventure tourism: Resolving the risk recreation paradox. *Tourism Management, 33*(4), 961–970. https://doi.org/10.1016/j.tourman.2011.10.002

Budker, P. & P. J. P. Whitehead (1971). *The life of sharks*. New York, Columbia University Press.

Bullingham, L., & Vasconcelos, A. C. (2013). The presentation of self in the online world: Goffman and the study of online identities. *Journal of Information Science, 39*(1), 101–112. https://doi.org/10.1177/0165551512470051

Buray, N., Mourier, J., Planes, S., & Clua, E. (2009). *Underwater photo-identification of sicklefin lemon sharks,* Negaprion acutidens, at Moorea (French Polynesia). 8.

Byard, R. W., Gilbert, J. D., & Brown, K. (2000). Pathologic features of fatal shark attacks. *The American Journal of Forensic Medicine and Pathology, 21*(3), 225.

Byard, Roger W., James, R. A., & Heath, K. J. (2006). Recovery of human remains after shark attack. *The American Journal of Forensic Medicine and Pathology, 27*(3), 256.

Cagua, E. F., Collins, N., Hancock, J., & Rees, R. (2014). Whale shark economics: A valuation of wildlife tourism in South Ari Atoll, Maldives. *PeerJ, 2,* 515. https://doi.org/10.7717/peerj.515

Cahill, S. E. (1998). Toward a Sociology of the Person. *Sociological Theory, 16*(2), 131–148. JSTOR.

Caillon, S., Cullman, G., Verschuuren, B., & Sterling, E. J. (2017). Moving beyond the human–nature dichotomy through biocultural approaches: Including ecological well-being in resilience indicators. *Ecology and Society, 22*(4), 27. https://doi.org/10.5751/ES-09746-220427

Calvet-Mir, L., & Salpeteur, M. (2016). Humans, Plants, and Networks: A Critical Review. *Environment & Society, 7*(1), 107–128. https://doi.org/10.3167/ares.2016.070107

Cama, N. (2012). *Object of the week: The horrors of the deep*. Australian National Maritime Museum. https://www.sea.museum/2012/05/01/object-of-the-week-the-horrors-of-the-deep

Camhi, M., et al. (1998). *Sharks and their relatives*. Gland, Switzerland, The IUCN Species Survival Commission.

Campbell, M. (2008). An animal geography of avian feeding habits in Peterborough, Ontario. *Area, 40*(4), 472–480. https://doi.org/10.1111/j.1475-4762.2008.00839.x

Candea, M. (2010). "I fell in love with Carlos the meerkat": Engagement and detachment in human-animal relations. *American Ethnologist, 37*(2), 241–258.

Candea, M. (2012). Different species, one theory: Reflections on anthropomorphism and anthropological comparison. *Cambridge Anthropology, 30*(2), 118–135.

Candea, M., & Alcayna-Stevens, L. (2012). Internal others: Ethnographies of naturalism. *Cambridge Anthropology, 30*(2), 36–47.

Carne, N. (2019). Sharks 'threatened by global fisheries.' *Cosmos Magazine*. https://cosmosmagazine.com/earth-sciences/sharks-threatened-by-global-fisheries

Carta, S. (2012). Visual anthropology and sensory ethnography in contemporary Sardinia: A film of a different kind. *Modern Italy, 17*(03), 305–324. https://doi.org/10.1080/13532944.2012.658154

Carter, G. L., Campbell, A. C., & Muncer, S. (2014). The Dark Triad personality: Attractiveness to women. *Personality and Individual Differences, 56*, 57–61. https://doi.org/10.1016/j.paid.2013.08.021

Cassel, S. H., & Maureira, T. M. (2017). Performing identity and culture in Indigenous tourism – a study of Indigenous communities in Québec, Canada. *Journal of Tourism and Cultural Change, 15*(1), 1–14. https://doi.org/10.1080/14766825.2015.1125910

Cassidy, R. (2012). Lives with others: Climate change and human-animal relations. *Annual Review of Anthropology, 41*, 21–36.

Cassidy, R., & Mullin, M. H. (2007). *Where the wild things are now: Domestication reconsidered*. Oxford: Berg.

Castree, N., Nash, C., Badmington, N., Braun, B., Murdoch, J., & Whatmore, S. (2004). Mapping posthumanism: An exchange. *Environment and Planning, 36*(8), 1341–1363.

Castro, J. I. (2016). The Origins and Rise of Shark Biology in the 20th Century. *Marine Fisheries Review, 78*(1–2), 20.

Catlin, J., Jones, T., & Jones, R. (2012). Balancing commercial and environmental needs: Licensing as a means of managing whale shark tourism on Ningaloo reef. *Journal of Sustainable Tourism, 20*(2), 163–178. https://doi.org/10.1080/09669582.2011.602686

Cempbell, E. (n.d.). *Psychological Issues in Diving II - Anxiety, Phobias in Diving*—DAN | Divers Alert Network—Medical Dive Article. Retrieved August 28, 2020, from https://www.diversalertnetwork.org/medical/articles/Psychological_Issues_in_Diving_II_Anxiety_Phobias_in_Diving

Cerullo, M. (n.d.). *Mary Cerullo*. Retrieved August 6, 2020, from https://marymcerullo.com/

Cerullo, M. M. (2015a). *Searching for great white sharks: a shark divers' quest for Mr. Big. Minnesota*, Compass point Books.

Cerullo, M. M. (2015b*). Journey to the shark island: a shark photographer's close encounters*. Minnesota, Compass point books.

Cerullo, M. M., & Rotman, J. L. (2000). *The Truth about Great White Sharks*. San Francisco: Chronicle Books.

Chabrak, N., & Craig, R. (2013). Student imaginings, cognitive dissonance and critical thinking. *Critical Perspectives on Accounting, 24*(2), 91–104. https://doi.org/10.1016/j.cpa.2011.07.008

Chao, S. (2018). In the shadow of the palm: Dispersed Ontologies among Marind, West Papua. *Cultural Anthropology, 33*(4), 621–649. https://doi.org/10.14506/ca33.4.08

Chapman, B. (2017). *Shark Attacks: Myths, Misunderstandings and Human Fear*. Retrieved from http://ebookcentral.proquest.com/lib/canterbury/detail.action?docID=5148911

Chapman, B. K., & McPhee, D. (2016). Global shark attack hotspots: Identifying underlying factors behind increased unprovoked shark bite incidence. *Ocean & Coastal Management, 133*, 72–84. https://doi.org/10.1016/j.ocecoaman.2016.09.010

Chapple Taylor K., Jorgensen Salvador J., Anderson Scot D., Kanive Paul E., Klimley A. Peter, Botsford Louis W., & Block Barbara A. (2011). A first estimate of white shark, Carcharodon carcharias, abundance off Central California. *Biology Letters, 7*(4), 581–583. https://doi.org/10.1098/rsbl.2011.0124

Cheng, E. H. (2015). *Underwater Photography - Introduction, Historical Background, Challenges and Opportunities*. from http://encyclopedia.jrank.org/articles/pages/1193/Underwater-Photography.html.

Cheyne, J. A. (2001). The Ominous Numinous Sensed Presence and 'Other' Hallucinations. In *Between Ourselves: Second-person Issues in the Study of Consciousness*. Thorverton: Imprint Academic.

Chin, A., Kyne, P. M., Walker, T. I., & McAULEY, R. B. (2010). An integrated risk assessment for climate change: Analysing the vulnerability of sharks and rays on Australia's Great Barrier Reef: Integrated risk assessment for climate change. *Global Change Biology, 16*(7), 1936–1953. https://doi.org/10.1111/j.1365-2486.2009.02128.x

Choi, C. Q. (2010). *"How 'Jaws' Forever Changed Our View of Great White sharks*.

Choy, T. K., Faier, L., Hathaway, M. J., Inoue, M. J., Satuska, S., & Tsing, A. (2009). A new form of collaboration in cultural anthropology: Matsutake worlds. *American Ethnologist, 36*(2), 380–403.

Chrisltft. (1993). *Baudrillard's Fatal Strategies* (Rick Roderick). Retrieved from https://www.youtube.com/watch?v=z0MUurcsVLs

Chriss, J. J. (2015). Goffman, Parsons, and the Negational Self. Academicus; *Albania, 11*, 11–31. http://dx.doi.org.ezproxy.canterbury.ac.nz/10.7336/academicus.2015.11.01

Christiansen, P. (2009). *Hey-Day to May-Day*. Bluff, NZ: Bluff History Group.

Christiansen, P. (2015). *Shipwrecks Bluff Area, 1845-1920*. Bluff, NZ: Bluff History Group.

Chrulew, M. (2011). Managing love and death at the zoo: The biopolitics of endangered species preservation. *Australasian Humanities Review, 50*, 137–157.

Cione, A. L. And Bonomo M. (2003). Great White Shark Teeth Used as Pendants and Possible Tools by Early-Middle Holocene Terrestrial Mammal Hunter-Gatherers in the Eastern Pampas (Southern South America). *International Journal of Osteoarchaeology*, 13.

Cisneros-Montemayor, A. M., Barnes-Mauthe, M., Al-Abdulrazzak, D., Navarro-Holm, E., & Sumaila, U. R. (2013). Global economic value of shark ecotourism: Implications for conservation. *Oryx, 47*(3), 381–388. https://doi.org/10.1017/S0030605312001718

Clarke, S. (2008). Culture and Identity. In the *SAGE Handbook of Cultural Analysis* (pp. 510–529). SAGE Publications Ltd. https://doi.org/10.4135/9781848608443

Clarke, V., & Braun, V. (2017). Thematic analysis. *The Journal of Positive Psychology, 12*(3), 297–298. https://doi.org/10.1080/17439760.2016.1262613

Clausen, R., & Longo, S. B. (2012). The tragedy of the commodity and the farce of AquAdvantage salmon. *Development and Change, 43*(1), 229–251.

Clua, E., & Séret, B. (2010). Unprovoked fatal shark attack in Lifou Island (Loyalty Islands, New Caledonia, South Pacific) by a great white shark, Carcharodon carcharias. *The American Journal of Forensic Medicine and Pathology, 31*(3), 281.

Clua, E., & Séret, B. (2016). Species identification of the shark involved in the 2007 Lifou fatal attack on a swimmer: A reply to Tirard et al. (2015). *Journal of Forensic and Legal Medicine*. https://doi.org/10.1016/j.jflm.2016.03.004

Clua, E., Buray, N., Legendre, P., Mourier, J., & Planes, S. (2010). Behavioural response of sicklefin lemon sharks Negaprion acutidens to underwater feeding for ecotourism purposes. *Marine Ecology Progress Series, 414*(Journal Article), 257–266. https://doi.org/10.3354/meps08746

Cocker, H., Piacentini, M., & Banister, E. (2018). Managing dramaturgical dilemmas: Youth drinking and multiple identities. *European Journal of Marketing, 52*(5/6), 1305–1328. https://doi.org/10.1108/EJM-01-2017-0045

Coghlan, S. (2016). Humanism, Anti-Humanism, and Nonhuman Animals. *Society & Animals, 24*(4), 403–419. https://doi.org/10.1163/15685306-12341416

Cohen, D. & Crabtree, B. (2006). *Qualitative Research Guidelines Project*. From http://www.qualres.org/HomeSemi-3629.html.

Cohen, D.E. (2011). The paradoxical politics of viral containment or, how scale undoes us one and all. *Social Text 29*(1): 15-35.

Cohen, E. (2019). Posthumanism and tourism. *Tourism Review, 74*(3), 416–427. https://doi.org/10.1108/TR-06-2018-0089

Cohen, S.M. (2004). *Identity, Persistence, and the Ship of Theseus*. Retrieved June 28, 2019, from https://faculty.washington.edu/smcohen/320/theseus.html

Colangelo, J. (2015). *Diving Beneath the Surface: A Phenomenological Exploration of Shark Ecotourism and Environmental Interpretation from the Perspective of Tourists*. Thesis, Université d'Ottawa / University of Ottawa. http://dx.doi.org/10.20381/ruor-4232

Collier, J. & Collier, M. (1992). *Visual anthropology- Photography as a research method*, University of New Mexico Press.

Collier, J. (1967). *Visual anthropology: photography as a research method*. New York Book, Holt, Rinehart and Winston.

Collier, J., & Collier, M. (1986). *Visual Anthropology: Photography as a Research Method*. Albuquerque: University of New Mexico Press.

Compagno, L. J. V. (1988). *Sharks of the order Carcharhiniformes*. Princeton, N.J, Princeton University Press.

Compagno, L. J. V., & Cook, S. F. (1995). The exploitation and conservation of freshwater elasmobranchs: Status of taxa and prospects for the future. *Journal of Aquariculture and Aquatic Sciences*.

Conrad, J. (1899). *Heart of Darkness*. CreateSpace Independent Publishing Platform.

Coote, T. (1994). *From the Bluff: A Social History*. Invercargill: Published by Invercargill City Council.

Corbett, J. (1944). *Man-eaters of Kumaon*. Oxford University Press, Amen House, London. Retrieved from http://archive.org/details/in.ernet.dli.2015.458957

Corbett, J. (1948). *The man-eating leopard of Rudraprayag*. Oxford University Press, Bombay Retrieved from http://archive.org/details/maneatingleopard00corb

Corbett, J. (1959). *Jungle lore*. Oxford University Press, Amen House, London Retrieved from http://archive.org/details/dli.bengal.10689.20127

Corbett, J., & Marczak, P. (1992). *The champawat man-eater*. Retrieved from http://archive.org/details/maneatersofkumao00jimc

Corbey, R., & Lanjouw, A. (2013). *The politics of species: Reshaping our relationships with other animals*. Cambridge, UK: Cambridge Univ. Press.

Cormier, L. (2011). *The ten thousand year fever: Rethinking human and wild-primate malarias*. Walnut Creek, CA: Left Coast.

Cousteau, J. Y. & F. Dumas (1954). *Le Monde du silence*. Paris, Éditions de Paris.

Cousteau, J. Y. & P. Diolé (1973). *Octopus and squid: the soft intelligence* [by] Jacques-Yves Cousteau and Philippe Diole. Translated from the French by J.F.Bernard. London Book, Cassell.

Cousteau, J.-M., & Richards, M. (1992). *Cousteau's Great White Shark* (First Edition edition). New York: Harry N Abrams Inc.

Cousteau, J.-Y. & F. Dumas (1953). *Silent World*, HarperCollins. New York.

Cousteau, J.-Y., & Malle, L. (1956, August 15). *Le monde du silence* [Documentary]. FSJYC Production, Requins Associés, Société Filmad.

Cox, G., & Francis, M. (1997). *Sharks and Rays of New Zealand*: An Illustrated Guide. Christchurch: Canterbury University Press.

Craven, J. (2018). *Form Follows Function" A Famous Phrase in Architecture*. Retrieved June 5, 2019, from: https://www.thoughtco.com/form-follows-function-177237

Crawford, S. (2017). *White Pointer Chronicles*. Retrieved April 23, 2019, from White pointer Chronicles website: https://www.whitepointer.cloud/

CreativecomP47 (N.D.). *"Underwater Photography."* from http://www.timetoast.com/timelines/underwater-photography--8.

White Shark Video (2015). *Hooking the Submarine: Full story with Craig Ferreira*: WSV. https://www.youtube.com/watch?v=j5VnDYVoH-w

Cresswell, L., McNicholas, S., & Newton, D. (2013). *Great White Shark*. Retrieved from http://www.imdb.com/title/tt2989326/

Crist, E. (2004). Can an insect speak? The case of the honeybee dance language. *Social Studies of Science 34* (1): 7-43.

Crystal, M., Riley, K., Bryan L. K., & Crickette M. S. (2017). Teaching Anthropology with Primate Documentaries: Investigating Instructors' Use of Films and Introducing the Primate Films Database. *American Anthropologist, 120*(1), 24–38. https://doi.org/10.1111/aman.12974

CSIRO Publishing (n.d.). *Marine and Freshwater Research*. Retrieved December 4, 2018, from http://www.publish.csiro.au.ezproxy.canterbury.ac.nz/mf/MF11049

Cubero-Pardo, P., Herrón, P., & González-Pérez, F. (2011). Shark reactions to scuba divers in two marine protected areas of the Eastern Tropical Pacific. *Aquatic Conservation: Marine and Freshwater Ecosystems, 21*(3), 239–246. https://doi.org/10.1002/aqc.1189

Cultural Anthropology. (n.d.). Retrieved from http://www.culanth.org/fieldsights/222–2011-culture-large-session-the-human-is-more-than-human

Cunningham-Day, R. (2001). *Sharks in Danger: Global shark conservation status with reference to management plans and legislation*. Parkland, FL USA, Universal Publishers.

Curtin, S. (2009). Wildlife tourism: The intangible, psychological benefits of human-wildlife encounters. *Current Issues in Tourism, 12*(5–6), 451–474. https://doi.org/10.1080/13683500903042857

Czerniak, C. M., & McDonald, J. (1993). Shark Attack. *Science Scope, 17*(2), 16–20.

Czyzewski, M. (1987). Erving Goffman on the Individual. A Reconstruction. *The Polish Sociological Bulletin, 79*, 31–41. JSTOR.

D'Costa, K. (n.d.). *How Our Love Affair with Reality Television Created Megalodon*. Retrieved March 5, 2019, from Scientific American Blog Network website: https://blogs.scientificamerican.com/anthropology-in-practice/how-our-love-affair-with-reality-television-created-megalodon/

Dahlke, S., Hall, W., & Phinney, A. (2015). Maximizing Theoretical Contributions of Participant Observation While Managing Challenges. *Qualitative Health Research, 25*(8), 1117–1122. https://doi.org/10.1177/1049732315578636

Dala-Corte, R. B., Moschetta, J. B., & Becker, F. G. (n.d.). Photo-identification as a technique for recognition of individual fish: A test with the freshwater armored catfish Rineloricaria aequalicuspis Reis & Cardoso, 2001 (Siluriformes: Loricariidae). *Neotropical Ichthyology, 14*(1). Retrieved from https://doaj.org

Daley, J. (2019). Great White Sharks Are Completely Terrified of Orcas. *Smart News, Smithsonian Magazine*. https://www.smithsonianmag.com/smart-news/great-white-sharks-are-completely-terrified-orcas-180972009/

DAN (n.d.). *How Diving Affects Your Health and Circulatory System* | Retrieved September 18, 2017, from https://www.diversalertnetwork.org/health/heart/how-diving-affects-health

Danby, P., Dashper, K., & Finkel, R. (2019). Multispecies leisure: Human-animal interactions in leisure landscapes. *Leisure Studies: Multispecies Leisure: Human-Animal Interactions in Leisure Landscapes, 38*(3), 291–302. https://doi.org/10.1080/02614367.2019.1628802

Dant, T. (1999). *Material culture in the social world: Values, activities, lifestyles. Buckingham* ; Philadelphia: Open University Press.

Dant, T. (2005). *Materiality and society. Maidenhead*: Open University Press.

Dashper, K. (2020). More-than-human emotions: Multispecies emotional labour in the tourism industry. *Gender, Work & Organization, 27*(1), 24–40. https://doi.org/10.1111/gwao.12344

Dashper, K., Dashper, K., Brymer, E., & Brymer, E. (2019). An ecological-phenomenological perspective on multispecies leisure and the horse-human relationship in events. *Leisure Studies: Multispecies Leisure: Human-Animal Interactions in Leisure Landscapes, 38*(3), 394–407. https://doi.org/10.1080/02614367.2019.1586981

Daston, L., & Mitman, G. (2005). *Thinking with animals: New perspectives on anthropomorphism*. New York: Columbia Univ. Press.

Davey, G. C. L. (1991). Characteristics of individuals with fear of spiders. *Anxiety Research, 4(4)*, 299–314. https://doi.org/10.1080/08917779208248798

Davies, O., & Riach, K. (2019). From manstream measuring to multispecies sustainability? A gendered reading of bee-ing sustainable. *Gender, Work & Organization, 26*(3), 246–266. https://doi.org/10.1111/gwao.12245

De Brigard, E. (1975). The History of Ethnographic Film. Principles of Visual Anthropology. *P. Hockings, The Hague: Mouton*.: 13-43.

De la Puente Jeri, S. (2017). *Characterizing the knowledge and attitudes towards sharks and the domestic use of shark meat and fins in Peru* (University of British Columbia). https://doi.org/10.14288/1.0355844

De Leon, J. P., & Cohen, J. H. (2005). Object and Walking Probes in Ethnographic Interviewing. *Field Methods, 17*(2), 200–204. https://doi.org/10.1177/1525822X05274733

De Wolff, K. (2017). Plastic Naturecultures: Multispecies Ethnography and the Dangers of Separating Living from Nonliving Bodies. *Body & Society, 23*(3), 23–47. https://doi.org/10.1177/1357034X17715074

Deckha, M. (2013). *Animal Advocacy, Feminism and Intersectionality*. 23, 18.

Deleuze, G., & Guattari, F. (1987). *A thousand plateaus: Capitalism and schizophrenia*. London: Athlone.

DeMello, M. (2012). *Animals and society: an introduction to human-animal studies*. New York, Columbia University Press.

Denzel, T. (2017). *St.Helenaâ, A remote island in the Atlantic*, (Travel Documentary) DW Documentaryâ YouTube. Retrieved from https://www.youtube.com/watch?v=Egzd7PsmNTo

Department of conservation (2013). *Commercial Great White Shark Cage Diving New Zealand*.

Department of Conservation (2015a). *Commercial great white shark cage diving New Zealand code of practice 2015* (p. 18).

Department of conservation (2015b). *Wildlife Act 1953*. D. o. conservation. New Zealand.

Department of conservation (N.D.). *White Sharks*. from http://www.doc.govt.nz/nature/native-animals/marine-fish-and-reptiles/sharks-mango/white-shark/.

Department of Conservation. (2019, December). *Shark cage diving*. https://www.doc.govt.nz/news/issues/shark-cage-diving/

Department of Conservation. (n.d.). *New Zealand is recognised as one of the world's hot spots for white sharks*. Retrieved July 22, 2020, from https://www.doc.govt.nz/nature/native-animals/marine-fish-and-reptiles/sharks-mango/white-shark/

Department of Conservation. (n.d.). *White sharks*. Retrieved August 6, 2020, from https://www.doc.govt.nz/nature/native-animals/marine-fish-and-reptiles/sharks-mango/white-shark/

Descola, P. (2013). *Beyond nature and culture*. Chicago: Univ. of Chicago Press.

Descola, P. (2014). Modes of being and forms of predication. *HAU: Journal of Ethnographic Theory, 4*(1), 271–280. https://doi.org/10.14318/hau4.1.012

Despret, V. (2013). Responding bodies and partial affinities in human-animal worlds. *Theory, Culture & Society, 30*(7–8), 51–76.

DeWalt, K. M., & DeWalt, B. R., (2002). *Participant Observation: A Guide for Fieldworkers*. Rowman, Altamira.

Dicken, M. L., & Hosking, S. G. (2009). Socio-economic aspects of the tiger shark diving industry within the Aliwal Shoal Marine Protected Area, South Africa. *African Journal of Marine Science, 31*(2), 227–232. https://doi.org/10.2989/AJMS.2009.31.2.10.882

Discovery. (2019, August 3). *Rare Shark Brothers Spotted in Australia* | Shark Week. https://www.youtube.com/watch?v=qxzLDzQOiCc&t=74s

Dickman, A. J. (2010). Complexities of conflict: The importance of considering social factors for effectively resolving human-wildlife conflict: Social factors affecting human-wildlife conflict resolution. *Animal Conservation, 13*(5), 458–466. https://doi.org/10.1111/j.1469-1795.2010.00368.x

Dive Worldwide. (n.d.). *Shark Diving Trips: Tailor-Made Ideas and Group Tours*. Retrieved July 15, 2020, from https://www.diveworldwide.com/discover/interests/shark-diving/trips

Diving with the sharks. (2017). *NewsRx Health & Science*. Retrieved from University of California - santa Barbara

Doak, W. (1975). *Sharks and other ancestors: patterns of survival in the South Seas*. Auckland, Hodder and Stoughton.

Dobson, J. (2007). *Jaws or Jawsome? Exploring the shaek-diving experience*. 5th International coastal and marine tourism congress., Auckland, New Zealand, New Zealand tourism research institute, AUT university.

DocumentaryTM. (n.d.). *Full Documentary Inside Nature's Giants Great White Shark*. Retrieved from https://www.youtube.com/watch?v=zsbXuakhGkQ

Dolan, R. J. (2002). Emotion, Cognition, and Behavior. *Science, 298*(5596), 1191–1194. Retrieved from JSTOR.

Domeier, M. L. (Ed.). (2012). *Global Perspectives on the Biology and Life History of the White Shark* (1 edition). Boca Raton, FL: CRC Press.

Domeier, M. L., & Nasby-Lucas, N. (2007). Annual re-sightings of photographically identified white sharks at an eastern Pacific aggregation site (Guadalupe Island, Mexico). *Marine Biology, 150*(5), 977–984. https://doi.org/10.1007/s00227-006-0380-7

Dowie, S. (2017). *Shark Diving and Paua Fishery*. Retrieved July 17, 2018, from Sarah Dowie. MP for Invercargill website: https://sarahdowie.national.org.nz/news/2014-12-06-shark-diving-and-paua-fishery

Down, S., & Reveley, J. (2009). Between narration and interaction: Situating first-line supervisor identity work. *Human Relations, 62*(3), 379–401. https://doi.org/10.1177/0018726708101043

Doyle, K. (2018, September 4). *Court ruling sinks teeth into shark cage diving*. RNZ. https://www.rnz.co.nz/news/national/365675/court-ruling-sinks-teeth-into-shark-cage-diving

Dransart, P. (2013). *Living beings: Perspectives on interspecies engagements*. London: Bloomsbury.

Dudley, S. F. J., & Simpfendorfer, C. A. (2006). Population status of 14 shark species caught in the protective gillnets off KwaZulu–Natal beaches, South Africa, 1978–2003. *Marine and Freshwater Research, 57*(2), 225–240. https://doi.org/10.1071/MF05156

Duffy, C. A. J., Francis, M. P., & Bonfil, R. (2012). Regional Population Connectivity, Oceanic Habitat, and Return Migration Revealed by Satellite Tagging of White Sharks, Carcharodon carcharias, at New Zealand Aggregation Sites. In *Global persectives on the biology and life history of the White shark*. Retrieved from https://www.researchgate.net/publication/234066813_Regional_Population_Connectivity_Oceanic_Habitat_and_Return_Migration_Revealed_by_Satellite_Tagging_of_White_Sharks_Carcharodon_carcharias_at_New_Zealand_Aggregation_Sites

Duim, van der, R., Ren, C., & Johannesson, G. T. (2013). Ordering, materiality, and multiplicity: Enacting Actor–Network Theory in tourism. *Tourist Studies, 13*(1), 3–20. https://doi.org/10.1177/1468797613476397

Dulvy, N. K., Baum, J. K., Clarke, S., Compagno, L. J. V., Cortés, E., Domingo, A., ... Valenti, S. (2008). You can swim but you can't hide: The global status and conservation of oceanic pelagic sharks and rays. Aquatic Conservation: *Marine and Freshwater Ecosystems, 18*(5), 459–482. https://doi.org/10.1002/aqc.975

Dulvy, N. K., Simpfendorfer, C. A., Davidson, L. N. K., Fordham, S. V., Bräutigam, A., Sant, G., & Welch, D. J. (2017). Challenges and Priorities in Shark and Ray Conservation. *Current Biology, 27*(11), R565–R572. https://doi.org/10.1016/j.cub.2017.04.038

Dunn, K. (2016). [Review] Animal Horror Cinema: Genre, History and Criticism, Katarina Gregersdotter, Johan Höglund and Nicklas Hållén (eds). Basingstoke and New York: Palgrave Macmillan, 2015. *Animal Studies Journal, 5*(2), 228–231.

Dutton, D. (2012). Being-with-Animals: Modes of Embodiment in Human-Animal Encounters. *Crossing Boundaries, 89*, 111. https://doi.org/10.1163/9789004233041_007
Ebert, D. A., et al. (2013). *Sharks of the world: a fully illustrated guide.* Plymouth, UK, Wild Nature Press.
Ebert, D. A., Fowler, S., & Dando, M. (2015). *A Pocket Guide to Sharks of the World* (Poc edition). Princeton: Princeton University Press.
eBuckley, R. (2015). Autoethnography helps analyse emotions. *Frontiers in Psychology, 6*. Retrieved from https://doaj.org
Eddy, M. D. (2013). The Shape of Knowledge: Children and the Visual Culture of Literacy and *Numeracy. Science in Context, 26*, 215-145.
Edensor, T. (2001). Haunting in the ruins: Matter and immateriality. *In Spatial Hauntings. Thousand Oaks,* Calif.: Space and Culture.
Edge of existence. (n.d.). *Great White Shark*. EDGE of Existence. Retrieved July 21, 2020, from https://www.edgeofexistence.org/species/great-white-shark/
Edge, M. (2010). *The Underwater Photographer*. Burlington, MA, Focal Press.
Editorial (2017). A debate on transparency, accessibility of data, and protecting confidential sources in qualitativeresearch.(2017*).Ethnography,18*(3),279–280.https://doi.org/10.1177/1466138117725090
Edwards, E. (2015). Anthropology and Photography: a long history of knowledge and affect. *Photographies: 1*, 18.
Edwards, E., et al. (1992). *Anthropology and photography, 1860-1920*. New Haven, Yale University Press in association with the Royal Anthropological Institute, London.
Elliott, R. (2014). *Shark Man: One Kiwi Man's Mission to Save Our Most Feared and Misunderstood Predator*. Random House New Zealand.
Ellis, E. C., Ellis, E. C., Magliocca, N. R., Magliocca, N. R., Stevens, C. J., Stevens, C. J., Fuller, D. Q., & Fuller, D. Q. (2018). Evolving the Anthropocene: Linking multi-level selection with long-term social–ecological change. *Sustainability Science, 13*(1), 119–128. https://doi.org/10.1007/s11625-017-0513-6
Ellwanger, A. L., & Lambert, J. E. (2018). Investigating Niche Construction in Dynamic Human-Animal Landscapes: Bridging Ecological and Evolutionary Timescales. *International Journal of Primatology, 39*(5), 797–816. https://doi.org/10.1007/s10764-018-0033-y
Elwin, V. (1964). *The Tribal World of Verrier Elwin: An Autobiography*. Oxford University Press.
Encyclopedia Mythica. (1997). *Miru*. https://pantheon.org/articles/m/miru.html
Engelmann, S. (2017). Social spiders and hybrid webs at Studio Tomás Saraceno. *Cultural Geographies, 24*(1), 161–169. https://doi.org/10.1177/1474474016647371
Eovaldi, B., Thompson, P., Eovaldi, K., & Eovaldi, R. (2016). Shark Fears and the Media. *Wilderness & Environmental Medicine, 27*(1), 184–185. https://doi.org/10.1016/j.wem.2015.10.012
Ercan, Z. (2016, March 20). *There is an animal in you, too* • TEDxVienna. Retrieved August 12, 2018, from TEDxVienna website: https://www.tedxvienna.at/blog/there-is-an-animal-in-you-too/
Erickson, J. (2017). Walking with Elephants: A Case for Trans-Species Ethnography. *The Trumpeter, 33*(1), 23–47.
Erlandson, D. A., et al. (1993). *Doing naturalistic inquiry: a guide to methods*. Newbury Park, CA., Sage.
Esler, L. (2017). *The Monica now a museum piece*. Retrieved May 16, 2019, from Stuff website: https://www.stuff.co.nz/southland-times/news/features/95363899/the-monica-now-a-museum-piece
Eugene L. Arva. (2008). Writing the Vanishing Real: Hyperreality and Magical Realism. *Journal of Narrative Theory, 38*(1), 60–85. https://doi.org/10.1353/jnt.0.0002
Everingham, C. R., Heading, G., & Connor, L. (2006). Couples' experiences of postnatal depression: A framing analysis of cultural identity, gender and communication. *Social Science & Medicine, 62*(7), 1745–1756. https://doi.org/10.1016/j.socscimed.2005.08.039
Ezeh, P. J. (2003). Participant Observation. *Qualitative Research, 3*(2), 191–205. https://doi.org/10.1177/14687941030032003
Fahrenkamp-Uppenbrink, J. (2015). Humans and animals vying for space in the air. *Science, 348*(6234), 536–536. https://doi.org/10.1126/science.348.6234.536-s

Fairhead, J. R. (2016). Termites, Mud Daubers and their Earths: A Multispecies Approach to Fertility and Power in West Africa. *Conservation and Society, 14*(4), 359–367. JSTOR.

Faier, L. (2010). *Cultural Anthropology*. Retrieved from http://www.culanth.org/fieldsights/276-thoughts-for-a-world-of-poaching

Faier, L. (2011). Fungi, trees, people, nematodes, beetles, and weather: Ecologies of vulnerability and ecologies of negotiation in matsutake commodity exchange. *Environment and Planning, 43*, 1079–1097.

Faier, L., & Rofel, L. (2014). Ethnographies of Encounter. *Annual Review of Anthropology, 43*(1), 363–377. https://doi.org/10.1146/annurev-anthro-102313-030210

Falzon, M.-A. (2009). Multi-sited ethnography: Theory, praxis and locality in contemporary research. In T. K. Faier Choy Lieba, Hathaway, Michael J., Inoue, Miyako J., Satuska, Shiho, Tsing, Anna (Ed.), *Strong collaboration as a method for multi-sited ethnography: On mycorrhizal relations* (pp. 380–403). Farnham, UK: Ashgate.

Farquhar, L. (2013). Performing and interpreting identity through Facebook imagery. *Convergence, 19*(4), 446–471. https://doi.org/10.1177/1354856512459838

Fawkes, J. (2015). Performance and Persona: Goffman and Jung's approaches to professional identity applied to public relations. *Public Relations Review, 41*(5), 675–680. https://doi.org/10.1016/j.pubrev.2014.02.011

FearOf (2014, April 15). *Galeophobia or Selachophobia; Fear of Sharks Phobia*. Retrieved June 20, 2017, from: http://www.fearof.net/fear-of-sharks-phobia-galeophobia-or-selachophobia/

Feinberg, R., Nason, P., & Sridharan, H. (2013). Introduction: Human-animal relations. *Environment and Society: Advances in Research, 4*(1), 1–4.

Fereday, J., & Muir-Cochrane, E. (2006). Demonstrating Rigor Using Thematic Analysis: A Hybrid Approach of Inductive and Deductive Coding and Theme Development. *International Journal of Qualitative Methods, 5*(1), 80–92. https://doi.org/10.1177/160940690600500107

Ferrante, A. (2013). *Sharknado*. Retrieved from https://www.rottentomatoes.com/m/sharknado_2013

Ferreira, C. A. (2011). *Great White sharks on their best behavior: An adventure and guide into the world of Jaws.* Amazon Digital Services, Inc, Middleton & Strauss.

Festinger, L. (1962). Cognitive Dissonance. *Scientific American, 207*(4), 93–106.

Fijn, N. (2012). A Multi-species etho-ethnographic approach to filmmaking. *Humanities Research, 18*(1), 71–88.

Fijn, N. (2013). Living with Crocodiles: Engagement with a Powerful Reptilian Being. *Animal Studies Journal, 2*(2), 1–27.

Fijn, N. (2019). The Multiple Being: Multispecies Ethnographic Filmmaking in Arnhem Land, Australia. *Visual Anthropology, 32*(5), 383–403. https://doi.org/10.1080/08949468.2019.1671747

Fijn, N., & Baynes-Rock, M. (2018). A Social Ecology of Stingless Bees. *Human Ecology, 46*(2), 207–216. https://doi.org/10.1007/s10745-018-9983-0

Findlay, R., Gennari, E., Cantor, M., & Tittensor, D. P. (2016). How solitary are white sharks: Social interactions or just spatial proximity? *Behavioral Ecology and Sociobiology, 70*(10), 1735–1744. https://doi.org/10.1007/s00265-016-2179-y

Fink, B., Neave, N., & Seydel, H. (2007). Male facial appearance signals physical strength to women. *American Journal of Human Biology: The Official Journal of the Human Biology Council, 19*(1), 82–87. https://doi.org/10.1002/ajhb.20583

Fitzgerald, D. (2013). *Somatosphere*. Retrieved from http://somatosphere.net/2013/10/philippe-descolas-beyond-nature-and-culture.html

Fitzpatrick, R., Abrantes, K. G., Seymour, J., & Barnett, A. (2011). Variation in depth of whitetip reef sharks: Does provisioning ecotourism change their behaviour? *Coral Reefs, 30*(3), 569–577. https://doi.org/10.1007/s00338-011-0769-8

Fleet, J. (2012, August 26). The Daring Spiritual Practice Of Shark Calling. *HuffPost India.* https://www.huffpost.com/entry/shark-callers-photos_n_1828134

Flick, U. (2009). *An introduction to qualitative research* (4th ed). Los Angeles: Sage Publications.

Florida Museum. (2018). *The First Documented Shark Attack in the Americas*. https://www.floridamuseum.ufl.edu/caribarch/education/sharks/

Florida Museum. (2020, January 21). *Yearly Worldwide Shark Attack Summary*. https://www.floridamuseum.ufl.edu/shark-attacks/yearly-worldwide-summary/

For Scuba Divers (2016, February 27). *She Rides Great White Sharks* [Video]. Retrieved June 16, 2017, from http://forscubadivers.com/photosvideos/she-rides-great-white-sharks-video/

Ford, A. (2019). Sport horse leisure and the phenomenology of interspecies embodiment. *Leisure Studies: Multispecies Leisure: Human-Animal Interactions in Leisure Landscapes, 38*(3), 329–340. https://doi.org/10.1080/02614367.2019.1584231

Fortwangler, C. (2013). Untangling introduced and invasive animals. *Environment and Society: Advances in Research, 4*(1), 41–59.

Fox, R. (2006). Animal behaviours, post-human lives: Everyday negotiations of the animal–human divide in pet-keeping. *Social & Cultural Geography, 7*(4), 525–537. https://doi.org/10.1080/14649360600825679

Francis, B. (2012). Before and after "Jaws": Changing representations of shark attacks. *Great Circle: Journal of the Australian Association for Maritime History, 34*(2), 44–64.

Francis, M. P., Duffy, C., & Lyon, W. (2015). *Spatial and temporal habitat use by white sharks (Carcharodon carcharias) at an aggregation site in southern New Zealand*. Retrieved January 28, 2019, from ResearchGate website: https://www.researchgate.net/publication/277621686_Spatial_and_temporal_habitat_use_by_white_sharks_Carcharodon_carcharias_at_an_aggregation_site_in_southern_New_Zealand

Freccero, C. (2011). Carnivorous virility; or, becoming-dog. *Social Text, 29*(1), 177–195.

Freeman, M. (2011). *The Photographer's Mind*. United Kingdom, The Ilex Press.

Friedrich, L. A., Jefferson, R., & Glegg, G. (2014). Public perceptions of sharks: Gathering support for shark conservation. *Marine Policy, 47*, 1–7. https://doi.org/10.1016/j.marpol.2014.02.003

Fu, J. (2018). Chinese youth performing identities and navigating belonging online. *Journal of Youth Studies, 21*(2), 129–143. https://doi.org/10.1080/13676261.2017.1355444

Fuentes, A. & Hockings, K. J. (2010). The ethnoprimatological approach in primatology. *American Journal of Primatology, 72*(10), 841–847. https://doi.org/10.1002/ajp.20844

Fuentes, A. & Kohn, E. (2012). Two proposals. *Cambridge Anthropology, 30*(2), 136–146.

Fuentes, A. (2010). "Naturalcultural encounters in Bali: Monkeys, temples, tourists and ethnoprimatology." *Cultural Anthropology 25*(4): 600-624.

Fuentes, A. (2012). Ethnoprimatology and the anthropology of the human-primate interface. *Annual Review of Anthropology, 41*, 101–117.

Galapagos Shark Diving. (n.d.). *Eco & Conservation Dive Liveaboard*., Retrieved July 15, 2020, from https://www.galapagossharkdiving.com/

Gallagher, A. J., & Hammerschlag, N. (2011). Global shark currency: The distribution, frequency, and economic value of shark ecotourism. *Current Issues in Tourism, 14*(8), 797–812. https://doi.org/10.1080/13683500.2011.585227

Gallagher, A. J., & Huveneers, C. P. M. (2018). Emerging challenges to shark-diving tourism. *Marine Policy, 96*, 9–12. https://doi.org/10.1016/j.marpol.2018.07.009

Gallagher, A. J., Vianna, G. M. S., Papastamatiou, Y. P., Macdonald, C., Guttridge, T. L., & Hammerschlag, N. (2015). Biological effects, conservation potential, and research priorities of shark diving tourism. *Biological Conservation, 184*, 365–379. https://doi.org/10.1016/j.biocon.2015.02.007

Galloway, A. (2018). *More-Than-Human Lab. » Epizoic media and multispecies ethnography*. Retrieved May 16, 2018, from http://morethanhumanlab.org/blog/2011/07/26/epizoic-media-and-multispecies-ethnography/

Galvin, S. (2018). Interspecies Relations and Agrarian Worlds. *Annual Review of Anthropology, 47*, 233–249. https://doi.org/10.1146/annurev-anthro-102317-050232

Gandhi, A. (2012). Catch me if you can: Monkey capture in Delhi. *Ethnography, 13*(1), 43–56.

Gane, M. (1991). *Baudrillard: Critical and fatal theory*. London, New York, N.Y: Routledge.

Gangestad, S. W., Garver-Apgar, C. E., Simpson, J. A., & Cousins, A. J. (2007). Changes in women's mate preferences across the ovulatory cycle. *Journal of Personality and Social Psychology, 92*(1), 151–163. https://doi.org/10.1037/0022-3514.92.1.151

Gannon, S., & Gannon, S. (2017). Saving squawk? Animal and human entanglement at the edge of the lagoon. *Environmental Education Research, 23*(1), 91–110. https://doi.org/10.1080/13504622.2015.1101752

Garla, R. C., Freitas, R. H. A., Calado, J. F., Paterno, G. B. C., & Carvalho, A. R. (2015). Public awareness of the economic potential and threats to sharks of a tropical oceanic archipelago in the western South Atlantic. *Marine Policy, 60*, 128–133. https://doi.org/10.1016/j.marpol.2015.06.012

Gaskill, F. (2016). *A Psychologist's Take on Our Fear of Sharks*. https://www.shrinktank.com/psychologists-take-fear-sharks/

Geiling, N. (2013). *The Worst Shark Attack in History*. Retrieved May 15, 2019, from Smithsonian website: https://www.smithsonianmag.com/history/the-worst-shark-attack-in-history-25715092/

Gelsleichter, J., Manire, C. A., Szabo, N. J., Cortés, E., Carlson, J., & Lombardi-Carlson, L. (2005). Organochlorine Concentrations in Bonnethead Sharks (Sphyrna tiburo) from Four Florida Estuaries. *Archives of Environmental Contamination and Toxicology, 48*(4), 474–483. https://doi.org/10.1007/s00244-003-0275-2

Getting-in (N.D.). *The Advantages and Disadvantages of Structured Interviews & Postal Questionaires*. from http://www.getting-in.com/guide/gcse-sociology-sampling-techniques-the-advantages-and-disadvantages-of-structured-interviews-postal-questionaires/.

Gibbs, L., & Warren, A. (2015). Transforming shark hazard policy: Learning from ocean-users and shark encounter in Western Australia. *Marine Policy, 58*, 116–124. https://doi.org/10.1016/j.marpol.2015.04.014

Gibbs, R. W. (2003). Embodied experience and linguistic meaning. *Brain and Language, 84*(1), 1–15. https://doi.org/10.1016/S0093-934X(02)00517-5

Gietler, S. (2013a). *Underwater lighting fundamentals*.

Gietler, S. (2013b). *Magic filters underwater*.

Gillespie, K., & Narayanan, Y. (2020). Animal Nationalisms: Multispecies Cultural Politics, Race, and the (Un)Making of the Settler Nation-State. *Journal of Intercultural Studies: Animal Nationalisms: Multispecies Cultural Politics, Race, and the (Un)Making of the Settler Nation-State, 41*(1), 1–7. https://doi.org/10.1080/07256868.2019.1704379

Ginn, F. (2013). Sticky lives: Slugs, detachment and more than human ethics in the garden. *Transactions of the Institute of British Geographers 39*(4): 532-544.

Glaus, K. B. J., Adrian-Kalchhauser, I., Piovano, S., Appleyard, S. A., Brunnschweiler, J. M., & Rico, C. (2019). Fishing for profit or food? Socio-economic drivers and fishers' attitudes towards sharks in Fiji. *Marine Policy, 100*, 249–257. https://doi.org/10.1016/j.marpol.2018.11.037

Global Shark Attack File home page. (n.d.). *Shark attack list*. Retrieved August 23, 2017, from http://www.sharkattackfile.net/

Glover, D. (2014). *Shark of Darkness: Wrath of Submarine*. http://www.imdb.com/title/tt4177960/

Gobo, G. (2008). *Doing Ethnography*. https://doi.org/10.4135/9780857028976

Goffman, E. (1959). *The Presentation of Self in Everyday Life*. New York, NY: Anchor.

Goldman, M. J., Nadasdy, P., & Turner, M. D. (2011). Knowing nature: Conversations at the intersection of political ecology and science studies. In M. Ingram (Ed.), *Fermentation, rot, and other human-microbial performances* (pp. 99–112). Chicago: Univ. of Chicago Press.

Gore, M. A., Frey, P. H., Ormond, R. F., Allan, H., & Gilkes, G. (2016). Use of Photo-Identification and Mark-Recapture Methodology to Assess Basking Shark (Cetorhinus maximus) Populations. *PLoS One; San Francisco, 11*(3). http://dx.doi.org.ezproxy.canterbury.ac.nz/10.1371/journal.pone.0150160

Gore, M. L., Muter, B. A., Lapinski, M. K., Neuberger, L., & Heide, B. van der. (2011). Risk frames on shark diving websites: Implications for global shark conservation. *Aquatic Conservation: Marine and Freshwater Ecosystems, 21*(2), 165–172. https://doi.org/10.1002/aqc.1171

Gouk, C., Pasricha, D., & Lingathas, S. (2015). *Shark attack: The emergency presentation and management.* BMJ Case Reports; London, 2015. http://dx.doi.org.ezproxy.canterbury.ac.nz/10.1136/bcr-2015-212380

Government of Canada (2016). *White Shark (Atlantic Population)*. Fisheries and Oceans Canada. https://www.dfo-mpo.gc.ca/species-especes/profiles-profils/whiteshark-requinblanc-eng.html

Gracie, A. (2016). Art, Space and Hyperreality: An Artistic Exploration of Artificiality, Meaning and Boundaries within Astrobiological Practice. *Leonardo, 49*(1), 6–13. https://doi.org/10.1162/LEON_a_00925

Graham, R. T., & Roberts, C. M. (2007). Assessing the size, growth rate and structure of a seasonal population of whale sharks (Rhincodon typus Smith 1828) using conventional tagging and photo identification. *Fisheries Research, 84*(1), 71–80. https://doi.org/10.1016/j.fishres.2006.11.026

Graizbord, D., Rodríguez-Muñiz, M., & Baiocchi, G. (2017). Expert for a day: Theory and the tailored craft of ethnography. *Ethnography, 18*(3), 322–344. https://doi.org/10.1177/1466138116680007

Granqvist, P., Fredrikson, M., Unge, P., Hagenfeldt, A., Valind, S., Larhammar, D., & Larsson, M. (2005). Sensed presence and mystical experiences are predicted by suggestibility, not by the application of transcranial weak complex magnetic fields. *Neuroscience Letters, 379*(1), 1–6. https://doi.org/10.1016/j.neulet.2004.10.057

Graves-Brown, P. (2000). *Matter, materiality and modern culture.* London; New York: Routledge.

Gray, G. M. E., & Gray, C. A. (2017). Beach-User Attitudes to Shark Bite Mitigation Strategies on Coastal Beaches; Sydney, Australia. *Human Dimensions of Wildlife, 22*(3), 282. https://doi.org/10.1080/10871209.2017.1295491

Greenhough, B. (2012). Where species meet and mingle: Endemic human virus relations, embodied communication and more than human agency at the Common Cold Unit 1946–90. *Cultural Geographies 19*(3): 281-301.

Greenspun, P. (2007). *History of photography timeline*. from http://photo.net/history/timeline.

Gregory, B., & Mead, M. (1952). *Trance and Dance in Bali*. Retrieved from http://www.imdb.com/title/tt0221658/

Griffiths, A. (2002). *Wondrous Difference: Cinema, Anthropology and Turn-of-the-century Visual Culture.* New York, Columbia University Press.

Grimshaw, A. & A. Ravetz (2005). *Visualizing anthropology*. Portland, OR;Bristol, UK, Intellect.

Grimshaw, A. (2001) *The Ethnographer's Eye*, Cambridge: Cambridge University Press.

Griselda, P. (1976, April 4). *Jaws.* pp. 41–42.

Gruber, D. M. & Krause, J. S. H. (2018). Are some sharks more social than others? Short- and long-term consistencies in the social behavior of juvenile lemon sharks. *Behavioral Ecology and Sociobiology; Heidelberg,* 72(1), 1–10. http://dx.doi.org.ezproxy.canterbury.ac.nz/10.1007/s00265-017-2431-0

Gruen, L., & Weil, K. (2012). Animal Others—Editors' Introduction. *Hypatia, 27*(3), 477–487. Retrieved from JSTOR.

Gubili, C., Johnson, R., Gennari, E., Oosthuizen, W. H., Kotze, D., Meÿer, M., ... Noble, L. R. (2009). Concordance of genetic and fin photo identification in the great white shark, off Mossel Bay, South Africa. *Marine Biology, 156*(10), 2199–2207. https://doi.org/10.1007/s00227-009-1233-y

Guindi, F. E. (2004). *Visual Anthropology: Essential Method and Theory*. Oxford UK, Altamira press.

Guttridge, T. L., Gruber, S. H., Krause, J., & Sims, D. W. (2010). Novel Acoustic Technology for Studying Free-Ranging Shark Social Behaviour by Recording Individuals' Interactions. PLoS One; *San Francisco, 5*(2), e9324. http://dx.doi.org.ezproxy.canterbury.ac.nz/10.1371/journal.pone.0009324

Guttridge, T. L., Myrberg, A. A., Porcher, I. F., Sims, D. W., & Krause, J. (2009). The role of learning in shark behaviour. *Fish and Fisheries, 10*(4), 450–469. https://doi.org/10.1111/j.1467-2979.2009.00339.x

Guttridge, T. L., van Dijk, S., Stamhuis, E. J., Krause, J., Gruber, S. H., & Brown, C. (2013). Social learning in juvenile lemon sharks, Negaprion brevirostris. *Animal Cognition, 16*(1), 55–64. https://doi.org/10.1007/s10071-012-0550-6

Haden, A. (2017, June 6). *Killer whales have been killing great white sharks in Cape waters*. Retrieved June 23, 2017, from The South African website: https://www.thesouthafrican.com/killer-whales-have-been-killing-great-white-sharks-in-cape-waters/

Hailemichael, A. (1995). A thematic analysis of the afar camel folk literature: an ethnography-of-communication approach. *Journal of Ethiopian Studies, 28*(1), 1–22.

Hammerschlag, N., Gallagher, A. J., Wester, J., Luo, J., & Ault, J. S. (2012). Don't bite the hand that feeds: Assessing ecological impacts of provisioning ecotourism on an apex marine predator: Ecological impacts of shark ecotourism. *Functional Ecology, 26*(3), 567–576. https://doi.org/10.1111/j.1365-2435.2012.01973.x

Hammerschlag, N., Gutowsky, L. F. G., Gallagher, A. J., Matich, P., & Cooke, S. J. (2017). Diel habitat use patterns of a marine apex predator (tiger shark, Galeocerdo cuvier) at a high use area exposed to dive tourism. *Journal of Experimental Marine Biology and Ecology, 495*, 24–34. https://doi.org/10.1016/j.jembe.2017.05.010

Hammerton, Z., & Ford, A. (2018). Decolonising the Waters: Interspecies Encounters Between Sharks and Humans. *Animal Studies Journal, 7*(1), 270–303.

Haraway, D. (1989). *Primate visions: Gender, race, and nature in the world of modern science*. London: Routledge.

Haraway, D. (2003). *The companion species manifesto: Dogs, people and significant otherness*. Chicago: Prickly Paradigm.

Haraway, D. (2008). *When species meet*. Minneapolis: Univ. of Minnesota Press.

Haraway, D. J. (2008). Training in the Contact Zone: Power, Play, and Invention in the Sport of Agility. In *Tactical Biopolitics*. The MIT Press. https://mitpress.universitypressscholarship.com/view/10.7551/mitpress/9780262042499.001.0001/upso-9780262042499-chapter-26

Hardt, M. & A. Negri (2004). *Multitude: War and Democracy in the Age of Empire*. New York, Penguin.

Hardy, K. C. (2019). Provincialising the Cow: Buffalo-Human Relationships. South Asia: *Journal of South Asian Studies, 42*(6), 1156–1172. https://doi.org/10.1080/00856401.2019.1680484

Hart, N. S., Theiss, S. M., Harahush, B. K., & Collin, S. P. (2011). Microspectrophotometric evidence for cone monochromacy in sharks. *Naturwissenschaften, 98*(3), 193–201. https://doi.org/10.1007/s00114-010-0758-8

Haselton, M. G., & Miller, G. F. (2006). Women's fertility across the cycle increases the short-term attractiveness of creative intelligence. *Human Nature (Hawthorne, N.Y.), 17*(1), 50–73. https://doi.org/10.1007/s12110-006-1020-0

Hastrup, K. (1992). Anhropological vision: some notes on visual and textual authority, in Crawford, P. I. and D. Turton (eds), *Film as Ethnography*. Manchester: Manchester University press.

Hathaway, M. J. (2018). Elusive fungus? Forms of Attraction in Multispecies World Making. *Social Analysis, 62*(4), 37. https://doi.org/10.3167/sa.2018.620403

Hays, S. (2014). *Stewart Island fishermen at odds with shark tourism operators*. http://www.3news.co.nz/Stewart-Island-fishermen-at-odds-with-shark-tourism-operators/tabid/1771/articleID/340377/Default.aspx?fbclid=IwAR0Rhf7kMH14aJqBff66IZUr9LvqmB23axhtuJYGP1hMDe7dibDNYFRdhOs

Hayward, E. (2010). Fingeryeyes: Impressions of cup corals. *Cultural Anthropology, 25*(4), 577–599.

Hazin, F. H. V., & Afonso, A. S. (2014). Response: A conservation approach to prevention of shark attacks off Recife, Brazil. *Animal Conservation, 17*(4), 301–302. https://doi.org/10.1111/acv.12160

Hazin, Fabio H. V., Burgess, G. H., & Carvalho, F. C. (2008). A Shark Attack Outbreak Off Recife, Pernambuco, Brazil: 19922006. *Bulletin of Marine Science, 82*(2), 199–212.

Heberlein, T. A. (2012). *Navigating Environmental Attitudes* (1 edition). New York: Oxford University Press.

Heider, K. G. (2006). *Ethnographic Film*. Austin, University of Texas Press,

Heist, B. (2012). *From Thematic Analysis to Grounded Theory*. Slides.

Helmreich, S. (2003). Trees and seas of information: Alien kinship and the biopolitics of gene transfer in marine biology and biotechnology. *American Ethnologist, 30*(3), 340–358.

Helmreich, S. (2009). *Alien ocean: Anthropological voyages in microbial seas*. Berkeley: Univ. of California Press.

Henare, A. J. M., Holbraad, M., & Wastell, S. (Eds.). (2007). *Thinking through things: Theorising artefacts ethnographically.* Milton Park, Abingdon, Oxon ; New York: Routledge.

Henshaw, S. (n.d.). *Bluff, South Island*. from http://www.newzealand.com/us/article/bluff-south-island

Herald, T. N. Z. (2015). *Calls for Stewart Island shark diving ban*.

Hetzel, P. (1998). When Hyperreality, Reality, Fiction and Non-Reality Are Brought Together: A Fragmented Vision of the Mall of America Through Personal Interpretation. *ACR European Advances, E-03*. Retrieved from http://acrwebsite.org/volumes/11709/volumes/e03/E-03

Hewitt, A. M., Kock, A. A., Booth, A. J., & Griffiths, C. L. (2018). Trends in sightings and population structure of White Sharks, Carcharodon carcharias, at Seal Island, False Bay, South Africa, and the emigration of subadult female sharks approaching maturity. *Environmental Biology of Fishes, 101*(1), 39–54. https://doi.org/10.1007/s10641-017-0679-x

Hicks, D., & Beaudry, M. C. (Eds.). (2010). *The Oxford handbook of material culture studies*. Oxford ; New York: Oxford University Press.

Hillary, R., Bradford, R., & Patterson, T. (2018). *World-first genetic analysis reveals Aussie white shark numbers*. (2018, February 8). CSIROscope. https://blog.csiro.au/world-first-genetic-analysis-reveals-aussie-white-shark-numbers/

Hillary, R., Bravington, M. V., Patterson, T. A., Grewe, P., Bradford, R. W., Feutry, P., ... Bruce, B. D. (n.d.). *Genetic relatedness reveals total population size of White Sharks in eastern Australia and New Zealand*. Retrieved January 28, 2019, from ResearchGate website: https://www.researchgate.net/publication/323016897_Genetic_relatedness_reveals_total_population_size_of_white_sharks_in_eastern_Australia_and_New_Zealand

Hillman, G. R., Würsig, B., Gailey, G. A., Kehtarnavaz, N., Drobyshevsky, A., Araabi, B. N., ... Weller, D. W. (2003). Computer-assisted photo-identification of individual marine vertebrates: A multi-species system. *Aquatic Mammals, 29*(1), 117–123. https://doi.org/10.1578/016754203101023960

Hinchliffe, S., Kearnes, M. B., Degen, M., & Whatmore, S. (2005). Urban Wild Things: A Cosmopolitical Experiment. Environment and Planning. *Society and Space, 23*(5), 643–658. https://doi.org/10.1068/d351t

Hird, M. J. (2009). *The origins of sociable life: Evolution after science studies*. New York: Palgrave Macmillan.

Hird, M. J. (2010). Meeting with the microcosmos. *Environment and Planning D: Society and Space, 28*(1), 36–39.

HistoryvsHollywood.Com. (n.d.). *Is Jaws a True Story? Learn the Real Inspiration for Jaws*. HistoryvsHollywood.Com. Retrieved August 28, 2020, from https://www.historyvshollywood.com/reelfaces/jaws/

Hockings, K. (2009). Living at the interface: Human-chimpanzee competition, coexistence and conflict in Africa. *Interaction Studies, 10*(2), 183–205.

Hodgetts, T., & Lorimer, J. (2015). Methodologies for animals' geographies: Cultures, communication and genomics. *Cultural Geographies, 22*(2), 285–295. https://doi.org/10.1177/1474474014525114

Hogan, K., & R. Stubbs (2003). *Can't get through 8 barriers to communication.* Grenta, LA, pelican publishing company.

Hohti, R., & Tammi, T. (2019). The greenhouse effect: Multispecies childhood and non-innocent relations of care. *Childhood, 26*(2), 169–185. https://doi.org/10.1177/0907568219826263

Holmberg, J., Norman, B., & Arzoumanian, Z. (2009). Estimating population size, structure, and residency time for whale sharks Rhincodon typus through collaborative photo-identification. *Endangered Species Research, 7*(1), 39–53. https://doi.org/10.3354/esr00186

Hotel Magazine. (n.d.). *Court bans shark cage diving*. Retrieved October 2, 2018, from http://hotelmagazine.co.nz/2018/09/05/court-bans-shark-cage-diving/

Houston Press (n.d.) *Shark Week Has Officially [Puts on Sunglasses] Jumped the Shark*. Retrieved June 19, 2017, from http://www.houstonpress.com/arts/shark-week-has-officially-puts-on-sunglasses-jumped-the-shark-6366123

Howard, B. (2015, July 20). *Surfer Attacked by Shark 'Did Everything Right.'* Retrieved May 8, 2019, from National Geographic News website: https://news.nationalgeographic.com/2015/07/150720-shark-attack-surfer-mick-fanning-south-africa/

Hoxsey, D. (2008). Mixed Communities Require Mixed Theories: Using Mills to Broaden Goffman's Exploration of Identity within the GBLT Communities. *Sociological Research Online, 13*(1), 215–231. https://doi.org/10.5153/sro.1683

Hromatka, B. S., Tung, J. Y., Kiefer, A. K., Do, C. B., Hinds, D. A., & Eriksson, N. (2015). Genetic variants associated with motion sickness point to roles for inner ear development, neurological processes and glucose homeostasis. *Human Molecular Genetics, 24*(9), 2700–2708. https://doi.org/10.1093/hmg/ddv028

Huffadine. (n.d.). *Shark cage diving an offence under the Wildlife Act. Here's why*. Retrieved September 5, 2018, from : https://www.stuff.co.nz/national/106827512/shark-cage-diving-an-offence-under-the-wildlife-act-heres-why

Hughes, H. (2011). Humans, sharks and the shared environment in the contemporary eco-doc. *Environmental Education Research 17*(6): 735-749.

Hunt, C. (1981). *Shark tooth and stone blade, Pacific island art from the University of Aberdeen*. Aberdeen, University of Aberdeen.

Hurn, S. (2012). *Humans and other animals: Cross-cultural perspectives on human-animal interactions*. London: Pluto.

Hustak, C., & Myers, N. (2012). Involutionary momentum: Affective ecologies and the sciences of plant/insect encounters. *Differences: A Journal of Feminist Cultural Studies, 23*(3), 74–118.

Hutching, G. (2012). Māori and sharks, *Te Ara - The Encyclopedia of New Zealand*. Retrieved 22.7.15, from http://www.teara.govt.nz/en/sharks-and-rays/page-2.

Huveneers, C., Apps, K., Becerril-García, E. E., Bruce, B., Butcher, P. A., Carlisle, A. B., ... Werry, J. M. (2018). Future Research Directions on the "Elusive" White Shark. *Frontiers in Marine Science, 5*. https://doi.org/10.3389/fmars.2018.00455

Huveneers, C., Meekan, M. G., Apps, K., Ferreira, L. C., Pannell, D., & Vianna, G. M. S. (2017). The economic value of shark-diving tourism in Australia. *Reviews in Fish Biology and Fisheries, 27*(3), 665–680. https://doi.org/10.1007/s11160-017-9486-x

Huveneers, C., Rogers, P. J., Beckmann, C., Semmens, J. M., Bruce, B. D., & Seuront, L. (2013). The effects of cage-diving activities on the fine-scale swimming behaviour and space use of White Sharks. *Marine Biology, 160*(11), 2863–2875. https://doi.org/10.1007/s00227-013-2277-6.

Huveneers, C., Watanabe, Y. Y., Payne, N. L., & Semmens, J. M. (2018). Interacting with wildlife tourism increases activity of white sharks. *Conservation Physiology, 6*(1), coy019. https://doi.org/10.1093/conphys/coy019

Ihama, Y., Ninomiya, K., Noguchi, M., & Fuke, C. (2009). Characteristic features of injuries due to shark attacks: A review of 12 cases. *Legal Medicine, 11*(5), 219–225. https://doi.org/10.1016/j.legalmed.2009.06.002

Industries, M. f. P. (2013). *This National Plan of Action for the Conservation and Management of Sharks*.

Industry, M. o. p. (2014). *Conservation and management of New Zealand Sharks*.

Ingold, T. (1994). *What is an animal?* London: Routledge.

Ingold, T. (2013). Anthropology Beyond Humanity. *Suomen Antropologi /Journal of the Finnish Anthropology Society, 38*(3), 15–23.

Ingold, T., & Pálsson, G. (2013). *Biosocial becomings: Integrating social and biological anthropology*. Cambridge, UK: Cambridge Univ. Press.

Ingram, M. (2011). Fermentation, rot, and other humanmicrobial performances. In *Knowing nature: Conversations at the intersection of political ecology and science studies*. Chicago, Univ. of Chicago Press: 99–112.

Insight Traction. (n.d.). *Environmental Cues*. Retrieved from http://insight-traction.com/environmental-cues/

IOL News. (n.d.). *Appeal court rules against 'shark cage widow'*. Retrieved October 2, 2018, from https://www.iol.co.za/news/appeal-court-rules-against-shark-cage-widow-2004200

Irvine, L. (2004). A Model of Animal Selfhood: Expanding Interactionist Possibilities. *Symbolic Interaction, 27*(1), 3–21. https://doi.org/10.1525/si.2004.27.1.3

Isaacs, J. R. (2019). The "bander's grip": Reading zones of human–shorebird contact. *Environment and Planning E: Nature and Space, 2*(4), 732–760. https://doi.org/10.1177/2514848619866331
Isaacs, J. R., & Gillespie, K. (2017). *CFP AAG 2017: The Contact Zone II & II: Where Species Meet – Political Geography Specialty Group.* https://www.politicalgeography.org/2016/10/18/cfp-aag-2017-the-contact-zone-ii-ii-where-species-meet/
Ivanov, O. A., Ivanova, V. V., & Saltan, A. A. (2018). Likert-scale questionnaires as an educational tool in teaching discrete mathematics. *International Journal of Mathematical Education in Science and Technology, 49*(7), 1110–1118. https://doi.org/10.1080/0020739X.2017.1423121
Jacknis, I. (1984). Franz Boas and Photography. *Studies in Visual Communication 10*(1).
Jackson, P. (2012). Situated Activities in a Dog Park: Identity and Conflict in Human-Animal Space. *Society & Animals, 20*(3), 254–272. https://doi.org/10.1163/15685306-12341237
Jaclin, D. (2013). In the (bleary) eye of the tiger: An anthropological journey into jungle backyards. *Social Science Information, 52*(2), 257–271.
Jacoby, D. M., P, Fear, L. N., Sims, D. W., & Croft, D. P. (2014). Shark personalities? Repeatability of social network traits in a widely distributed predatory fish. *Behavioral Ecology and Sociobiology; Heidelberg, 68*(12), 1995–2003. http://dx.doi.org.ezproxy.canterbury.ac.nz/10.1007/s00265-014-1805-9
Jadhav, S., & Barua, M. (2012). The elephant vanishes: Impact of human-elephant conflict on people's wellbeing. *Health & Place, 18*(6), 1356–1365.
Jambura, P. L., Kindlimann, R., López-Romero, F., Marramà, G., Pfaff, C., Stumpf, S., ... Kriwet, J. (2019). Micro-computed tomography imaging reveals the development of a unique tooth mineralization pattern in mackerel sharks (Chondrichthyes; Lamniformes) in deep time. *Scientific Reports, 9*(1), 9652. https://doi.org/10.1038/s41598-019-46081-3
James, P. (2015). Shark attacks in Sydney estuary. *Journal of the Royal Australian Historical Society, 101*(1), 45.
Johnson, R., & Kock, A. (2006). South Africa's White Shark cage-diving industry Is their cause for concern? *WWF South Africa Report Series, 20.*
Johnston, C. (2008). Beyond the clearing: Towards a dwelt animal geography. *Progress in Human Geography, 32*(5), 633–649. https://doi.org/10.1177/0309132508089825
Jøn. A, & Aich, R. S. (2015). Southern Shark Lore Forty Years after Jaws: The Positioning of Sharks Within Murihiku, New Zealand. *Australian Folklore, 30*, 169–192.
Jonason, P. K., Valentine, K. A., Li, N. P., & Harbeson, C. L. (2011). Mate-selection and the Dark Triad: Facilitating a short-term mating strategy and creating a volatile environment. *Personality and Individual Differences, 51*(6), 759–763. https://doi.org/10.1016/j.paid.2011.06.025
Jones, A. M., & Boivin, N. (2010). The Malice of Inanimate Objects. *The Oxford Handbook of Material Culture Studies*. https://doi.org/10.1093/oxfordhb/9780199218714.013.0014
Jones, B. (2019). Bloom/Split/Dissolve: Jellyfish, H. D., and Multispecies Justice in Anthropocene Seas. *Configurations, 27*(4), 483–499. https://doi.org/10.1353/con.2019.0032
Jorgensen, D. (1989). *Participant Observation*. https://doi.org/10.4135/9781412985376
Jørgensen, I. K. H., & Wirman, H. (2016). Multispecies methods, technologies for play. *Digital Creativity, 27*(1), 37–51. https://doi.org/10.1080/14626268.2016.1144617
Jorgensen, S. J., Anderson, S., Ferretti, F., Tietz, J. R., Chapple, T., Kanive, P., ... Block, B. A. (2019). Killer whales redistribute white shark foraging pressure on seals. *Scientific Reports,* 9(1), 6153. https://doi.org/10.1038/s41598-019-39356-2
Jorgensen, S. J., Arnoldi, N. S., Estess, E. E., Chapple, T. K., Rückert, M., Anderson, S. D., & Block, B. A. (2012). Eating or Meeting? Cluster Analysis Reveals Intricacies of White Shark (Carcharodon carcharias) Migration and Offshore Behavior. *PLoS One; San Francisco, 7*(10), e47819. http://dx.doi.org.ezproxy.canterbury.ac.nz/10.1371/journal.pone.0047819
Jost Robinson, C. A., & Remis, M. J. (2018). Engaging Holism: Exploring Multispecies Approaches in Ethnoprimatology. *International Journal of Primatology, 39*(5), 776–796. https://doi.org/10.1007/s10764-018-0036-8

Jowaheer, R., & Jowaheer, R. (n.d.-a). *Michael Phelps to race a great white shark*. Retrieved June 23, 2017, from http://www.aol.co.uk/travel/2017/06/18/michael-phelps-race-great-white-shark/

JR., M. (2005). Biodiversity conservation and the extinction of experience. *Trends Ecol Evol 20*(8): 430-434.

Just Gotta Dive (n.d.). *How To Maintain Your Body Temperature*. Retrieved September 18, 2017, http://www.justgottadive.com/dive_resources/technical_articles/dive-essentials-how-maintain-your-body-temperature

Kahler, B. (n.d.). *Jaws*. Retrieved from https://www.academia.edu/32842114/Jaws

Kalof, L. & Montgomery, G. M. (2011). *Making animal meaning*. East Lansing, Michigan State University Press.

Kansky, R., & Knight, A. T. (2014). Key factors driving attitudes towards large mammals in conflict with humans. *Biological Conservation, 179*, 93–105. https://doi.org/10.1016/j.biocon.2014.09.008

Kasprak, A. (2017). *Fact Check: Do Falling Coconuts Kill More People Than Sharks Each Year?* https://www.snopes.com/fact-check/coconuts-kill-more-sharks/

Kawulich, B. B. (2005). Participant Observation as a Data Collection Method. *Forum: Qualitative Social Research, 6*(2), 1–22.

Kayange, D. S. (2015). *Evaluating the specific learning mechanisms and a longitudinal transformation output by using the performance indicators*. 13.

Keane, K. (2007). 5. – *Taniwha – Te Ara Encyclopedia of New Zealand*. Retrieved January 7, 2019, from https://teara.govt.nz/en/5833

Keck, F. (2013). *Somatosphere*. Retrieved from http://somatosphere.net/2013/09/eduardo-kohns-how-forests-think.html

Keeton, C. (2018). The return of the great white sharks. *Times LIVE*. https://www.timeslive.co.za/news/sci-tech/2018-04-16-the-return-of-the-great-white-sharks/

Kellert, S. R., et al. (1996). Human culture and large carnivore, conservation in North America. *Conserv Biol. 10*(4): 977-990.

Kellner, D. (Ed.). (1994). *Baudrillard: A critical reader*. Oxford; Cambridge, Mass: Blackwell.

Kelly Tarlton's Sea Life Aquarium. (n.d.). *Snorkel with sharks in Auckland | Shark Dive |*. Retrieved July 15, 2020, from https://www.kellytarltons.co.nz/tickets/shark-cage-adventure/

Kelly, A. H., & Lezaun, J. (2014). Urban mosquitoes, situational publics, and the pursuit of interspecies separation in Dar es Salaam. *American Ethnologist, 41*(2), 368–383. https://doi.org/10.1111/amet.12081

Kelly, C. R., & Hoerl, K. E. (2017). *Genesis in Hyperreality: Legitimizing Disingenuous Controversy at the Creation Museum*. Argumentation and Advocacy. Retrieved from http://www.tandfonline.com/doi/abs/10.1080/00028533.2012.11821759

Kemper, C. (2017, March 16). *How Cage Diving with Great White Sharks Began P1*. Retrieved July 17, 2018, from Tracking Sharks website: https://www.trackingsharks.com/cage-diving-great-white-sharks-began-part-1/

Kenyon, G. (n.d.). *Australia's ancient language shaped by sharks*. Retrieved April 30, 2018, from http://www.bbc.com/travel/story/20180429-australias-ancient-language-shaped-by-sharks

Kermode, M. (2015). *Jaws, 40 years on: 'One of the truly great and lasting classics of American cinema' ,Film*, The Guardian. https://www.theguardian.com/film/2015/may/31/jaws-40-years-on-truly-great-lasting-classics-of-america-cinema

Kerr, J. H., & Houge Mackenzie, S. (2012). Multiple motives for participating in adventure sports. *Psychology of Sport and Exercise, 13*(5), 649–657. https://doi.org/10.1016/j.psychsport.2012.04.002

Keul, A. (2013). Embodied encounters between humans and gators. *Social & Cultural Geography, 14(*8), 930–953. https://doi.org/10.1080/14649365.2013.837190

Khan, S. (2017). Beyond action. *Ethnography, 18*(1), 88–96. https://doi.org/10.1177/1466138115592420

Kharel, D. (2015). Visual Ethnography, Thick Description and Cultural Representation. Dhaulagiri *Journal of Sociology and Anthropology, 9*(0), 147–160. https://doi.org/10.3126/dsaj.v9i0.14026

Kiik, L. (2018). Wild-ing the Ethnography of Conservation: Writing Nature's Value and Agency In. *Anthropological Forum, 28*(3), 217–235. https://doi.org/10.1080/00664677.2018.1476222

Kikuchi, R. (2012). Captive Bears in Human-Animal Welfare Conflict: A Case Study of Bile Extraction on Asia's Bear Farms. Journal of Agricultural and Environmental Ethics; *Dordrecht, 25*(1), 55–77. http://dx.doi.org.ezproxy.canterbury.ac.nz/10.1007/s10806-010-9290-2

Kinder, J. (2012). *The Spectres of Simulacra: Hyperreality, Consumption as Ideology, and the (Im)Possible Future of Radical Politics*. https://www.academia.edu/1595967/The_Spectres_of_Simulacra_Hyperreality_Consumption_as_Ideology_and_the_Im_Possible_Future_of_Radical_Politics

Kingery, W. D. (Ed.). (1996*). Learning from things: Method and theory of material culture studies.* Washington, D.C: Smithsonian Institution Press.

Kinzey, W. (1997). New world primates, ecology, evolution and behavior. In L. E. Sponsel (Ed.), *The human niche in Amazonia: Explorations in ethnoprimatology* (pp. 143–165). New York: Aldine de Gruyter.

Kirk, R. (n.d.). *Stop the killing of Great White Sharks in NZ. Petitions.Net*. Retrieved July 15, 2020, from https://www.petitions.net/stop_the_killing_of_great_white_sharks_in_nz

Kirksey, E. (2014). *The Multispecies Salon*. United states of America, Duke University Press Books.

Kirksey, S. E., & Helmreich, S. (2010). The Emergence of Multispecies Ethnography. *Cultural Anthropology, 25*(4), 545–576. https://doi.org/10.1111/j.1548-1360.2010.01069.x

Kirksey, S. E. (2012a). Living with parasites in Palo Verde National Park. *Environmental Humanities, 1*, 23–55.

Kirksey, S. E. (2012b). Thneeds reseeds: Figures of biocultural hope in the Anthropocene. *Rachel Carson Perspectives, 9*, 89–94.

Kirksey, S. E., Schuetze, C., & Shapiro, N. (2011). Poaching at the Multispecies Salon. *Kroeber Anthropological Society, 100*(1), 129–153.

Kirksey, S. E., Shapiro, N., & Brodine, M. (2013). Hope in blasted landscapes. *Social Science Information, 52*(2), 228–256.

Klein, S. (2015). *We Are All Stardust: Leading Scientists Talk About Their Work, Their Lives, and the Mysteries of Our Existence by Stefan Klein*. Retrieved from https://www.goodreads.com/book/show/16238031-we-are-all-stardust

Klimley, A. P. (2013). *The Biology of Sharks and Rays*. Chicago; London: University of Chicago Press.

Klugers, D. (2010). *Keeping It Green: New York Jaws: The Biology of the Great White Shark* (part 5).

Knight, J. (2005). Animals in person: Cross-cultural perspectives on human-animal intimacies. In John Knight (Ed.), *Feeding Mr. Monkey: Cross species food "exchange" in Japanese monkey parks* (pp. 231–253). Oxford: Berg.

Knight, S., & Barnett, L. (2008). Justifying Attitudes toward Animal Use: A Qualitative Study of People's Views and Beliefs. *Anthrozoös, 21*(1), 31–42. https://doi.org/10.2752/089279308X274047

Knudsen, S. (2014). Multiple sea snails: The uncertain becoming of an alien species. *Anthropological Quarterly, 87*(1), 59–92.

Kock, A. (n.d.). *Gaping Jaws of a Great White Shark. Smithsonian Ocean*. Retrieved August 20, 2020, from http://ocean.si.edu/ocean-life/sharks-rays/gaping-jaws-great-white-shark

Kohn, E. (2007). How Dogs Dream: Amazonian Natures and the Politics of Transspecies Engagement. *American Ethnologist, 34*(1), 3–24. https://doi.org/10.1525/ae.2007.34.1.3

Kohn, E. (2013). *How forests think: Toward an anthropology beyond the human*. Berkeley: Univ. of California Press.

Kopnina, H. (2016). Wild Animals and Justice: The Case of the Dead Elephant in the Room. *Journal of International Wildlife Law & Policy: Wild Animals and Justice, 19*(3), 219–235. https://doi.org/10.1080/13880292.2016.1204882

Kopnina, H. (2017). Beyond multispecies ethnography: Engaging with violence and animal rights in anthropology. *Critique of Anthropology, 37*(3), 333–357. https://doi.org/10.1177/0308275X17723973

Kosek, J. (2010). Ecologies of empire: On the new uses of the honeybee. *Cultural Anthropology, 25*(4), 650–678.

Kosuch, L. (1993). *Sharks and shark products in Prehistoric South Florida. Monographs of the Institute of Archaeology and Paleoenvironmental Studies*. Gainesville, University of Florida. 2: 1–52.

Kroeber, A. L. (1948). *Anthropology: race, language, culture, psychology, prehistory*. London Book, Harrap.

Kulick, D. (2017). Human–Animal Communication. *Annual Review of Anthropology, 46*(1), 357–378. https://doi.org/10.1146/annurev-anthro-102116-041723

Kuper, A. (1996). *Anthropology and Anthropologists*. London, Routledge.

Labaree, R. V. (n.d.). *Research Guides: Organizing Your Social Sciences Research Paper:* 6. The Methodology [Research Guide]. Retrieved May 4, 2018, from http://libguides.usc.edu/writingguide/methodology

Lama, R. L. de la, Puente, S. D. la, & Riveros, J. C. (2018). Attitudes and misconceptions towards sharks and shark meat consumption along the Peruvian coast. *PLOS ONE, 13*(8), e0202971. https://doi.org/10.1371/journal.pone.0202971

Lampen, C. (2019). *Great White Shark Deep Blue Possibly Spotted Near Hawaii.* https://www.thecut.com/2019/01/great-white-shark-deep-blue-possibly-spotted-near-hawaii.html

Lane, R. J. (2009*). Jean Baudrillard (2nd ed*). London; New York: Routledge.

Laroche, R. K., Kock, A. A., Dill, L. M., & Oosthuizen, W. H. (2007). Effects of provisioning ecotourism activity on the behaviour of White Sharks Carcharodon carcharias. *Marine Ecology Progress Series, 338,* 199–209. https://doi.org/10.3354/meps338199

Larson, C. L., Aronoff, J., Sarinopoulos, I. C., & Zhu, D. C. (2009). Recognizing Threat: A Simple Geometric Shape Activates Neural Circuitry for Threat Detection. *Journal of Cognitive Neuroscience, 21*(8), 1523–1535. https://doi.org/10.1162/jocn.2009.21111

Latimer, J., & Miele, M. (2013). Naturecultures? Science, affect and the non-human. *Theory, Culture & Society, 30*(7), 5–31.

Latour, B. (1996). On Interobjectivity. Mind, Culture and Activity. *An International Journal of undregraduate research 3*: 228-245.

Latour, B. (2004). *Politics of nature: How to bring the sciences into democracy.* Cambridge, MA: Harvard Univ. Press.

Latour, B. (2005). *Reassembling the Social - An Introduction to Actor. Network-Theory,* Oxford University Press.

Law, J. (1986). Power, action and belief: A new sociology of knowledge. In M. Callon (Ed.*), Some elements of a sociology of translation: Domestication of the scallops and the fishermen of St Brieuc Bay* (pp. 196–233). London: Routledge and Kegan Paul.

Law, J. (2004). *After Method: Mess in Social Science Research*. https://doi.org/10.4324/9780203481141

Law, J. (2009). Actor Network Theory and Material Semiotics. In the *New Blackwell Companion to Social Theory* (pp. 141–158). https://doi.org/10.1002/9781444304992.ch7

Law, J., & Hassard, J. (1999). *Actor Network Theory and After*. Wiley.

Leather, M., & Gibson, K. (2019). The consumption and hyperreality of nature: Greater affordances for outdoor learning. *Curriculum Perspectives, 39*(1), 79–83. https://doi.org/10.1007/s41297-019-00063-7

Lee, P. C. (2010). Sharing space: Can ethnoprimatology contribute to the survival of nonhuman primates in human-dominated globalized landscapes? *American Journal of Primatology, 72(10)*, 925–931.

Lemahieu, A., Blaison, A., Crochelet, E., Bertrand, G., Pennober, G., & Soria, M. (2017). Human-shark interactions: The case study of Reunion island in the south-west Indian Ocean. *Ocean & Coastal Management, 136*, 73–82. https://doi.org/10.1016/j.ocecoaman.2016.11.020

Lemke, J. L. (2001). Articulating communities: Sociocultural perspectives on science education. *Journal of Research in Science Teaching, 38*(3), 296–316. https://doi.org/10.1002/1098-2736(200103)38:3<296::AID-TEA1007>3.0.CO;2-R

Lentz, A. K., Burgess, G. H., Perrin, K., & Brown, J. A. (2010). Mortality and Management of 96 Shark Attacks and Development of a Shark Bite Severity Scoring System. *The American Surgeon, 76*(1), 101.

Leopold, L. B. (n.d.). *The Conservation Attitude*. 7.

Lestel, D. (2013). The withering of shared life through the loss of biodiversity. *Social Science Information, 52*(2), 307–325.

Lestel, D., & Taylor, H. (2013). Shared life: An introduction. *Social Science Information, 52*(2), 183–186.

Lestel, D., Brunois, F., & Gaunet, F. (2006). Etho-ethnology and ethno-ethology. *Social Science Information, 45*(2), 155–177.

Lev, M., & Barkai, R. (2016). Elephants are people, people are elephants: Human–proboscideans similarities as a case for cross cultural animal humanization in recent and Paleolithic times. *Quaternary International, 406*, 239–245. https://doi.org/10.1016/j.quaint.2015.07.005

Levin, C. M., & Cruz, A. R. (2008). Behind the Scenes of a Visual Ethnography: A Dialogue between Anthropology and Film. *Journal of Film and Video, 60*(2), 59–68. JSTOR.

Levine, M., Collier, R. S., Ritter, E., & Fouda, M. (2014). Shark Cognition and a Human Mediated Driver of a Spate of Shark Attacks. *Open Journal of Animal Sciences, 4*(5), 263–269. https://doi.org/10.4236/ojas.2014.45033

Lewis, K. (2011). Shark attack survivor. *Scholastic Scope, 59*(15), 4.

Lichterman, P. (2017). Interpretive reflexivity in ethnography. *Ethnography, 18*(1), 35–45. https://doi.org/10.1177/1466138115592418

Lien, M. E., & Law, J. (2011). "Emergent Aliens": On salmon, nature, and their enactment. *Ethnos, 76*(1), 65–87.

Lima, T. A. (1999–2000). Em busca dos frutos do mar: os pescadores-coletores do litoral centro-sul do Brasil. *Revista USP 44*, 270–327.

Linch, A., & Holland, B. (2017). Cultural Killing and Human–Animal Capability Conflict. *Journal of Human Development and Capabilities, 18*(3), 322–336. https://doi.org/10.1080/19452829.2017.1342383

Lindgren, N., & Öhman, J. (2019). A posthuman approach to human-animal relationships: Advocating critical pluralism. *Environmental Education Research, 25*(8), 1200–1215. https://doi.org/10.1080/13504622.2018.1450848

Linton, K. (n.d.). *On Eating and Killing Multispecies Entanglements and Implications for Ecology* . Retrieved April 12, 2019, from https://www.academia.edu/38756063/On_Eating_and_Killing_Multispecies_Entanglements_and_Implications_for_Ecology?email_work_card=title

Liu, L. (2009). *Performing place and ethnic identity within ethnic tourism among the Miao of Dehang village Western Hunan China* [Ph.D., The University of Utah]. http://search.proquest.com/docview/305013073/abstract/ECAF6775D1454E74PQ/1

Livingston, J., & Puar, J. K. (2011). Interspecies. *Social Text, 29*(1), 3–14.

Lloro-Bidart, T. (2018). A Feminist Posthumanist Multispecies Ethnography for Educational Studies. *Educational Studies, 54*(3), 253–270. https://doi.org/10.1080/00131946.2017.1413370

Lloro-Bidart, T. (2018). Cultivating affects: A feminist posthumanist analysis of invertebrate and human performativity in an urban community garden. *Emotion, Space and Society, 27*(Journal Article), 23–30. https://doi.org/10.1016/j.emospa.2018.02.006

Lo´ pez Mazz, J. M. (1994–1995). "Cabo Polonio: sitio arqueolo´ gico del litoral atla´ntico uruguayo." *Anais da VII Reunia˜o Cientı´fica da SAB, Revista de Arqueologı´a 8*(2): 239-265.

Locke, P. (2013). Explorations in ethnoelephantology: Social, historical, and ecological intersections between Asian elephants and humans. *Environment and Society: Advances in Research 4*(1): 79-97.

Locke, P. (2017). Elephants as persons, affective apprenticeship, and fieldwork with nonhuman informants in Nepal. *HAU: Journal of Ethnographic Theory. 7*, (1), 353-376.

Locke, P. (n.d.). *Humans, Elephants, and Interspecies Intimacy in The Chitwan National Park, Nepal*. 20.

Locke, P. & P. Keil (2015). Multispecies Methodologies and Human-Elephant Relations, *Anthropology and Environment Society*.

Locke, P. & U. Muenster (2015). *Multispecies Ethnography, oxford Bibliography online*. Retrieved 7/12/2015. https://doi.org/10.1093/OBO/9780199766567-0130

Lockwood, A. (2018). Bodily Encounter, Bearing Witness and the Engaged Activism of the Global Save Movement. *Animal Studies Journal,* 7(1), 104–126.

Löe, J., & Röskaft, E. (2004). Large Carnivores and Human Safety: A Review. *Ambio, 33*(6), 283–288. Retrieved from JSTOR.

Loizos, P. (1993). *Innovation in Ethnographic Film: From Innocence to Self-consciousness, 1955-85.* Manchester University Press.

Long, C., & Laughren, P. (1993). Australia's first films: Facts and fables [Series of parts]: Part 6: Surprising survivals from colonial Queensland. *Cinema Papers, (96)*, 32-37,59-61.

Lorimer, H. (2006). Herding Memories of Humans and Animals. Environment and Planning. *Society and Space, 24*(4), 497–518. https://doi.org/10.1068/d381t

Lorimer, J. (2007). Nonhuman charisma. *Environment and Planning D: Society and Space 25*(5): 911-932. https://doi.org/10.1068/d71j

Lorimer, J. (2010a). Elephants as Companion Species: the Lively Biogeographies of Asian Elephant Conservation in Sri Lanka. *Transactions of the Institute of the British Geographers 35*(4): 491-506.

Lorimer, J. (2010b). Moving image methodologies for more-than-human geographies. *Cultural Geographies, 17(*2), 237–258.

Lorimer, J. (2012a). Aesthetics for post-human worlds: difference, expertise and ethics. *Dialogues in Human Geography 2*(3): 284-287.

Lorimer, J. (2012b). Multinatural geographies for the Anthropocene. *Progress in Human Geography, 36(*5), 593–612.

Lorimer, J. (2017). Parasites, ghosts and mutualists: A relational geography of microbes for global health. *Transactions of the Institute of British Geographers, 42*(4), 544–558. https://doi.org/10.1111/tran.12189

Lorimer, J. (N.D.). *Charisma*. The Multispecies salon, Retrieved 12/6/2015.

Lowe, C. (2004). Making the monkey: How the Togean Macaque went from "new form" to "endemic species" in Indonesians' conservation biology. *Cultural Anthropology, 19*(4), 491–516.

Lowe, C. (2010). Viral clouds: Becoming H5N1 in Indonesia. *Cultural Anthropology, 25*(4), 625–649.

Lowe, R. L., Stevenson, R. J., & Bothwell, M. L. (1996). *Algal ecology: Freshwater benthic ecosystems*. Retrieved from https://trove.nla.gov.au/version/46512170

Lück, M. (2003). Education on marine mammal tours as agent for conservation—but do tourists want to be educated? *Ocean & Coastal Management, 46*(9), 943–956. https://doi.org/10.1016/S0964-5691(03)00071-1

Lulka, D. (2008). Embodying Anthropomorphism: Contextualizing Commonality in the Material Landscape. *Anthrozoös, 21*(2), 181–196. https://doi.org/10.2752/175303708X305828

Lulka, D. (2009). The residual humanism of hybridity: Retaining a sense of the earth. *Transactions of the Institute of British Geographers, 34*(3), 378–393.

Luo, Y., Liu, Y., & Zhang, D. Y. (2015). Prediction the variation of shark scale's attack angles in swimming. *Indian Journal of Animal Research, 49*(3), 295–302. https://doi.org/10.5958/0976-0555.2015.00088.6

Lyes, L. (2013, August 8). *Big Fish, Little Person*. Retrieved April 5, 2019, from History Geek website: https://historygeek.co.nz/2013/08/08/big-fish-little-person/

Macara, G. R. (2013). *The climate and weather of southland, 2nd edition. NIWA.*

MacDougall, D. (1998). *Transcultural Cinema*, Princeton University Press.

MacDougall, D. (2006). *The corporeal image: film, ethnography, and the senses.* Princeton, N.J, Princeton University Press.

MacDougall, J. & D. MacDougall (1992). *Photo Wallahs*, Berkeley Media LLC: 60 Min.

Macintosh, N. B., Shearer, T., Thornton, D. B., & Welker, M. (2000). Accounting as simulacrum and hyperreality: Perspectives on income and capital. *Accounting, Organizations and Society, 25*(1), 13–50. https://doi.org/10.1016/S0361-3682(99)00010-0

Mack, B. (2016). No luck for DOC in search for injured shark near Stewart Island | Stuff.co.nz. Retrieved September 5, 2018, from https://www.stuff.co.nz/environment/76575711/no-luck-for-doc-in-search-for-injured-shark-near-stewart-island?rm=m

MacLean, D. (2017, August 27). Shark bites canoe near Tutukaka. NZ Herald. Retrieved from http://www.nzherald.co.nz/nz/news/article.cfm?c_id=1&objectid=11912072

MacLeod, A., Cameron, P., Ajjawi, R., Kits, O., & Tummons, J. (2019). Actor-network theory and ethnography: Sociomaterial approaches to researching medical education. *Perspectives on Medical Education, 8*(3), 177–186. https://doi.org/10.1007/s40037-019-0513-6

MacQuitty M. (1993). *Tiburones*. Altea: Madrid.

Madden, F. (2010). Creating Coexistence between Humans and Wildlife: Global Perspectives on Local Efforts to Address Human–Wildlife Conflict. *Human Dimensions of Wildlife, 9*(4), 247–257.

Mahtab, M., & Jalais, A. (n.d.). *The land of tales and tigers*. Retrieved from https://www.academia.edu/39631252/The_land_of_tales_and_tigers_--_by_Moyukh_Mahtab

MakeMyTrip Blog. (n.d.). *Diving with the Sharks – An Enthralling Ride Under Water*. Retrieved July 15, 2020, from https://www.makemytrip.com/blog/diving-with-sharks-adventure

Man attacked by great white shark at Baylys Beach, Northland. (2018, October 19). Retrieved from https://www.nzherald.co.nz/nz/news/article.cfm?c_id=1&objectid=12145577

Marriam-Webster. (n.d.). *Definition of Avatar.* Retrieved August 28, 2020, from https://www.merriam-webster.com/dictionary/avatar

Marshall, A. D., & Pierce, S. J. (2012). The use and abuse of photographic identification in sharks and rays. *Journal of Fish Biology, 80*(5), 1361–1379. https://doi.org/10.1111/j.1095-8649.2012.03244.x

Marshall, C. & G. B. Rossman (1989). *Designing qualitative research.* Newbury Park. CA, Sage.

Marshall, G., & Jonker, L. (2010). An introduction to descriptive statistics: A review and practical guide. *Radiography, 16*(4), e1–e7. https://doi.org/10.1016/j.radi.2010.01.001

Marshall, G., & Jonker, L. (2011). An introduction to inferential statistics: A review and practical guide. *Radiography, 17*(1), e1–e6. https://doi.org/10.1016/j.radi.2009.12.006

Martin, J. A. (2011). When Sharks (Don't) Attack: Wild Animal Agency in Historical Narratives. *Environmental History, 16*(3), 451–455. https://doi.org/10.1093/envhis/emr051

Martin, R. A. (N. D.). *Mental Processes of the White Shark.* Retrieved May 12, 2018, from http://www.elasmo-research.org/education/white_shark/mental_process.htm

Martin, R. A. (n.d.). *Fossil History of the White Shark. Biology of Sharks and Rays.* Retrieved February 10, 2020, from http://www.elasmo-research.org/education/white_shark/carcharodon.htm

Marvin, G. & S. McHugh (2015*). In it together: An introduction to human animal studies. In The handbook of human animal studies*. London, Routledge.

Marvin, Garry, & McHugh, S. (2014). The handbook of human-animal studies. In Garry McHugh Marvin Susan (Ed.), *In it together: An introduction to human-animal studies* (pg. 1–9). London: Routledge.

Mason, P. H. (2016). The anthropology of multispecies assemblages. *The Australian Journal of Anthropology, 27*(3), 398–404. https://doi.org/10.1111/taja.12189

Maurstad, A., Davis, D., & Cowles, S. (2013). *Co-being and intra-action in horse–human relationships: A multi-species ethnography of be(com)ing human and be(com)ing horse. Social Anthropology, 21*(3), 322–335.

Maxwell, G. (1952). *Harpoon at a Venture. London*: Rupert Hart- Davis.

McAndrew, F. T. (2015). *Why Some People See Ghosts and Other Presences*. Retrieved from https://www.psychologytoday.com/us/blog/out-the-ooze/201507/why-some-people-see-ghosts-and-other-presences

McCagh, C., Sneddon, J., & Blache, D. (2015). Killing sharks: The media's role in public and political response to fatal human–shark interactions. *Marine Policy, 62,* 271–278. https://doi.org/10.1016/j.marpol.2015.09.016

McCarthy, P. (2015). *Stewart Island Shark diving breaches confirmed "Department of Conservation.* Retrieved September 5, 2018, from Stuff website: https://www.stuff.co.nz/southland-times/news/70230966/stewart-island-shark-diving-breaches-confirmed--department-of-conservation

Mcclellan Press, K., Mandelman, J., Burgess, E., Cooke, S. J., Nguyen, V. M., & Danylchuk, A. J. (2016). Catching sharks: Recreational saltwater angler behaviours and attitudes regarding shark encounters and conservation: Saltwater angler behaviours and attitudes regarding shark encounters. *Aquatic Conservation: Marine and Freshwater Ecosystems, 26*(4), 689–702. https://doi.org/10.1002/aqc.2581

McClellan, K. (2015). Envisioning Multispecies Encounters: Photographing Fish in Illinois and Birds in Qatar. *Visual Anthropology Review, 31*(1), 87–93. https://doi.org/10.1111/var.12065

McCracken, H. (2015) Fears cage diving is increasing shark risk. The New Zealand Herald McFarland, S. E., & Hediger, R. (2009). *Animals and agency: An interdisciplinary exploration*. Retrieved from https://ebookcentral.proquest.com/lib/canterbury/reader.action?docID=467653&query=

McGinity, R., & Salokangas, M. (2014). Introduction: 'Embedded research' as an approach into academia for emerging researchers. *Management in Education, 28*(1), 3–5. https://doi.org/10.1177/0892020613508863

McIntosh, D., & Wright, P. A. (2017). Emotional processing as an important part of the wildlife viewing experience. *Journal of Outdoor Recreation and Tourism, 18*, 1–9. https://doi.org/10.1016/j.jort.2017.01.004

McKechnie, C. C. (2012). Spiders, Horror, and Animal Others in Late Victorian Empire Fiction. *Journal of Victorian Culture, 17*(4), 505–516. https://doi.org/10.1080/13555502.2012.733065

McKeon, M. G., & Drew, J. A. (2019). Community dynamics in Fijian coral reef fish communities vary with conservation and shark-based tourism. *Pacific Conservation Biology, 25*(4), 363. https://doi.org/10.1071/PC18045

McKinney, J. A., Hoffmayer, E. R., Holmberg, J., Graham, R. T., Driggers, W. B., de la Parra-Venegas, R., ... Dove, A. D. M. (2017). Long-term assessment of whale shark population demography and connectivity using photo-identification in the Western Atlantic Ocean. PLoS One; *San Francisco, 12*(8), http://dx.doi.org.ezproxy.canterbury.ac.nz/10.1371/journal.pone.0180495

McLeod, H. (n.d*.). Department of Conservation gives Stewart Island shark cage-diving a tick "For now*. Retrieved September 5, 2018, from Stuff website: https://www.stuff.co.nz/national/87913847/doc-gives-shark-cage-diving-a-tick--for-now

McLintock, A. H., William Henry Dawbin, M. S., & Taonga, N. Z. M. for C. and H. T. M. (n.d*.). Early Whaling Operations* [Web page]. Retrieved May 16, 2019, from An encyclopaedia of New Zealand, edited by A. H. McLintock, 1966. website: https://teara.govt.nz/en/26893

McLoughlin, E. (2019). Knowing cows: Transformative mobilizations of human and non-human bodies in an emotionography of the slaughterhouse. *Gender, Work & Organization, 26(3)*, 322–342. https://doi.org/10.1111/gwao.12247

McNamara, P. (2013). The Folly of Dream Interpretation. Retrieved March 10, 2019, from Psychology Today website: http://www.psychologytoday.com/blog/dream-catcher/201307/the-folly-dream-interpretation

Mcneilly, H. (2018). *Below the Surface: Dunedin shark attacks killed three in 1960s*. Retrieved January 6, 2019, from Stuff website: https://www.stuff.co.nz/national/109444407/below-the-surface-dunedins-shark-attacks-chapter-1-keep-out-of-the-water

Mcneilly, H. (2019a). *Below the Surface: Dunedin shark attack claimed life of William Black in 1967*. Retrieved January 6, 2019, from Stuff website: https://www.stuff.co.nz/national/109495764/below-the-surface-dunedins-shark-attacks-chapter-2-i-thought-i-was-dead

Mcneilly, H. (2019b). *Below the Surface: Dunedin shark attacks men spearfishing*. Retrieved January 6, 2019, from Stuff website: https://www.stuff.co.nz/national/109497460/below-the-surface-dunedins-shark-attacks-chapter-3-get-me-ashore

Mcneilly, H. (2019c). *Below the Surface: Dunedin shark's "Huge bloody eyes*." Retrieved January 6, 2019, from Stuff website: https://www.stuff.co.nz/national/109510098/below-the-surface-dunedins-shark-attacks-chapter-4-huge-bloody-eyes

Mcneilly, H. (2019d). *Below the surface: Dunedin's "rogue shark" theory*. Retrieved January 6, 2019, from Stuff website: https://www.stuff.co.nz/national/109525105/below-the-surface-dunedins-shark-attacks-chapter-5-i-have-been-bitten

Mead, M. & F. C. Macgregor (1951). *Growth and culture*. New york, Putnam.

Meekan, M., & Lowe, J. (2019). Does provisioning for tourism harm whale sharks at Oslob? A review of the evidence and reply to Ziegler et al. (2018). *Tourism Management, 75*(Journal Article), 626–629. https://doi.org/10.1016/j.tourman.2019.02.003

Meel, B. L. (2009). Shark attacks on the Transkei Coast of South Africa: A case report. *African Journal of Primary Health Care & Family Medicine, 1*(1), 135–136.

Mehmet, M., & Simmons, P. (n.d.). Feeding frenzy: Public accuse the media of deliberately fuelling shark fear. Retrieved June 23, 2018, from The Conversation website: http://theconversation.com/feeding-frenzy-public-accuse-the-media-of-deliberately-fuelling-shark-fear-95858

Memorandum of Understanding on the Conservation of Migratory Sharks. (2010). *Journal of International Wildlife Law & Policy, 13*(2), 188–199. https://doi.org/10.1080/13880292.2010.486737

Merriam-Webster. (n.d.). Definition of Holism. Retrieved August 5, 2020, from https://www.merriam-webster.com/dictionary/holism

Merriam-webster. (n.d.). *Definition of Xenophobia*. Retrieved January 16, 2021, from https://www.merriam-webster.com/dictionary/xenophobia

Mexico Unexplained. (2018). *The Black Demon, Gigantic Shark of Mexico: Mexico Unexplained*. Retrieved from https://www.youtube.com/watch?v=Dsw3cq2r1Hg

Meyer, M. (2012). Placing and tracing absence: A material culture of the immaterial. *Journal of Material Culture, 17*(1), 103–110. https://doi.org/10.1177/1359183511433259

Meyer, M., & Woodthorpe, K. (2008). The Material Presence of Absence: A dialogue between museums and cemeteries. *Sociological Research Online, 13.* Retrieved from http://www.socresonline.org.uk/13/5/1.html

Michelle, C., Davis, C. H., Hight, C., & Hardy, A. L. (2015). *The Hobbit hyperreality paradox: Polarization among audiences for a 3D high frame rate film.* Convergence. https://doi.org/10.1177/1354856515584880

Miller, D. (2010). *Stuff.* Cambridge: Polity Press.

Miller, D. (Ed.). (1998). *Material cultures: Why some things matter*. Chicago: University of Chicago Press.

Miller, D. (Ed.). (2005). *Materiality*. Durham, N.C: Duke University Press.

Miller, G. (2001). *The Mating Mind: How Sexual Choice Shaped the Evolution of Human Nature* (Reprint edition). New York: Anchor.

Miller, G. F. (2001). Aesthetic fitness: How sexual selection shaped artistic virtuosity as a fitness indicator and aesthetic preferences as mate choice criteria. *Bulletin of Psychology and the Arts: Special Issue on Evolution, Creativity and Aesethetics*, 2(1), 20–25.

Miller, J. C., and Del Casino, V. J. (2018). Negative simulation, spectacle and the embodied geopolitics of tourism. *Transactions of the Institute of British Geographers, 43*(4), 661–673. https://doi.org/10.1111/tran.12244

Miller, J. R. (2005). Biodiversity conservation and the extinction of experience. *Trends in Ecology & Evolution, 20*(8), 430–434. https://doi.org/10.1016/j.tree.2005.05.013

Minas, M. (2016). Shark investigation. *Prime Number, 31*(4), 4–5.

Miner, Q. (n.d.). *Qualitative Data Analysis Software for Mixed Methods Research*. Retrieved May 14, 2018, from Provalis Research website: https://provalisresearch.com/products/qualitative-data-analysis-software/

Mishra, S. (2009). Camera lucida. *Literary Review: an international journal of contemporary writing 52*(3): 198-198.

Mislinski, P. (N.D.). *Why do sharks attack?* Retrieved 19.9.2015, from http://www.discovery.com/tv-shows/shark-week/about-this-show/why-do-sharks-attack/.

Mitchell, C. (n.d.). *Shark-diving conflict in southern seas continues with High Court ruling.* Retrieved September 5, 2018, from Stuff website: https://www.stuff.co.nz/environment/93303304/sharkdiving-conflict-in-southern-seas-continues-with-high-court-ruling

Mitri, S., Xavier, J. B., & Foster, K. R. (2011). Social evolution in multispecies biofilms. *Proceedings of the National Academy of Sciences of the United States of America, 108,* 10839–10846.

Mizejewski, D. (2013). *Inspiring Wildlife Conservation with Photography.* Retrieved 10.9.2015, from http://www.huffingtonpost.com/david-mizejewski/inspiring-wildlife_b_3534478.html.

Mol, A., & Law, J. (2001). Complexities. In C. Thompson (Ed.), *When elephants stand for competing models of nature* (pp. 166–189). Durham, NC: Duke Univ. Press.

Moore, A. (2012). The aquatic invaders: Marine management figuring fishermen, fisheries, and lionfish in the Bahamas. *Cultural Anthropology, 27*(4), 667–688.https://doi.org/10.1111/j.1548-1360.2012.01166.x

Moore, L. J. (2015). A Day at the Beach: Rising Sea Levels, Horseshoe Crabs, and Traffic Jams. *Sociology, 49*(5), 886–902. https://doi.org/10.1177/0038038515573474

Moore, L. J., & Kosut, M. (2014). Among the colony: Ethnographic fieldwork, urban bees and intra-species mindfulness. *Ethnography, 15*(4), 516–539. JSTOR.

Moors, A. (2017). On autoethnography. *Ethnography, 18*(3), 387–389. https://doi.org/10.1177/1466138117723354

Morey, S. (2002). The shark in modern culture: Beauty and the beast. *Journal of undergraduate research 4.*

Morgan, L. H. (1868). *The American Beaver and His Works*. J.B. Lippincott & Company.

Morris, D., & Beaven, K. (2012). *A Pool to be Proud Of.* http://inspiringcommunities.org.nz/wp-content/uploads/2015/03/A-pool-to-be-proud-of.pdf

Morton, B. (2013). The Beginners Guide to shark diving. *Sport Diver, 21*(7), 36–47.

Mourier, J., Vercelloni, J., & Planes, S. (2012). Evidence of social communities in a spatially structured network of a free-ranging shark species. *Animal Behaviour, 83*(2), 389–401. https://doi.org/10.1016/j.anbehav.2011.11.008

Muhly, T. B., Semeniuk, C., Massolo, A., Hickman, L., & Musiani, M. (2011). Human Activity Helps Prey Win the Predator-Prey Space Race. *Plos One, 6*(3), e17050. https://doi.org/10.1371/journal.pone.0017050

Müller, M. (2015). Assemblages and Actor-networks: Rethinking Socio-material Power, Politics and Space. *Geography Compass, 9*(1), 27–41. https://doi.org/10.1111/gec3.12192

Multispecies salon. (n.d.). *The Multispecies Salon*. Retrieved May 16, 2018, from The Multispecies Salon website: http://www.multispecies-salon.org/

Munster, U. (2014). Working for the forest: The ambivalent intimacies of human-elephant collaboration in south Indian wildlife conservation. *Ethnos: Journal of Anthropology*. Retrieved from http://www.tandfonline.com/doi/abs/10.1080/00141844.2014.969292

Murdoch, J. (1997). Inhuman/Nonhuman/Human: Actor-Network Theory and the Prospects for a Nondualistic and Symmetrical Perspective on Nature and Society. *Environment and Planning. Society and Space, 15*(6), 731–756. https://doi.org/10.1068/d150731

Muter, B. A., Gore, M. L., Gledhill, K. S., Lamont, C., & Huveneers, C. (2013). Australian and U.S. News Media Portrayal of Sharks and Their Conservation: Media Portrayal of Sharks. *Conservation Biology, 27*(1), 187–196. https://doi.org/10.1111/j.1523-1739.2012.01952.x

Myers, R. A., et al. (2007). "Cascading effects of the loss of apex predatory sharks from coastal ocean." *Science. 315:* 1846-1850.

Myrick, J. G., & Evans, S. D. (2014). Do PSAs Take a Bite Out of Shark Week? The Effects of Juxtaposing Environmental Messages with Violent Images of Shark Attacks. *Science Communication, 36*(5), 544–569. https://doi.org/10.1177/1075547014547159

Nading, A. M. (2012). Dengue mosquitoes are single mothers: Biopolitics meets ecological aesthetics in Nicaraguan community health work. *Cultural Anthropology, 27*(4), 572–596.

Nading, A. M. (2013). Humans, animals, and health: From ecology to entanglement. *Environment and Society: Advances in Research, 4*(1), 60–78.

Nagel, T. (1974). What Is It Like to Be a Bat? *The Philosophical Review, 83*(4), 435–450.

Nakamura, K. (2013). Making Sense of Sensory Ethnography: The Sensual and the Multisensory. *American Anthropologist, 115*(1), 132–135. https://doi.org/10.1111/j.1548-1433.2012.01544.x

Nance, S. (2013*). Entertaining Elephants: Animal Agency and the Business of the American Circus.* Baltimore: Johns Hopkins Univ. Press.

Nasby- Lucas, N., & Domeier, M. L. (2012). Global Perspectives on the Biology and Life History of the White Shark. In *Use of Photo Identification to Describe a White Shark Aggregation at Guadalupe Island, Mexico*. Boca Raton, FL: CRC Press.

Nason, P. (2016, April 21). *The Shark Callers of Papua New Guinea*. Retrieved March 5, 2019, from Shark Research Institute website: https://www.sharks.org/blogs/shark-culture/shark-calling-papua-new-guinea

National Geographic. (2014, January 30). *Shark Tagging & Tracking: Separating Fact from Fiction*. https://blog.nationalgeographic.org/2014/01/30/shark-tagging-tracking-separating-fact-from-fiction/

Nazimi, L., Robbins, W. D., Schilds, A., & Huveneers, C. (2018). Comparison of industry-based data to monitor white shark cage-dive tourism. *Tourism Management, 66*, 263–273. https://doi.org/10.1016/j.tourman.2017.12.002

NBC News website (n.d.) *Great white hype: What are the odds of being killed by a shark*. Retrieved June 12, 2018,: https://www.nbcnews.com/news/us-news/great-white-hype-what-are-odds-being-killed-shark-n776431

Neff, C. (2014). *Human Perceptions and Attitudes Toward Sharks: Examining the predator policy paradox*. Retrieved from https://www.academia.edu/8295663/Human_Perceptions_and_Attitudes_Toward_Sharks_Examining_the_predator_policy_paradox

Neff, C. (2015). The Jaws Effect: How movie narratives are used to influence policy responses to shark bites in Western Australia. *Australian Journal of Political Science, 50*(1), 114–127. https://doi.org/10.1080/10361146.2014.989385

Neff, C. (2019). *Governing Emotion: How to Analyze Emotional Political Situations. Flaws: Shark Bites and Emotional Public Policymaking*. Retrieved from https://www.academia.edu/39348023/Governing_Emotion_How_to_Analyze_Emotional_Political_Situations

Neff, C. L., & Yang, J. Y. H. (2013). Shark bites and public attitudes: Policy implications from the first before and after shark bite survey. *Marine Policy, 38*, 545–547. https://doi.org/10.1016/j.marpol.2012.06.017

Neff, C., & Hueter, R. (2013). Science, policy, and the public discourse of shark "attack": A proposal for reclassifying human–shark interactions. *Journal of Environmental Studies and Sciences, 3*(1), 65–73. https://doi.org/10.1007/s13412-013-0107-2

Nelson, L. H., & Nelson, J. (1996). Feminism, Science, and the Philosophy of Science. In K. Barad (Ed.), *Meeting the universe halfway: Realism and social constructivism without contradiction* (pp. 161–194). Dordecht, The Netherlands: Kluwer Academic.

Nemoto, T., & Beglar, D. (n.d.). *Developing Likert-Scale Questionnaires*. 8.

Nevres, M. Ö. (2018, October 6). *Largest great white sharks ever recorded.* Our Planet. https://ourplnt.com/largest-great-white-sharks-ever-recorded/

New Scientist (n.d.) *Sharks now protected no matter whose waters they swim in*. Retrieved November 3, 2017, from : https://www.newscientist.com/article/2152073-sharks-now-protected-no-matter-whose-waters-they-swim-in/

New Zealand Geographic (n.d.) *Shark diver faces major loss after court ruling*. Retrieved October 2, 2018, from : https://www.nzgeo.com/audio/shark-diver-faces-major-loss-after-court-ruling/

New Zealand Ministry of Foreign Affairs and trade. (n.d.). *Protecting whales. New Zealand Ministry of Foreign Affairs and Trade*. Retrieved July 22, 2020, from https://www.mfat.govt.nz/en/environment/biodiversity-and-species-conservation/protecting-whales/

New Zealand, R. (2015). *Government signs migratory shark agreement*.

New Zealand, S. (N. D.). *QuickStats About Bluff*.

Australasian Leisure Management. (n.d.). *"New Zealand's Tourism Industry Association call for law change to permit shark cage diving"* Retrieved October 2, 2018, from https://www.ausleisure.com.au/news/new-zealands-tourism-industry-association-call-for-law-change-to-permit-shark-cage-diving/

Newsbeat (n.d.). *Is banning cage diving actually bad for sharks*? Retrieved October 2, 2018, from http://www.newsbeat.kiwi/2018/09/18/is-banning-cage-diving-actually-bad-for-sharks/

Newshub (2018, May 9). *Court of Appeal shark cage diving ruling major victory for pāua diving community*. Retrieved from https://www.newshub.co.nz/home/new-zealand/2018/09/court-of-appeal-shark-cage-diving-ruling-major-victory-for-p-ua-diving-community.html

Newshub. (2019). *"Huge relief" for businesses as Supreme Court overturns shark cage diving decision*. https://www.newshub.co.nz/home/new-zealand/2019/10/huge-relief-for-businesses-as-supreme-court-overturns-shark-cage-diving-decision.html

Nicoll, D. (2016). *Shark Dive NZ operator says he was not run off Stewart Island* | Stuff.co.nz. Retrieved September 5, 2018, from https://www.stuff.co.nz/national/76732699/Shark-Dive-NZ-operator-says-he-was-not-run-off-Stewart-Island?rm=m

Nicoll, D. (2018). *Court finds shark diving illegal but says law could be changed*. Retrieved May 9, 2019, from Stuff website: https://www.stuff.co.nz/southland-times/news/106798859/court-finds-shark-diving-illegal-but-says-law-could-be-changed

Nicoll, D. (n.d.). *PressReader.com, Connecting People Through News*. Retrieved July 17, 2018, from https://www.pressreader.com/new-zealand/the-southland-times/20151212/281560879735233

Nielsen, B. F. (2014). Out of Context: Ethnographic Interviewing, Empathy, and Humanitarian Design. *Design Philosophy Papers; Crows Nest, 12*(1), 51–64.

Nietzsche, F. (1907). *Beyond Good and Evil*. Macmillan Co.

Niwa (2011, 1.5.15). *White sharks*. Retrieved 17.6.15, from https://www.niwa.co.nz/coasts-and-oceans/research-projects/white-sharks.

Niwa (2013). *Shark Tagging*. Retrieved 3.8.2015, from http://www.niwa.co.nz/gallery/shark-tagging.

Norouzitallab, P., Baruah, K., Vanrompay, D., & Bossier, P. (2019). Can epigenetics translate environmental cues into phenotypes? *Science of The Total Environment, 647*, 1281–1293. https://doi.org/10.1016/j.scitotenv.2018.08.063

North Shore Shark Adventures (n.d.). *Shark Diving Tours - Honolulu, Oahu*. Retrieved July 15, 2020, from https://sharktourshawaii.com/

Nosal, A. P., Keenan, E. A., Hastings, P. A., & Gneezy, A. (2016). The Effect of Background Music in Shark Documentaries on Viewers' Perceptions of Sharks. *Plos One, 11*(8), e0159279. https://doi.org/10.1371/journal.pone.0159279

Nottle, C., Nottle, C., Young, J., & Young, J. (2019). Individuals, instinct and moralities: Exploring multi-species leisure using the serious leisure perspective. *Leisure Studies: Multispecies Leisure: Human-Animal Interactions in Leisure Landscapes, 38*(3), 303–316. https://doi.org/10.1080/02614367.2019.1572777

Nxumalo, F., & Pacini-Ketchabaw, V. (2017). "Staying with the trouble" in child-insect-educator common worlds. *Environmental Education Research, 23*(10), 1414–1426. https://doi.org/10.1080/13504622.2017.1325447

NZ Herald. (2006, March 12). *Great white shark could become protected species*. https://www.nzherald.co.nz/nz/news/article.cfm?c_id=1&objectid=10372415

NZ herald. (2018, October 19). *Man attacked by great white shark at Baylys Beach,* Northland. NZ Herald. https://www.nzherald.co.nz/nz/news/article.cfm?c_id=1&objectid=12145577

NZ Herald. (n.d.). *Court ruling another twist in shark cage stoush-* Retrieved June 7, 2017, from http://www.nzherald.co.nz/nz/news/article.cfm?c_id=1&objectid=11871254

O'Bryhim, J. R., & Parsons, E. C. M. (2015). Increased knowledge about sharks increases public concern about their conservation. *Marine Policy, 56*, 43–47. https://doi.org/10.1016/j.marpol.2015.02.007

O'Connell, C. P., & Leurs, G. (2016). A minimally invasive technique to assess several life-history characteristics of the endangered great hammerhead shark Sphyrna mokarran. *Journal of Fish Biology, 88*(3), 1257–1264. https://doi.org/10.1111/jfb.12900

O'Connell, C. P., Andreotti, S., Rutzen, M., Meÿer, M., Matthee, C. A., & He, P. (2014). Effects of the Sharksafe barrier on white shark (Carcharodon carcharias) behavior and its implications for future conservation technologies. *Journal of Experimental Marine Biology and Ecology, 460*, 37–46. https://doi.org/10.1016/j.jembe.2014.06.004

O'Leary, K., & Murphy, S. (2019). Moving beyond Goffman: The performativity of anonymity on SNS. *European Journal of Marketing, 53*(1), 83–107. https://doi.org/10.1108/EJM-01-2017-0016

O'Sullivan, S. (2001). The aesthetics of affect: Thinking art beyond representation. *Angelaki, 6*(3), 125–135. https://doi.org/10.1080/09697250120087987

O'Bryhim, J. (2011). *The effects of public opinion on sharks: knowledge, attitudes, and behaviors towards sharks and shark conservation*. Saarbrücken, LAP, Lambert Academic Publishing.

Office of the Minister of Conservation. (2018). *New science report lists sharks as threatened*. https://www.doc.govt.nz/news/media-releases/2018/new-science-report-lists-sharks-as-threatened/

Ogden, L. (2011). *Swamplife: people, gators, and mangroves entangled in the Everglades*. Minneapolis, University of Minnesota Press.

Ogden, L. A., Hall, B., & Tanita, K. (2013). Animals, Plants, People, and Things: A Review of Multispecies Ethnography. *Environment & Society, 4*, 5–24.

Oma, K. A. (2013). Human-Animal Meeting Points: Use of Space in the Household Arena in Past Societies. *Society & Animals, 21*(2), 162–177. https://doi.org/10.1163/15685306-12341300

Oppenheim, R. (2007). Actor-network theory and anthropology after science, technology, and society. *Anthropological Theory, 7*(4), 471–493. https://doi.org/10.1177/1463499607083430

Orne, J. (2013). Queers in the Line of Fire: Goffman's Stigma Revisited. *Sociological Quarterly, 54*(2), 229–253. https://doi.org/10.1111/tsq.12001

Otago Daily Times Online News (2018) *Shark cage diving deemed offence under Wildlife Act.* Retrieved October 2, 2018, from website: https://www.odt.co.nz/regions/southland/shark-cage-diving-deemed-offence-under-wildlife-act

Otago Daily Times Online News. (n.d.). *A long haul 100 years of fishing at Port*. Retrieved September 19, 2017, from https://www.odt.co.nz/lifestyle/magazine/long-haul-100-years-fishing-port.

Otis, M. (2006). Visual Anthropology. *Current Anthropology, 47*(6), 889–889. https://doi.org/10.1086/508693

Ovid, & Feeney, D. (2004). *Metamorphoses* (Reprint edition; D. Raeburn, Trans.). London: Penguin Classics.

Oxford Reference, O. (2016). *"Contact zone."* from http://www.oxfordreference.com/view/10.1093/oi/authority.20110803095634533.

Packer, C., Ikanda, D., Kissui, B., & Kushnir, H. (2005). Conservation biology: Lion attacks on humans in Tanzania. *Nature, 436*(7053), 927–928. https://doi.org/10.1038/436927a

Paddenburg, T. (2017). *No explosion of great whites: Authorities*. Retrieved October 29, 2017, from http://www.perthnow.com.au/news/western-australia/csiro-says-no-evidence-to-support-claims-of-explosion-in-great-white-shark-numbers/news-story/f74d9ff2d6484366c4e03e50b7f39e1c

Padi (2015). *Underwater Photography and Video Equipment*. Retrieved 9.10.2015, from http://www.padi.com/scuba-diving/padi-courses/course-catalog/about-scuba-gear/photo-video-equipment.

Paleček, M., & Risjord, M. (2013). Relativism and the ontological turn within anthropology. *Philosophy of the Social Sciences, 43*(1), 3–23.

Palermo, E. (n.d.). *Why Are People So Afraid of Sharks?* Retrieved June 20, 2017, from Live Science website: https://www.livescience.com/51579-fear-of-sharks-psychology.html

Palmer, A., Malone, N., & Park, J. (2015). Accessing Orangutans' Perspectives: Interdisciplinary Methods at the Human/Animal Interface. *Current Anthropology, 56*(4), 571–578. https://doi.org/10.1086/682053

Panelli, R. (2010). More-than-human social geographies: Posthuman and other possibilities. *Progress in Human Geography, 34*(1), 79–87.

Parathian, H. E., McLennan, M. R., Hill, C. M., Frazão-Moreira, A., & Hockings, K. J. (2018). Breaking Through Disciplinary Barriers: Human-Wildlife Interactions and Multispecies Ethnography. *International Journal of Primatology, 39*(5), 749–775. https://doi.org/10.1007/s10764-018-0027-9

Parker, A., & Manley, A. (2017). Goffman, Identity and Organizational Control: Elite Sports Academies and Social Theory. *Sociology of Sport Journal, 34*(3), 211–222. https://doi.org/10.1123/ssj.2016-0150

Parker, L. (N.D,) Studying Sharks' Social Lives to Expose Their Friendly Side. The New York Times.

Parreñas, J. S. (2018). *Decolonizing Extinction: The Work of Care in Orangutan Rehabilitation by Juno Salazar Parreñas*. Retrieved from https://www.goodreads.com/book/show/36419812-decolonizing-extinction

Parreñas, R. (2012). Producing affect: Transnational volunteerism in a Malaysian orangutan rehabilitation center. *American Ethnologist, 39*(4), 673–687.

Patton, M. Q., & Cochran, M. (n.d.). *Reproduced with kind permission of Michael Quinn Patton*. 36.

Pavsek, C. (2015). Leviathan and the Experience of Sensory Ethnography: Leviathan and Sensory Ethnography. *Visual Anthropology Review, 31*(1), 4–11. https://doi.org/10.1111/var.12056

Paxson, H. (2008). Post-Pasteurian cultures: The microbiopolitics of raw-milk cheese in the United States. *Cultural Anthropology, 231*(1), 15–47.

Paxson, H., & Helmreich, S. (2013). The perils and promises of microbial abundance: Novel natures and model ecosystems, from artisanal cheese to alien seas. *Social Studies of Science, 44*(2), 165–193.

Peace, A. (2015). Shark attack! A cultural approach. *Anthropology Today, 31*(5), 3–7. https://doi.org/10.1111/1467-8322.12197

Pedro Barreiros, J., Gadig, O. B. F., & Haddad, V. (2015). In Reply to Shark Attacks and Shark Diving. *Wilderness & Environmental Medicine, 26*(2), 277–278. https://doi.org/10.1016/j.wem.2014.11.018

Penberthy, N. (2015). *Great White Sharks prefer rock music*. Retrieved June 17, 2017, from Australian Geographic website: http://www.australiangeographic.com.au/topics/science-environment/2015/11/great-white-sharks-prefer-rock-music

Pepin-Neff, C. L. (2019). *Flaws: Shark Bites and Emotional Public Policymaking* (1st ed. 2019 edition). Palgrave Macmillan.

Pepin-Neff, C. L., & Wynter, T. (2018). Reducing fear to influence policy preferences: An experiment with sharks and beach safety policy options. *Marine Policy, 88*, 222–229. https://doi.org/10.1016/j.marpol.2017.11.023

Persson, K., Elger, B. S., & Shaw, D. M. (2017). The Indignity of Relative Concepts of Animal Dignity: A Qualitative Study of People Working with Nonhuman Animals. *Anthrozoös, 30*(2), 237–247. https://doi.org/10.1080/08927936.2017.1311004

Perthnow (2016). *Shark tags may trigger fatal attacks*. Retrieved June 23, 2017, from http://www.perthnow.com.au/news/western-australia/wa-shark-attacks-evidence-reveals-possible-link-between-shark-tagging-and-attacks/news-story/0f305e86f30a16a2346cdb14dfcfa076.

Petropavlovsky, M.-N. (2014). Hugh Raffles, 2010, Insectopedia, Pantheon Books, New York, 465 pages. *Développement durable et territoires. Économie, géographie, politique, droit, sociologie*, (Vol. 5, n°2). Retrieved from http://journals.openedition.org/developpementdurable/10308

Pettit, M., Serykh, D., & Green, C. D. (2015). Multispecies Networks: Visualizing the Psychological Research of the Committee for Research in Problems of Sex. *Isis, 106*(1), 121–149. https://doi.org/10.1086/681039

Pettitt, J. (2016). Sartre, Goffman, and Fictional Nazis: Homogeneity as Identity in Martin Amis's Time's Arrow (1991) and Edgar Hilsenrath's The Nazi and the Barber (1971). *Interdisciplinary Literary Studies, 18*(4), 441–460. JSTOR. https://doi.org/10.5325/intelitestud.18.4.0441

Philo, C., & Wilbert, C. (2000). *Animal Spaces, Beastly Places: New Geographies of Human-animal Relations*. Psychology Press.

Photovoice (N.D.). *Vission and Mission*. Retrieved 12.8.20151, from http://www.photovoice.org/vision-and-mission/.

Pickering, A. (2010). *Material Culture and the Dance of Agency*. The Oxford Handbook of Material Culture Studies. https://doi.org/10.1093/oxfordhb/9780199218714.013.0007

Pickett, B. (2015). *Stewart Island shark diving operators permits safe for now*. Retrieved September 5, 2018, from Stuff website: https://www.stuff.co.nz/southland-times/news/70486582/Stewart-Island-shark-diving-operators-permits-safe-for-now

Pink, S. (2001). *Doing visual ethnography: images, media and representation in research*. London, SAGE.

Pink, S. (2006). *The Future of Visual Anthropology: Engaging the senses*. Usa and canada, Routledge.

Pink, S. (2007). *Doing visual ethnography: images, media, and representation in research*. Thousand Oaks, Calif; London, Sage Publications.

Pink, S. (2014). *Digital–visual–sensory-design anthropology: Ethnography, imagination and intervention: Arts and Humanities in Higher Education*. https://doi.org/10.1177/1474022214542353

Pink, S. (n.d.). *What is Sensory Ethnography*. Retrieved August 30, 2017, from https://www.ncrm.ac.uk/resources/video/RMF2010/pages/18_Sensory.php

Pink, S., et al. (2004). *Working images: visual research and representation in ethnography*. New York, Routledge.

Pinney, C. (1992). *The Parallel Histories of Anthropology and Photography*. New Haven CT, Yale University Press.

Pinney, C., & C. Ebooks (2011). *Photography and anthropology*. London, Reaktion Books.

Pitt-Rivers, J. (1963). *The People of the Sierra,Chicago and London*. Chicago and London, Chicago University Press.

Poggie, J. J. (1980). Maritime Anthropology: Socio-Cultural Analysis of Small-Scale Fishermen's Cooperatives: Introduction. *Anthropological Quarterly, 53*(1), 1–3. Retrieved from JSTOR.

Pomfret, G. (2012). Personal emotional journeys associated with adventure activities on packaged mountaineering holidays. *Tourism Management Perspectives, 4*, 145–154. https://doi.org/10.1016/j.tmp.2012.08.003

Pooley, S. (2016). A Cultural Herpetology of Nile Crocodiles in Africa. *Conservation and Society, 14*(4). Retrieved from https://www.researchgate.net/publication/307575977_A_Cultural_Herpetology_of_Nile_Crocodiles_in_Africa

Pooley, S., Barua, M., Beinart, W., Dickman, A., Holmes, G., Lorimer, J., Loveridge, A. J., Macdonald, D. W., Marvin, G., Redpath, S., Sillero-Zubiri, C., Zimmermann, A., & Milner-Gulland, E. J. (2017). *An interdisciplinary review of current and future approaches to improving human–predator relations. Conservation Biology, 31*(3), 513–523. https://doi.org/10.1111/cobi.12859

Popular Mechanics website (2016) *Scientists Discover New, Mysterious Prehistoric Shark.* Retrieved June 23, 2017, from: http://www.popularmechanics.com/science/animals/a23229/scientists-discover-prehistoric-shark/

Porter, N. (2019). Training Dogs to Feel Good: Embodying Well-being in Multispecies Relations. *Medical Anthropology Quarterly, 33*(1), 101–119. https://doi.org/10.1111/maq.12459

Pouliot, A., & Ryan, J. C. (2013). Fungi: An entangled exploration. *PAN: Philosophy Activism Nature, 10,* 1–5.

Hakai Magazine (n.d.-a). Poverty, Poaching, and Death by Great White Shark. Retrieved October 28, 2017, from https://www.hakaimagazine.com/article-short/poverty-poaching-and-death-great-white-shark

Pownall, A. (2014). *Scientists test bubbles to scare sharks*. Retrieved from https://thewest.com.au/news/australia/scientists-test-bubbles-to-scare-sharks-ng-ya-372733

Powter, D. M., & Gladstone, W. (2008). *Habitat preferences of Port Jackson sharks, Heterodontus portusjacksoni, in the coastal waters of Eastern Australia*. Retrieved from https://opus.lib.uts.edu.au/handle/10453/14860

Pratt, M. L. (1991). Arts of the Contact Zone. *Profession, 33–40*. Retrieved from JSTOR.

Pratt, S., Pratt, S., & Suntikul, W. (2016). Can Marine Wildlife Tourism Provide an "Edutaining" Experience? *Journal of Travel & Tourism Marketing, 33*(6), 867–884. https://doi.org/10.1080/10548408.2015.1069778

Preez, M. D., Dicken, M., & Hosking, S. G. (2012). The Value of Tiger Shark Diving Within the Aliwal Shoal Marine Protected Area: A Travel Cost Analysis. *South African Journal of Economics, 80*(3), 387–399. https://doi.org/10.1111/j.1813-6982.2011.01292.x

Press, A. (2017, March 30). *Real-life Sharknado? Cyclone Debbie washes up shark on Australian street.* Retrieved from https://www.theguardian.com/australia-news/2017/mar/30/shark-washes-up-puddle-australia-cylone-debbie-sharknado

Press, A. A. (2016, October 9). *Drones to monitor shark activity off Western Australia coastline.* Retrieved from https://www.theguardian.com/environment/2016/oct/09/drones-to-monitor-shark-activity-off-western-australia-coastline

Press, S. V. (2019). Terrestrial Cosmopolitanism, Posthumanism, and Multispecies Modes of Being in Cereus Blooms at Night. *Humanities, 8*(2), 92. https://doi.org/10.3390/h8020092

Press Reader (n.d.). *Connecting People Through News*. Retrieved October 2, 2018, from https://www.pressreader.com/new-zealand/the-dominion-post/20180328/281732680034248

Probyn, E. (2014). Ethnographic worldviews. In E. Probyn (Ed.), *Sustaining fish-human communities? A more-than-human question* (pg. 155–171). Dordrecht, The Netherlands: Springer.

DAN, Divers Alert Network — Medical Dive Article. (n.d.). *Psychological Issues in Diving II - Anxiety, Phobias in Diving*. Retrieved October 24, 2018, from https://www.diversalertnetwork.org/medical/articles/Psychological_Issues_in_Diving_II_Anxiety_Phobias_in_Diving

Queiroz, N., Humphries, N. E., Couto, A., Vedor, M., Costa, I. da, Sequeira, A. M. M., ... Sims, D. W. (2019). Global spatial risk assessment of sharks under the footprint of fisheries. *Nature, 1*. https://doi.org/10.1038/s41586-019-1444-4

Question Time. (2018). *Hon Maggie Barry, shark cage diving*. Retrieved July 15, 2020, from https://maggiebarry.national.org.nz/question_time_18_june_shark_cage_diving

Radhakrishna, S. (2018). Primate Tales: Using Literature to Understand Changes in Human–Primate Relations. *International Journal of Primatology, 39*(5), 878–894. https://doi.org/10.1007/s10764-018-0035-9

Radio New Zealand (2017, June 2). *Court decision muddies waters on shark-cage tourism*. Retrieved June 23, 2017, from website: http://www.radionz.co.nz/news/national/332190/court-decision-muddies-waters-on-shark-cage-tourism

Raffles, H. (2010). *Insectopedia.* New York: Pantheon.

Ratelle, A. (2015). Contact Zones, Becoming and the Wild Animal Body. In A. Ratelle (Ed.), *Animality and Children's Literature and Film* (pp. 41–64). https://doi.org/10.1057/9781137373168_3

Redpath, S. M., Bhatia, S., & Young, J. (2015). Tilting at wildlife: Reconsidering human–wildlife conflict. *Oryx, 49*(2), 222–225. https://doi.org/10.1017/S0030605314000799

Reed, I. A. (2017). Ethnography, theory, and sociology as a human science: An interlocution. *Ethnography, 18*(1), 107–129. https://doi.org/10.1177/1466138115592417

Reisland, M. A., & Lambert, J. E. (2016). Sympatric Apes in Sacred Forests: Shared Space and Habitat Use by Humans and Endangered Javan Gibbons (Hylobates moloch). *PLoS ONE, 11*(1). https://doi.org/10.1371/journal.pone.0146891

Remis, M. J., & Hardin, R. (2009). Transvalued Species in an African Forest. *Conservation Biology, 23*(6), 1588–1596.

Reynolds, J. (2017). 'Missing out': Reflections on the positioning of ethnographic research within an evaluative framing. *Ethnography, 18*(3), 345–365. https://doi.org/10.1177/1466138116664106

Richards, K., O'Leary, B. C., Roberts, C. M., Ormond, R., Gore, M., & Hawkins, J. P. (2015). Sharks and people: Insight into the global practices of tourism operators and their attitudes to Shark behaviour. *Marine Pollution Bulletin, 91*(1), 200–210. https://doi.org/10.1016/j.marpolbul.2014.12.004

Rigney, E. (2019). *Orcas eat great white sharks—New insights into rare behavior revealed.* https://www.nationalgeographic.com/animals/2019/07/killer-whales-orcas-eat-great-white-sharks/

Riley, E. P. (2008). Wild Profusion: Biodiversity Conservation in an Indonesian Archipelago by Celia Lowe. *American Anthropologist, 110*(1), 124–125. https://doi.org/10.1111/j.1548-1433.2008.00018_48.x

Robers, A. (2019). *Shark Week: Discovery Releases 2019 Schedule, Including First Original Film Starring Josh Duhamel. TV.* https://popculture.com/tv-shows/news/shark-week-discovery-releases-2019-schedule-including-first-original-film-josh-duhamel/

Rodney Fox Shark Expeditions (n.d*) In our blood*. Retrieved October 11, 2017, from: http://www.rodneyfox.com.au/about-us/

Rodney Fox. (n.d.). *Amazing Man. Amazing Life. Rodney Fox Shark Expeditions.* Retrieved August 6, 2020, from https://www.rodneyfox.com.au/about-us/rodney-fox/

Roff, G., Brown, C. J., Priest, M. A., & Mumby, P. J. (2018). Decline of coastal apex shark populations over the past half century. *Communications Biology, 1*(1), 223. https://doi.org/10.1038/s42003-018-0233-1

Rojek, C. (2006). After popular culture: Hyperreality and leisure. *Leisure Studies*. https://doi.org/10.1080/02614369300390261

Rojek, C., & Turner, B. S. (Eds.). (1993). *Forget Baudrillard?* London; New York: Routledge.

Rolland, M., De Melo, E. H., Vieira, É. L. R., & De Oliveira, M. D. C. A. (2012). Environmental conditions associated with shark attacks against humans: A case report about territoriality. *The American Journal of Forensic Medicine and Pathology, 33*(3), 265.

Romania, V. (n.d.). Goffman in Dixon. Ethnographer or Performer? *Italian Sociological Review*. Retrieved from https://www.academia.edu/39162398/Goffman_in_Dixon._Ethnographer_or_Performer

Roney, J. R. (2003). Effects of visual exposure to the opposite sex: Cognitive aspects of mate attraction in human males. *Personality & Social Psychology Bulletin, 29*(3), 393–404. https://doi.org/10.1177/0146167202250221

Ropeik, D. (2010*). How Risky Is It, Really? Why Our Fears Don't Always Match the Facts* (1 edition). New York: McGraw-Hill Education.

Rose, D. B. (2011). *Wild dog dreaming: Love and extinction*. Charlottesville: Univ. of Virginia Press.

Rose, D. B., & Van Dooren, T. (2011). Unloved others: Death of the disregarded in the time of extinctions. *Australian Humanities Review, 50*. Retrieved from http://www.australianhumanitiesreview.org/archive/Issue-May-2011/home.html

Rose, J., Rose, J., & Wilson, J. (2019). Assembling Homelessness: A Posthumanist Political Ecology Approach to Urban Nature, Wildlife, and Actor-Networks. *Leisure Sciences, 41*(5), 402–422. https://doi.org/10.1080/01490400.2019.1627964

Roy, E. A. (2016, January 29). *"It's like Jaws": New Zealand town fears attack by great white sharks lured by cage divers.* https://www.theguardian.com/world/2016/jan/29/jaws-new-zealand-town-attack-great-white-sharks-cage-divers

Rtshiladze, M. A., Andersen, S. P., Nguyen, D. Q. A., & Grabs, A. (2011). The 2009 Sydney shark attacks: Case series and literature review. *ANZ Journal of Surgery, 81*(5), 345–351. https://doi.org/10.1111/j.1445-2197.2010.05640.x

Ruby, J. & R. Chalfen (1974). *The Teaching of Visual Anthropology at Temple*.

Ruby, J. (2000). *Picturing culture: explorations of film & anthropology*. Chicago, University of Chicago Press.

Ruby, J. (2001). The Professionalization of Visual Anthropology in the United States: The 1960s and 1970s. *Visual Anthropology Review, 17*(2), 5–12. https://doi.org/10.1525/var.2001.17.2.5

Ruby, J. (2004). The Professionalization of Visual Anthropology in the United States -- The 1960s and 1970s. *Revista Brasileira de Sociologia da Emocao 3*(9): 352-366.

Rutherford, D. (2016). Affect Theory and the Empirical. *Annual Review of Anthropology, 45*(1), 285–300. https://doi.org/10.1146/annurev-anthro-102215-095843

Sachdeva, S. (2015). *Fatal shark attack likely if Stewart Island cage diving not stopped, MPs told*. Retrieved September 5, 2018, from Stuff website: https://www.stuff.co.nz/environment/74695792/Fatal-shark-attack-likely-if-Stewart-Island-cage-diving-not-stopped-MPs-told

Sagan, D. (2011). *The human is more than human: Interspecies communities and the new "facts of life."* Retrieved from http://www.culanth.org/fieldsights/228-the-human-is-more-than-human-interspecies-communities-and-the-new-facts-of-life

Sage, D., Justesen, L., Dainty, A., Tryggestad, K., & Mouritsen, J. (2016). Organizing space and time through relational human–animal boundary work: *Exclusion, invitation and disturbance. Organization, 23*(3), 434–450. https://doi.org/10.1177/1350508416629449

Salgado, S. V. A., Tonon, G., & Barrera, M. D. M. (2012). Qualitative Research: The Thematic Analysis for the Treatment of Information from the Approach of the Social Phenomenology. *Universitas Humanística, 74*, 196–226.

Sancho, D. (2017). Escaping India's culture of education: Migration desires among aspiring middle-class young men. *Ethnography, 18*(4), 515–534. https://doi.org/10.1177/1466138116687591

Sanders, C. R. (1990). The Animal "Other": Self Definition, Social Identity and Companion Animals. *ACR North American Advances, NA-17*. Retrieved from http://acrwebsite.org/volumes/7082/volumes/v17/NA-17

Sandoz, D. (2003). *Simulation/ simulacrum (1)*. The Chicago School of Media Theory. Retrieved August 10, 2019, from https://lucian.uchicago.edu/blogs/mediatheory/keywords/simulation-simulacrum/

Sands, K. L., & Sands, K. L. (2019). Shared spaces on the street: A multispecies ethnography of ex-racing greyhound street collections in South Wales, UK. *Leisure Studies: Multispecies Leisure: Human-Animal Interactions in Leisure Landscapes, 38*(3), 367–380. https://doi.org/10.1080/02614367.2019.1577904

Satama, S., & Huopalainen, A. (2019). 'Please tell me when you are in pain': A heartbreaking story of care, grief and female–canine companionship. *Gender, Work & Organization, 26*(3), 358–376. https://doi.org/10.1111/gwao.12294

Satsuka, S. (2018). Sensing Multispecies Entanglements. *Social Analysis, 62(*4), 78–101. https://doi.org/10.3167/sa.2018.620405

Save our seas magazine (2015). *The Politics of Shark Attacks*. Retrieved June 9, 2017, from http://www.saveourseasmagazine.com/the-politics-of-sharks-attacks/

Sax, B. (2009). Who patrols the human animal divide? *Minnesota Review 73-74*: 165-169.

Sayes, E. (2014). Actor-network theory and methodology: Just what does it mean to say that nonhumans have agency? *Social Studies of Science, 44*(1), 134–149.

Scheibe, S. (2018). Predicting real-world behaviour: Cognition-emotion links across adulthood and everyday functioning at work. *Cognition and Emotion*, 1–7. https://doi.org/10.1080/02699931.2018.1500446

Schensul, S. L., Schensul, J. J., LeCompte, M. D., & Ph.D, M. D. L., M. A. (1999). *Essential Ethnographic Methods: Observations, Interviews, and Questionnaires.* Rowman Altamira. Walnut Creek, Calif, AltaMira Press

Schiffer, M. B., & Miller, A. R. (1999). *The material life of human beings: Artifacts, behavior, and communication.* London; New York: Routledge.

Schlagloth, R., Santamaria, F., Golding, B., & Thomson, H. (2018). Why is it Important to Use Flagship Species in Community Education? The Koala as a Case Study. *Animal Studies Journal, 7*(1), 127–148.

Schneider, K. (2013). Pigs, Fish, and Birds: Toward Multispecies Ethnography in Melanesia. *Environment & Society, 4*, 25–40. Retrieved from JSTOR.

School, C. B. (2012, September 21). *Virtual Boundaries: How Environmental Cues Affect Motivation and Task–Oriented Behavior.* Retrieved June 6, 2019, from Newsroom website: https://www8.gsb.columbia.edu/newsroom/newsn/2144/virtual-boundaries-how-environmental-cues-affect-motivation-and-taskoriented-behavior

Schrader, A. (2010). Responding to Pfiesteria piscicida (the fish killer) phantomatic ontologies, indeterminacy, and responsibility in toxic microbiology. *Social Studies of Science, 40*(2), 275–306.

Schultze, U. (2014). Performing embodied identity in virtual worlds. *European Journal of Information Systems, 23*(1), 84–95. https://doi.org/10.1057/ejis.2012.52

Schwanebeck, W. (2017). Jaws: Directed by Alfred Hitchcock. In W. Schwanebeck (Ed.), *Reassessing the Hitchcock Touch: Industry, Collaboration, and Filmmaking* (pp. 241–257). https://doi.org/10.1007/978-3-319-60008-6_14

Schwarz, N. (2000). Emotion, cognition, and decision making. *Cognition and Emotion, 14*(4), 433–440. https://doi.org/10.1080/026999300402745

ScienceDirect Topics. (n.d.). *Environmental Cues An overview.* Retrieved June 6, 2019, from https://www.sciencedirect.com/topics/earth-and-planetary-sciences/environmental-cue

Scoop (2015). *DOC tightens controls on shark cage diving.* Retrieved May 12, 2019, from : http://www.scoop.co.nz/stories/PO1512/S00011/doc-tightens-controls-on-shark-cage-diving.htm

Scoop News. (2018) *Shark cage diving operations closed down by Court of Appeal.* Retrieved October 2, 2018, from http://www.scoop.co.nz/stories/BU1809/S00093/shark-cage-diving-operations-closed-down-by-court-of-appeal.htm

Scuba Diving. (n.d.). *25 Best Shark Diving Destinations, Shark Photos*. Retrieved July 15, 2020, from https://www.scubadiving.com/photos/25-best-destinations-sharks-and-adventure#page-2

Segur, J. (2011). *The shark god or demon.* Tribal Animals, ZED.

Seifertrebecca. (2015, April 30). *Shark Anthropologist*. Retrieved March 5, 2019, from What can you do with anthropology? website: https://whatcanyoudowithanthropology.wordpress.com/2015/04/30/shark-anthropologist/

Seligman, M. E. P., Walker, E. F., & Rosenhan, D. L. (2001). *Abnormal Psychology*. New York: W W Norton & Co Inc.

Sepie, A. (2017). More than Stories, More than Myths: Animal/Human/Nature(s) in Traditional Ecological Worldviews. *Humanities, 6*(4), 78. https://doi.org/10.3390/h6040078

Sernert, H. (2011). *Attacks on humans in Sweden by bear, wolf, lynx, wolverine, moose and wild boar in relation to Swedes' fear for these animals. 25.*

Shah, A. (2017). Ethnography? Participant observation, a potentially revolutionary praxis. HAU: *Journal of Ethnographic Theory, 7*(1), 45–59. https://doi.org/10.14318/hau7.1.008

Shark Attack Data (n.d.). *Shark attack at Waihere Bay in Chatham Islands, New Zealand.* Retrieved August 23, 2017, from: http://www.sharkattackdata.com/attack/new_zealand/chatham_islands/1996.09.06.b

Shark Attack Data (n.d.). New Zealand. Retrieved January 7, 2019. http://www.sharkattackdata.com/place/new_zealand

Shark Business. (2017). *The economic value of sharks*. Retrieved April 23, 2019. http://www.sharkbusiness.org/portfolio_page/the-economic-value-of-sharks/

New Zealand Parliament. (n.d.). *Shark Cage Diving (Permitting and Safety) Bill 2018: Bills Digest 2572*. Retrieved October 2, 2018, from https://www.parliament.nz/en/pb/bills-and-laws/bills-digests/document/52PLLaw25721/shark-cage-diving-permitting-and-safety-bill-2018-bills

SharkNewz website (n.d.). *Shark cage diving Archives*. Retrieved October 2, 2018, from: https://www.sharknewz.com/tag/shark-cage-diving/

Sharkcagediving (n.d) *Shark Cage Diving with Great White Shark Tours*. Retrieved July 15, 2020, from https://www.sharkcagediving.net/

Department of conservation (n.d.). *Shark cage diving*. Retrieved October 2, 2018, from https://www.doc.govt.nz/news/issues/shark-cage-diving/

Facebook. (2018, September). *Shark Dive New Zealand*. https://www.facebook.com/sharkdivenz/photos/a.164858540227348/1960054594041058/

Sharkdivenz.com (n.d.). *Shark Dive NZ – Cage Diving with Great White Sharks » Book Online*. Retrieved November 5, 2017, from http://www.sharkdivenz.com/book-online/

Shark divers. (2014, February 27). *Is tagging sharks bad?* https://www.sharkdiver.com/what-does-tagging-sharks-do-to-their-health/

Shark Experience Ltd. (n.d.). Reservations. Retrieved November 5, 2017, from https://sharkexperience.rezdy.com/catalog/155871/tours

Shark Experiences. (n.d.). About us. Retrieved July 15, 2020, from http://www.sharkexperience.co.nz/about-us/

MSN (n.d.) Shark Mythology: Maori. Retrieved August 13, 2018, from website: http://a.msn.com/09/en-gb/AAAipGQ

Sharkattacks (n.d.-a). *A detailed look at Shark History, from Myth about the Great White Shark Attacks Victim stories photo gallery of Bull, Tiger Attack on Divers & swimmers*. Retrieved July 21, 2018, from http://www.sharkattacks.com/historical.htm

Sharkey, S. (2019, March 14). *Study Proves Sharks Are an Important Part of The Ecosystem*. Retrieved March 16, 2019, from Awesome Ocean website: http://awesomeocean.com/conservation/study-proves-sharks-important-part-ecosystem/

Smithsonian Ocean website (n.d.) Sharks. Retrieved January 15, 2019, from: http://ocean.si.edu/ocean-life/sharks-rays/sharks

Sharp, L. A. (2011). Monkey business: Interspecies longing and scientific prophecy in experimental xenotransplantation. *Social Text, 29*(1), 43–69.

Shepard, P. (1996). *The others: how animals made us human*. Washington, D.C, Island Press.

Shepherd, T. D., & Myers, R. A. (2005). Direct and indirect fishery effects on small coastal elasmobranchs in the northern Gulf of Mexico: Fishery effects on Gulf of Mexico elasmobranchs. *Ecology Letters, 8*(10), 1095–1104. https://doi.org/10.1111/j.1461-0248.2005.00807.x

Shermer, M. (2010). *The Sensed-Presence Effect*. Retrieved June 18, 2017, from Scientific American website: https://www.scientificamerican.com/article/the-sensed-presence-effect/

Shir-Vertesh, D. (2012). "Flexible personhood": Loving animals as family members in Israel. *American Anthropologist, 114*(3), 420–432.

Shiv, B., & Fedorikhin, A. (1999). Heart and Mind in Conflict: The Interplay of Affect and Cognition in Consumer Decision Making. *Journal of Consumer Research, 26*(3), 278–292. https://doi.org/10.1086/209563

Simon, G. (2014, April 15). *Fear of Sharks Phobia – Galeophobia or Selachophobia*. http://www.fearof.net/fear-of-sharks-phobia-galeophobia-or-selachophobia/

Simpfendorfer, C. A. (2000). Predicting Population Recovery Rates for Endangered Western Atlantic Sawfishes Using Demographic Analysis. *Environmental Biology of Fishes, 58*(4), 371–377. https://doi.org/10.1023/A:1007675111597

Simpfendorfer, C. A., & Kyne, p. M. (2009). Limited potential to recover from overfishing raises concerns for deep-sea sharks, rays and chimaeras. *Environmental Conservation; Cambridge, 36*(2), 97–103. http://dx.doi.org.ezproxy.canterbury.ac.nz/10.1017/S0376892909990191

Simpfendorfer, C. A., Heupel, M. R., White, W. T., & Dulvy, N. K. (2011). The importance of research and public opinion to conservation management of sharks and rays: A synthesis. *Marine and Freshwater Research, 62*(6), 518–527. https://doi.org/10.1071/MF11086

Simpson, J. A., & Gangestad, S. W. (1992). Sociosexuality and Romantic Partner Choice. *Journal of Personality, 60*(1), 31–51. https://doi.org/10.1111/j.1467-6494.1992.tb00264.x

Simpson, P. (2013). Ecologies of Experience: Materiality, Sociality, and the Embodied Experience of (Street) Performing. *Environment and Planning A: Economy and Space, 45*(1), 180–196. https://doi.org/10.1068/a4566

Sims, D., Nash J., & Morritt, D. (2001). Movements and activity of male and female dogfish in a tidal sea lough: alternative behavioural strategies and apparent sexual segregation. *Mar Biol 139:*1165–1175

Sims, D.W., Southall, E.J., Wearmouth, V.J., Hutchinson, N., Budd, G.C., & Morritt, D. (2005) *Refuging behaviour in the nursehound Scyliorhinus stellaris (Chondrichthyes: Elasmobranchii)*: prelim

Singer, M. (2014). Zoonotic Ecosyndemics and Multispecies Ethnography. *Anthropological Quarterly, 87(*4), 1279–1309. https://doi.org/10.1353/anq.2014.0060

Skomal, D. G., & Caloyianis, N. (2008). *The Shark Handbook: The Essential Guide for Understanding the Sharks of the World (1 edition)*. Kennelbunkport, Maine: Cider Mill Press.

Smart, A. (2014). Critical perspectives on multispecies ethnography. *Critique of Anthropology, 34*(1), 3–7. https://doi.org/10.1177/0308275X13510749

Smith, K. R., Scarpaci, C., Scarr, M. J., & Otway, N. M. (2014). Scuba diving tourism with critically endangered grey nurse sharks (Carcharias taurus) off eastern Australia: Tourist demographics, shark behaviour and diver compliance. *Tourism Management, 45*(Journal Article), 211–225. https://doi.org/10.1016/j.tourman.2014.05.002

Smith, K., Scarr, M., & Scarpaci, C. (2010). Grey Nurse Shark Diving Tourism: Tourist Compliance and Shark Behaviour at Fish Rock, Australia. *Environmental Management, 46*(5), 699–710. https://doi.org/10.1007/s00267-010-9561-8

Smith, R. G. (Ed.). (2010). *The Baudrillard dictionary.* Edinburgh: Edinburgh University Press.

Smith, R. G., & Clarke, D. B. (Eds.). (2015). *Jean Baudrillard: From Hyperreality to Disappearance: Uncollected Interviews, 1986 to 2007 (1 edition).* Edinburgh: Edinburgh University Press.

Smith, R. J., et al. (2012). "Identifying Cinderella species: uncovering mammals with conservation flagship appeal." *Conservation Letters, 5*(3): 205-212.

Smith, W. R. (2003). *Myths and Legends of the Australian Aborigines*. Mineola, N.Y: Dover Publications.

Sodikoff, G. M. (2011). *The anthropology of extinction: Essays on culture and species death.* Bloomington: Indiana Univ. Press.

Song, H. (2010). *Pigeon trouble: Bestiary biopolitics in a deindustrialized America*. Philadelphia: Univ. of Pennsylvania Press.

Soniak, M. (2014, March 3). *Why you want to save the whales, but not the crickets*. https://theweek.com/articles/450037/why-want-save-whales-but-not-crickets

Sontag, S. (1973). *On photography*. New York, farer, Straus & Giroux.

Speed, C. W., Meekan, M. G., & Bradshaw, C. J. (2007). Spot the match – wildlife photo-identification using information theory. *Frontiers in Zoology, 4*, 2. https://doi.org/10.1186/1742-9994-4-2

Springsteen, B. (2016). *Born to Run (First Edition*). Simon & Schuster.

Spielberg, S. (1975, June 20). *Jaws* [Adventure, Thriller]. Zanuck/Brown Productions, Universal Pictures.

Spindler, G. (2008). Using Visual Stimuli in Ethnography. *Anthropology & Education Quarterly, 39*(2), 127–140. JSTOR.

Shark Spotters. (n.d.). *Safety*. Retrieved May 10, 2019, from https://sharkspotters.org.za/safety/spotting/

Spradley, J. P. (1979). *The Ethnographic Interview*. New York: Harcourt, Brace, Jovanovich.

SpringerLink (n.d.) *The effects of shark cage-diving operations on the behaviour and movements of white sharks,Carcharodon carcharias, at the Neptune Islands, South Australia*. Retrieved June 13, 2017, from https://link-springer-com.ezproxy.canterbury.ac.nz/article/10.1007%2Fs00227-012-2142-z

Squirrell, T. (n.d.). *"All the world's a stage"? Understanding Goffman's Presentation of the Self in Everyday Life*. Retrieved June 25, 2019, from Tim Squirrell website: https://www.timsquirrell.com/blog/2017/5/27/all-the-worlds-a-stage-understanding-goffmans-presentation-of-the-self-in-everyday-life

Staff, N. Z. (n.d.-a). *Outrage after shark cage diving ruled an offence*. Retrieved October 2, 2018, from ZB website: http://www.newstalkzb.co.nz/on-air/larry-williams-drive/audio/steve-hanrahan-shark-cage-diving-brings-tourists-to-south-island/

Stanford (n.d.). *Baudrillard Simulacra and Simulations*. Retrieved June 11, 2017, from http://web.stanford.edu/class/history34q/readings/Baudrillard/Baudrillard_Simulacra.html

Star, F. (2018). *Pocket Eyewitness Sharks: Facts at Your Fingertips*. DK Children.

Statistics New Zealand. (n.d.). *Bluff (Southland, South Island, New Zealand) Population Statistics, Charts, Map, Location, Weather and Web Information*. Retrieved July 2, 2018, from http://citypopulation.info/php/newzealand-southisland.php?cityid=297

Stearns, A. (n.d.). *Shark Week & the Thrill of Deviant Behavior*. Retrieved March 5, 2019, from Sociology in Focus website: http://sociologyinfocus.com/2013/09/shark-week-the-thrill-of-deviant-behavior/

Steward, H. (2009). Animal Agency. *Inquiry, 52*(3), 217–231. https://doi.org/10.1080/00201740902917119

Stock, M. K., Winburn, A. P., & Burgess, G. H. (n.d.). Skeletal Indicators of Shark Feeding on Human Remains: Evidence from Florida Forensic Anthropology Cases. *Journal of Forensic Sciences*. Retrieved from https://www.academia.edu/10997833/Skeletal_Indicators_of_Shark_Feeding_on_Human_Remains_A_Case_Study_from_the_Eastern_Coast_of_Florida

Stone, D. (2018, September 27). *Meet the Underwater Photographer Who Chases Sperm Whales*. Retrieved March 6, 2019, from National geographic website: https://www.nationalgeographic.com/magazine/2018/10/explore-field-notes-brian-skerry-underwater-photographer-sperm-whales-dominica/

Stone, E. (2019). What's in it for the cats?: Cat shows as serious leisure from a multispecies perspective. *Leisure Studies, 38*(3), 381–393. https://doi.org/10.1080/02614367.2019.1572776

Storr, J. (2009). Décalage: A Thematic Interpretation of Cultural Differences in the African Diaspora. *Journal of Black Studies, 39*(5), 665–688.

Strege, D. (2016, January 29). Fear of a great white shark attack grips New Zealand island; it's like 'Jaws.' *Men's Journal*. https://www.mensjournal.com/adventure/fear-of-a-great-white-shark-attack-grips-new-zealand-island-its-like-jaws/

Strike. (2014, April 21). *Dive 'talk' – manual signals*. Retrieved September 18, 2017, from NEKTONIX website: https://nektonix.com/2014/04/21/dive-talk-manual-signals/

Studniarz, S. (n.d.). *Ontology, simulacra and hyperreality. Philip K. Dick's Ubik and the question of postmodernist canon*. Retrieved July 1, 2020, from https://www.academia.edu/4457852/_Ontology_simulacra_and_hyperreality._Philip_K._Dick_s_Ubik_and_the_question_of_postmodernist_canon_

Stuff (2009). *Bluff to replicate anchor chain sculpture*. Retrieved May 16, 2019, from website: https://www.stuff.co.nz/southland-times/news/466612/Bluff-to-replicate-anchor-chain-sculpture.

Stuff (2013). *Bluff's calm voice in a storm*. Retrieved January 1, 2019, website: https://www.stuff.co.nz/southland-times/news/9564825/Bluffs-calm-voice-in-a-storm

Stuff (n.d.- c). *Southland shark cage diving operators not baited by latest controversy*. September 5, 2018, from website: https://www.stuff.co.nz/environment/74784056/Southland-shark-cage-diving-operators-not-baited-by-latest-controversy

Stuff (n.d.). *Department of Conservation review finds shark cage diving does not increase risks to water users*. Retrieved September 5, 2018, from : https://www.stuff.co.nz/environment/74600479/Department-of-Conservation-review-finds-shark-cage-diving-does-not-increase-risks-to-water-users

Stuff (n.d.-d). *Stewart Island Shark diving breaches confirmed Department of Conservation*. Retrieved September 5, 2018, from website: https://www.stuff.co.nz/southland-times/news/70230966/stewart-island-shark-diving-breaches-confirmed--department-of-conservation

Stuff (n.d.-e). *Great white numbers low, basking sharks rapidly decline*. Retrieved July 1, 2018, https://www.stuff.co.nz/science/105112605/great-white-numbers-low-basking-sharks-rapidly-decline

Stuff (2015). *Paua divers take Department of Conservation to court on shark cage diving*. Retrieved September 5, 2018, https://www.stuff.co.nz/national/75022978/paua-divers-take-department-of-conservation-to-court-on-shark-cage-diving

Stuff (2018). *Paua divers should steer clear of shark infested waters, cage diving operators say*. Retrieved September 5: https://www.stuff.co.nz/business/102617638/paua-divers-say-theres-a-duty-to-consider-them-in-shark-cage-diving-consents

Stuff (n.d.-b). *Paua group calls for halt to shark cage diving*. Retrieved September 5, 2018. https://www.stuff.co.nz/business/farming/aquaculture/63863066/paua-group-calls-for-halt-to-shark-cage-diving

Stuff (2019). *"Offence" of shark cage diving off Stewart Island thrashed out in Supreme Court*. https://www.stuff.co.nz/business/farming/agribusiness/111547770/offence-of-shark-cage-diving-off-stewart-island-thrashed-out-in-supreme-court

Sukumaran, J., & Knowles, L. L. (2017). Multispecies coalescent delimits structure, not species. *Proceedings of the National Academy of Sciences, 114*(7), 1607–1612. https://doi.org/10.1073/pnas.1607921114

Sutcliffe, S. R., & Barnes, M. L. (2018). The role of shark ecotourism in conservation behaviour: Evidence from Hawaii. *Marine Policy, 97*, 27–33. https://doi.org/10.1016/j.marpol.2018.08.022

Sutton, M. (2015, February 9). *Sharks flee Neptune Islands following killer whale attack: Tourism operators*. https://www.abc.net.au/news/2015-02-09/sharks-evacuate-neptune-islands-after-killer-whales-attack/6080788

Swanson, H. A. (2017). Methods for multispecies anthropology: Thinking with Salmon Otoliths and Scales. *Social Analysis, 61*(2), 81. https://doi.org/10.3167/sa.2017.610206

Tallman, S. (n.d.). *Shark Cage-Diving Safety Regulations Desperately Needed in South Africa*. Retrieved October 2, 2018, from GoPetition website: https://www.gopetition.com/petitions/shark-cage-diving-safety-regulations-desperately-needed-in-south-africa.html

Taonga, N. Z. M. for C., & H. T. M. (2006). *Te whānau puha – whales – Te Ara Encyclopedia of New Zealand*. Retrieved July 29, 2018, from /en/te-whanau-puha-whales/page-3

Taonga, N. Z. M. for C., & H. T. M. (n.d.). *Shark catch*. Retrieved August 27, 2017, from /en/photograph/7825/shark-catch

Tapper. J. (2013). *Why "Shark Week" is so successful*. Retrieved August 10, 2019, from http://thelead.blogs.cnn.com/2013/08/05/why-shark-week-is-so-successful/

Taylor, A., & Pacini-Ketchabaw, V. (2017). Kids, raccoons, and roos: Awkward encounters and mixed affects. *Children's Geographies, 15*(2), 131–145. https://doi.org/10.1080/14733285.2016.1199849

Taylor, M. (2012). *Scuba diving action cameras*.

Taylor, N. (2010). Animal Shelter Emotion Management: A Case of in situ Hegemonic Resistance? *Smociology, 44*(1), 85–101. Retrieved from JSTOR.

Teara (N.D.). *"Story: sharks and rays."* Retrieved 11.12.15, from http://www.teara.govt.nz/en/sharks-and-rays/page-2.

Techera, E. J., & Klein, N. (2013). The role of law in shark-based eco-tourism: Lessons from Australia. *Marine Policy, 39*, 21–28. https://doi.org/10.1016/j.marpol.2012.10.003

Tedeschi, E. (2016). Animals, Humans and Sociability. *Italian Sociological Review, 6*(2), 151–184.

TEDx Talks. (2015). *Seeing Deeper into the White Shark's World, Greg Skomal, TEDxNewBedford*. Retrieved from https://www.youtube.com/watch?v=03Ex3obOl1Q

The Blue Planet. (2001, September 12). [Documentary]. *Arbeitsgemeinschaft der öffentlich-rechtlichen Rundfunkanstalten der Bundesrepublik Deutschland* (ARD), British Broadcasting Corporation (BBC).

The Daily California. (2017). *14 things that kill more people than sharks.* https://www.dailycal.org/2017/06/30/14-things-kill-people-sharks/

The Economist (2017). *Fatal shore; Shark attacks in Australia.* 423(9041), 34.

The Inquisitr (2016). *Australian Great White Shark Attack Spurs Intense Debate Over Public Safety.* Retrieved June 23, 2017, from: http://www.inquisitr.com/3557384/australian-great-white-shark-attack-spurs-intense-debate-over-public-safety/

The IUCN Red List of Threatened Species; (n.d.). *IUCN Red List of Threatened Species. White Sharks.* Retrieved February 11, 2020, from https://www.iucnredlist.org/en

The Partially Examined Life. (2012, January 25). *Rick Roderick on Baudrillard—Fatal Strategies.* https://www.youtube.com/watch?v=2U9WMftV40c

The Paua Industrial Council (n.d.) *About Paua Mac's.* Retrieved July 17, 2018, from http://www.paua.org.nz/about-paua-macs/

Proquest (n.d) The role of embedded research in quality improvement. Retrieved May 10, 2019, from https://search- -com.ezproxy.canterbury.ac.nz/docview/1883802756?pq-origsite=summon&accountid=14499

The SAGE Encyclopedia of Communication Research (2017) *Ethnographic Interview.* https://doi.org/10.4135/9781483381411.n168

Jacques-Yves Cousteau (1956). The Silent World. Retrieved from http://www.imdb.com/title/tt0049518/

Thomas, J., & Harden, A. (2008). Methods for the thematic synthesis of qualitative research in systematic reviews. *BMC Medical Research Methodology, 8,* 45. https://doi.org/10.1186/1471-2288-8-45

Thompson, K. (2016, January 12). *The Presentation of the Self in Everyday Life – A Summary.* Retrieved June 25, 2019, from ReviseSociology website: https://revisesociology.com/2016/01/12/the-presentation-of-the-self-in-everyday-life-a-summary/

Thompson, T. L. & J. J. Mintzes (2002). Cognitive structure and the affective domain: on knowing and feeling in biology. *International journal of science education 24*: 645-660.

Thompson, T. L., & Mintzes. J. J. (2002). Cognitive structure and the affective domain: on knowing and feeling in biology. *International Journal of Science Education, 24*:645-660.

Thomson, J. A., Araujo, G., Labaja, J., McCoy, E., Murray, R., & Ponzo, A. (2017). Feeding the world's largest fish: Highly variable whale shark residency patterns at a provisioning site in the Philippines. *Royal Society Open Science, 4*(9), 170394. https://doi.org/10.1098/rsos.170394

Thorson, T. B. (1982). The Impact of Commercial Exploitation on Sawfish and Shark Populations in Lake Nicaragua. *Fisheries, 7*(2), 2–10. https://doi.org/10.1577/1548-8446(1982)007<0002:TIOCEO>2.0.CO;2

Tilley, C. (Ed.). (2013). *Handbook of material culture.* London: SAGE.

Tirard, P., Maillaud, C., & Borsa, P. (2015). Fatal tiger shark, Galeocerdo cuvier attack in New Caledonia erroneously ascribed to great white shark, Carcharodon carcharias. *Journal of Forensic and Legal Medicine, 33,* 68–70. https://doi.org/10.1016/j.jflm.2015.04.011

Topelko, K. N., & Dearden, P. (2005). The Shark Watching Industry and its Potential Contribution to Shark Conservation. *Journal of Ecotourism, 4*(2), 108–128. https://doi.org/10.1080/14724040409480343

Torres, P., Bolhão, N., Tristão da Cunha, R., Vieira, J. A. C., & Rodrigues, A. dos S. (2017). Dead or alive: The growing importance of shark diving in the Mid-Atlantic region. *Journal for Nature Conservation, 36,* 20–28. https://doi.org/10.1016/j.jnc.2017.01.005

Tourism Industry Association. (2018). *Call for law change to permit shark cage diving.* Scoop News. https://www.scoop.co.nz/stories/BU1809/S00115/call-for-law-change-to-permit-shark-cage-diving.htm

Accomnews NZ (2018, September 12). *Tourism outcry as shark cage ban bites.* Retrieved October 2, 2018, from website: https://www.accomnews.co.nz/2018/09/12/tourism-outcry-shark-cage-ban-bites/

Towner, A. V., Underhill, L. G., Jewell, O. J. D., & Smale, M. J. (2013). Environmental Influences on the Abundance and Sexual Composition of White Sharks Carcharodon carcharias in Gansbaai, South

Africa. *PLoS One; San Francisco, 8*(8), e71197. http://dx.doi.org.ezproxy.canterbury.ac.nz/10.1371/journal.pone.0071197

Towner, A. V., Wcisel, M. A., Reisinger, R. R., Edwards, D., & Jewell, O. J. D. (2013). Gauging the Threat: The First Population Estimate for White Sharks in South Africa Using Photo Identification and Automated Software. *PLOS ONE, 8*(6), e66035. https://doi.org/10.1371/journal.pone.0066035

Townsend, A. (2011). *Top 10 Unforgettable Shark Moments*. Time. Retrieved from http://content.time.com/time/specials/packages/article/0,28804,2085822_2085823_2086095,00.html

Travel Trade Gazette UK & Ireland (2002) *Bahamas bitten by shark-diving bug.* 2535, 44.

Treisman, M. (1977). Motion Sickness: An Evolutionary Hypothesis. *Science, 197*(4302), 493–495. Retrieved from JSTOR.

Tremblay, P. (2002). Tourism wildlife icons: Attractions or marketing symbols? *CAUTHE 2002: Tourism and Hospitality on the Edge.*

Tresch, J., & Latour, B. (2013). Another turn after ANT:An interview with Bruno Latour. *Social Studies of Science, 43(2)*, 302–313. Retrieved from JSTOR.

Tricas, T. C., et al. (1997). *Sharks and Rays*. London, Harper Collins Publishers.

Trip Advisor. (B) (n.d.). *Unfortunately no great white sharks in Gansbaai—Reviews, Photos—Marine Dynamics. Tripadvisor*. Retrieved August 10, 2020, from http://www.tripadvisor.in/ShowUserReviews-g665841-d1985812-r631617316-Marine_Dynamics-Kleinbaai_Overstrand_Overberg_District_Western_Cape.html

Trondman, M. (2017). Taking normative sense seriously: Ethnography in the light of a utopian referent. *Ethnography, 18*(1), 10–23. https://doi.org/10.1177/1466138115592421

Trout, P. A., & Ehrenreich, B. (2011*). Deadly Powers: Animal Predators and the Mythic Imagination*. Amherst, NY: Prometheus Books.

Tsing, A. (2010). Arts of inclusion, or how to love a mushroom. *Manoa, 22*(2), 191–203.

Tsing, A. (2012). Unruly edges: Mushrooms as companion species. *Environmental Humanities*, 1, 141–154.

Tsing, A. (2013). Dancing the mushroom forest. *PAN: Philosophy Activism Nature, 10*, 6–14.

Tsing, A., & Satsuka, S. (2008). Diverging understandings of forest management in matsutake science. *Economic Botany, 62*(3), 244–253.

Tsoi, K. H., Chan, S. Y., Lee, Y. C., Ip, B. H. Y., & Cheang, C. C. (2016). Shark Conservation: An Educational Approach Based on Children's Knowledge and Perceptions toward Sharks. *PLoS ONE, 11(*9). https://doi.org/10.1371/journal.pone.0163406

Tummons, J. (2010). Institutional ethnography and actor-network theory: A framework for researching the assessment of trainee teachers. *Ethnography & Education, 5*(3), 345–357. https://doi.org/10.1080/17457823.2010.511444

Tummons, J., Fournier, C., Kits, O., & MacLeod, A. (2018). Using technology to accomplish comparability of provision in distributed medical education in Canada: An actor-network theory ethnography. *Studies in Higher Education, 43*(11), 1912–1922. https://doi.org/10.1080/03075079.2017.1290063

Turner, L. (2018). *The Edinburgh Companion to Animal Studies (U. Sellbach & R. Broglio, Eds*.). Edinburgh: Edinburgh University Press.

TVNZ (n.d.) *Court of Appeal ruling sinks teeth into shark cage diving*. Retrieved from https://www.tvnz.co.nz/one-news/new-zealand/court-appeal-ruling-sinks-teeth-into-shark-cage-diving

Tyler, I. (2018). Resituating Erving Goffman: From Stigma Power to Black Power. *Sociological Review, 66(*4), 744–765. https://doi.org/10.1177/0038026118777450

Uexküll, J. von, & Uexküll, J. von. (2010). A foray into the worlds of animals and humans: *With A theory of meaning* (1st University of Minnesota Press ed). Minneapolis: University of Minnesota Press.

Ullrich, J. (2018). Contact Zones—Where Dogs and Humans Meet: Dog-Human Metamorphoses in Contemporary Art. *Animals and Their People*, 53–68. https://doi.org/10.1163/9789004386228_005

Ulmer, J. B., & Ulmer, J. B. (2017). Posthumanism as research methodology: Inquiry in the Anthropocene. *International Journal of Qualitative Studies in Education, 30*(9), 832–848. https://doi.org/10.1080/09518398.2017.1336806

University of Miami (n.d.) Shark Research. Animal Welfare – Shark Research & Conservation Program (SRC). Retrieved July 24, 2020, from //sharkresearch.rsmas.miami.edu/research/animal-welfare/

Urbanik, J. (2012). *Placing animals: An introduction to the geography of human-animal relations.* Lanham, MD: Rowman & Littlefield.

Vaismoradi, M., Turunen, H., & Bondas, T. (2013). Content analysis and thematic analysis: Implications for conducting a qualitative descriptive study. *Nursing & Health Sciences, 15*(3), 398–405. https://doi.org/10.1111/nhs.12048

Van de Port, M. (2017). The verification of ethnographic data: A response. *Ethnography, 18(*3), 295–299. https://doi.org/10.1177/1466138117722799

Van Dijck, J. (2013). 'You have one identity': Performing the self on Facebook and LinkedIn. *Media, Culture & Society, 35*(2), 199–215. https://doi.org/10.1177/0163443712468605

Van Dooremalen, T. (2017). The pros and cons of researching events ethnographically. *Ethnography, 18(*3), 415–424. https://doi.org/10.1177/1466138117709293

Van Dooren, T. (2011a). Invasive species in penguin worlds: An ethical taxonomy of killing for conservation. *Conservation and Society, 9*(4), 286–298.

Van Dooren, T. (2011b). Vultures and their people in India: Equity and entanglements in a time of extinctions. *Australian Humanities Review, 59*, 45–61.

Van Dooren, T. (2014). *Flight ways: Life and loss at the edge of extinction*. New York: Columbia Univ. Press.

Van Duppen, Z. (2016). The phenomenology of hypo- and hyperreality in psychopathology. *Phenomenology and the Cognitive Sciences, 15*(3), 423–441. https://doi.org/10.1007/s11097-015-9429-8

Van Ginkel, R. (2015). Humpback Johannes (a.k.a. Johanna): A Dutch Tragicomedy Featuring a Hyperreal Whale. *Etnofoor, 27*(1), 123–141. Retrieved from JSTOR.

Vanutelli, M. E., & Balconi, M. (2015). Perceiving emotions in human–human and human–animal interactions: Hemodynamic prefrontal activity (fNIRS) and empathic concern. *Neuroscience Letters, 605*, 1–6. https://doi.org/10.1016/j.neulet.2015.07.020

Vaughn, P., & Turner, C. (2016). Decoding via Coding: Analyzing Qualitative Text Data Through Thematic Coding and Survey Methodologies. *Journal of Library Administration, 56*(1), 41–51. https://doi.org/10.1080/01930826.2015.1105035

Vianna, G. M. S., Meekan, M. G., Pannell, D. J., Marsh, S. P., & Meeuwig, J. J. (2012). Socio-economic value and community benefits from shark-diving tourism in Palau: A sustainable use of reef shark populations. *Biological Conservation, 145*(1), 267–277. https://doi.org/10.1016/j.biocon.2011.11.022

News24 (2014). Victory for widow in shark cage dive case. Retrieved October 2, 2018. https://www.news24.com/SouthAfrica/News/Victory-for-widow-in-shark-cage-dive-case-20141228

Vidon, E. S., Rickly, J. M., & Knudsen, D. C. (2018). Wilderness state of mind: Expanding authenticity. *Annals of Tourism Research, 73*, 62–70. https://doi.org/10.1016/j.annals.2018.09.006

Viegas, J. (2014, July 14). *Submarine: Any Truth to the Legend*? Retrieved May 16, 2019, from Seeker website: https://www.seeker.com/submarine-any-truth-to-the-legend-1768787852.html

Vindrola-Padros, C., Pape, T., Utley, M., & Fulop, N. J. (2017). The role of embedded research in quality improvement: A narrative review. *BMJ Quality & Safety, 26*(1), 70–80. https://doi.org/10.1136/bmjqs-2015-004877

Viveiros de Castro, E. (2012). Cosmological perspectivism in Amazonia and elsewhere. *HAU: Journal of Ethnographic Theory Masterclass Series*. Retrieved from http://www.haujournal.org/index.php/masterclass/article/view/72

Vladic, L. V. & S. (n.d.). USS Indianapolis: Survivor Accounts from the Worst Sea Disaster in U.S. *Naval History. HISTORY*. Retrieved August 11, 2020, from https://www.history.com/news/uss-indianapolis-sinking-survivor-stories-sharks

Vokes, R. (2012). *Photography in Africa: ethnographic perspectives.* Woodbridge, Suffolk;New York;, James Currey.

Von Uexküll, J. (1992). A stroll through the worlds of animals and men: A picture book of invisible worlds. *Semiotica, 89*(4), 319–391. https://doi.org/10.1515/semi.1992.89.4.319

Voyer, A. (2017). Meanings, motives and action. *Ethnography, 18*(1), 97–106. https://doi.org/10.1177/1466138115592414

Voyer, A., & Trondman, M. (2017). Between theory and social reality: Ethnography and Interpretation and Social Knowledge: Introduction to the special issue. *Ethnography, 18*(1), 3–9. https://doi.org/10.1177/1466138115592415

Vytal, K. (2007). Learning to Fear. *APS Observer, 20*(1). Retrieved from https://www.psychologicalscience.org/observer/learning-to-fear

Walford, G. (2007). Classification and framing of interviews in ethnographic interviewing. *Ethnography and Education, 2*(2), 145–157. https://doi.org/10.1080/17457820701350491

Walsham, G. (1997). Actor-Network Theory and IS Research: Current Status and Future Prospects. In A. S. Lee, J. Liebenau, & J. I. DeGross (Eds.), *Information Systems and Qualitative Research: Proceedings of the IFIP TC8 WG 8.2 International Conference on Information Systems and Qualitative Research*, 31st May–3rd June 1997, Philadelphia, Pennsylvania, USA (pp. 466–480). https://doi.org/10.1007/978-0-387-35309-8_23Wanderer, E. (2018). THE AXOLOTL IN GLOBAL CIRCUITS OF KNOWLEDGE PRODUCTION: Producing Multispecies Potentiality. Cultural Anthropology, 33(4), 650–679. https://doi.org/10.14506/ca33.4.09

Wang, N., & Wang, N. (2019). The Rise of Posthumanism: Challenge to and Prospect for Mankind. *Fudan Journal of the Humanities and Social Sciences, 12*(1), 1–13. https://doi.org/10.1007/s40647-018-0242-y

Ward, T. (n.d.). *The smell of dead sharks is helping to keep surfers safe from attack*. Retrieved June 23, 2018, from WIRED UK website: http://www.wired.co.uk/article/shark-attacks-2018-smell-podi

Ward-Paige, C. A. (2017). A global overview of shark sanctuary regulations and their impact on shark fisheries. *Marine Policy, 82*, 87–97. https://doi.org/10.1016/j.marpol.2017.05.004

Washington Post (n.d. a). *'My whole arm was in its mouth': Woman loses arm in shark attack in the Bahamas*. Retrieved June 23, 2017. https://www.washingtonpost.com/news/early-lead/wp/2017/06/12/my-whole-arm-was-in-its-mouth-woman-loses-arm-in-shark-attack-in-the-bahamas/

Washington Post website (n.d.b). *Why we love to fear sharks*. Retrieved June 20, 2017. https://www.washingtonpost.com/news/morning-mix/wp/2015/07/02/why-were-terrified-of-sharks/

Waters, S., El Harrad, A., Bell, S., & Setchell, J. M. (2019). Interpreting People's Behavior Toward Primates Using Qualitative Data: A Case Study from North Morocco. *International Journal of Primatology, 40*(3), 316–330. https://doi.org/10.1007/s10764-019-00087-w

Watson Life Resources. (n.d.). *Environmental Cues to Help Transitions*. Retrieved June 6, 2019, from The Watson Institute website: https://www.thewatsoninstitute.org/watson-life-resources/situation/environmental-cues-to-help-transitions/

Watson, M. C. (2016). On Multispecies Mythology: A Critique of Animal Anthropology. *Theory, Culture & Society, 33*(5), 159–172. https://doi.org/10.1177/0263276416637128

Webber, S., Carter, M., Smith, W., & Vetere, F. (2017). Interactive technology and human–animal encounters at the zoo. *International Journal of Human-Computer Studies, 98*, 150–168. https://doi.org/10.1016/j.ijhcs.2016.05.003

Weigert, A. J. (2003). Terrorism, Identity, and Public Order: A Perspective From Goffman. *Identity, 3*(2), 93–113. https://doi.org/10.1207/S1532706XID030201

Welovesharks. (2017, April 6). *Shark Tagging: Understanding the risks involved. We Love Sharks*! https://www.welovesharks.club/shark-tagging-understanding-the-risks-involved/

Whatmore, S. (2002). *Hybrid Geographies: Natures, cultures, spaces*. London: SAGE.

Whatmore, S. (2006). Materialist returns: Practising cultural geography in and for a more-than-human world. *Cultural Geographies, 13*(4), 600–609.

Whatmore, S., & Thorne, L. (1998). Wild(er)ness: Reconfiguring the Geographies of Wildlife. *Transactions of the Institute of British Geographers, 23*(4), 435–454.

Whatmore, S., & Thorne, L. (2000). Elephants on the move: Spatial formations of wildlife exchange. Environment and Planning. *Society and Space, 18*(2), 185–203.

Whatmough, S., Putten, I. V., & Chin, A. (2011). From hunters to nature observers: A record of 53 years of diver attitudes towards sharks and rays and marine protected areas. *Marine and Freshwater Research, 62*(6), 755–763. https://doi.org/10.1071/MF10142

Wherry, F. F. (2017). Fragments from an ethnographer's field guide: Skepticism, thick minimalism, and big theory. *Ethnography, 18*(1), 46–56. https://doi.org/10.1177/1466138115592422

White Shark Video. (2015). *Hooking the Submarine: Full story with Craig Ferreira*: WSV. Retrieved from https://www.youtube.com/watch?v=j5VnDYVoH-w

White Shark Video. (n.d.). *Ocearch damaging sharks while profiting*. Retrieved July 24, 2020, from http://www.whitesharkvideo.com/ocearch-the-enemy.html

White, M. (2016). *Great white fight*. [Pressreader]. Retrieved May 15, 2019, from https://www.pressreader.com/

White, P. D. (2014). *Gay for the Great White*. White Climax Publishing.

Earthrace conservation (2018). *Why "Shark Cage Diving" ban matters*. Retrieved October 2, 2018. http://earthraceconservation.org/why-shark-cage-diving-ban-matters/

Wilde, N. (n.d.). *Environmental Cues*. Retrieved June 6, 2019, from Victoria Stilwell Positively website: https://positively.com/contributors/2203/

Wilkie, R. (2015). Multispecies Scholarship and Encounters: Changing Assumptions at the Human-Animal Nexus. *Sociology, 49*(2), 323–339. https://doi.org/10.1177/0038038513490356

Williams, R. W., & Bois, W. E. B. D. (2012). "The Sacred Unity in All the Diversity": The Text and a Thematic Analysis of W.E.B. Du Bois' "The Individual and Social Conscience" (1905). *Journal of African American Studies, 16*(3), 456–497. https://doi.org/10.1007/s12111-011-9171-4

Wicker, H.-R. (2001). Xenophobia. In N. J. Smelser & P. B. Baltes (Eds.), *International Encyclopedia of the Social & Behavioral Sciences* (pp. 16649–16652). Pergamon. https://doi.org/10.1016/B0-08-043076-7/00980-3

Wilson, M., & Daly, M. (2004). Do pretty women inspire men to discount the future? Proceedings of the Royal Society. *Biological Sciences, 271*(Suppl 4), S177–S179.

Wincer, S. (1993, July 16). *Free Willy* [Adventure, Drama, Family]. Warner Bros., Canal+, Regency Enterprises.

Winchatz, M. R. (2006). Fieldworker or Foreigner? Ethnographic Interviewing in Nonnative Languages. *Field Methods, 18*(1), 83–97. https://doi.org/10.1177/1525822X05279902

Wolfe, C. (2003). *Animal rites: American culture, the discourse of species, and posthumanist theory*. University of Chicago Press.

Wolfe, C. (2009). *On a Certain Blindness in Human Beings. In The Death of the Animal* (Vol. 1–Book, Section, p. 123). Columbia University Press. https://doi.org/10.7312/cava14544.14

Wolfe, C. (2010). *What is posthumanism?* Minneapolis: Univ. of Minnesota Press.

Wolny, R. W. (2017). Hyperreality and Simulacrum: Jean Baudrillard and European Postmodernism. *European Journal of Interdisciplinary Studies,* 8(1), 76. https://doi.org/10.26417/ejis.v8i1.p76-80

Wolper, J. (2014, August 13). *The Shark Attack that Shook the 1700s*. Retrieved May 29, 2019, from Narratively website: https://narratively.com/the-shark-attack-that-shook-the-1700s/

Woodward, I. (2007). *Understanding material culture*. Los Angeles; London: Sage Publications.

Woodward, W. (2018). [Review] Creatural Fictions David Herman, editor. Creatural Fictions: Human-Animal Relationships in Twentieth- and Twenty-First-Century Literature. *Animal Studies Journal, 7*(1), 319–321.

Woolgar, S., & Lezaun, J. (2013). The wrong bin bag: A turn to ontology in science and technology studies? *Social Studies of Science, 43*(3), 321–340.

Worley, P. (2011). *The If Machine: Philosophical Enquiry in the Classroom*. London ; New York: Continuum.

Wrenn, C. L. (n.d.). *Discriminating Spirits: Cultural Source Theory and the Human-Nonhuman Boundary*. Mortality. Retrieved from https://www.academia.edu/39600898/Discriminating_Spirits_Cultural_Source_Theory_and_the_Human-Nonhuman_Boundary

Wright, K. (2014). Becoming-with. *Environmental Humanities, 5*(1), 277–281. https://doi.org/10.1215/22011919-3615514

Wright, M. P. (2013, February 14). *Ernst Karel*. Retrieved August 4, 2018, from EAR ROOM website: https://earroom.wordpress.com/2013/02/14/ernst-karel/

Wright, T. (2008). *Visual impact: culture and the meaning of images*. Oxford; New York, Berg.

Dissertation Editing & Writing (2016). *Writing the methodology Chapter*. Retrieved May 4, 2018. https://dissertationwriting.com/2016/12/05/write-dissertation-methodology-help/

Wroe, S., et al. (2008). Three-dimensional computer analysis of white shark jaw mechanics: how hard can a great white bite? *Journal of Zoology 276*(4).

Yeo, J.-H., & Neo, H. (2010). Monkey business: Human–animal conflicts in urban Singapore. *Social & Cultural Geography, 11*(7), 681–699. https://doi.org/10.1080/14649365.2010.508565

York, R., & Mancus, P. (2013). The Invisible Animal: Anthrozoology and Macrosociology. *Sociological Theory, 31*(1), 75–91. https://doi.org/10.1177/0735275113477085

Young, B., & Hren, D. (n.d.). *Introduction to qualitative research methods*. 48.

Young, M. W. (1998). *Malinowski's Kiriwana: Fieldwork Photography*. Chicago and London, University of Chicago Press.

Zavattaro, S. M. (2013). Expanding Goffman's Theater Metaphor to an Identity-Based View of Place Branding. *Administrative Theory & Praxis (M.E. Sharpe), 35*(4), 510–528. https://doi.org/10.2753/ATP1084-1806350403

ZB. (n.d.). *Clayton Mitchell: Stewart Island sharks*. ZB. Retrieved July 15, 2020, from https://www.newstalkzb.co.nz/on-air/mike-hosking-breakfast/audio/clayton-mitchell-stewart-island-sharks/

Zemah Shamir, Z., Zemah Shamir, S., Tchernov, D., Scheinin, A., & Becker, N. (2019). Shark aggregation and tourism: Opportunities and challenges of an emerging phenomenon. *International Journal of Sustainable Development & World Ecology, 26*(5), 406–414. https://doi.org/10.1080/13504509.2019.1573769

Zhang, N. (2014). Performing identities: Women in rural–urban migration in contemporary China. *Geoforum, 54*, 17–27. https://doi.org/10.1016/j.geoforum.2014.03.006

Zhu, J., Adli, M., Zou, J. Y., Verstappen, G., Coyne, M., Zhang, X., ...Bernstein, B. E. (2013). Genome-wide Chromatin State Transitions Associated with Developmental and Environmental Cues. *Cell, 152*(3), 642–654. https://doi.org/10.1016/j.cell.2012.12.033

Ziegler, J. A., Silberg, J. N., Araujo, G., Labaja, J., Ponzo, A., Rollins, R., & Dearden, P. (2018). A guilty pleasure: Tourist perspectives on the ethics of feeding whale sharks in Oslob, Philippines. *Tourism Management, 68*(Generic), 264–274. https://doi.org/10.1016/j.tourman.2018.04.001

Ziegler, J., Dearden, P., & Rollins, R. (2012). But are tourists satisfied? Importance-performance analysis of the whale shark tourism industry on Isla Holbox, Mexico. *Tourism Management, 33(*3), 692–701. https://doi.org/10.1016/j.tourman.2011.08.004

Žikić, B. (2016). Qualitative Field Research in Anthropology: An Overview of Basic Research Methodology. *Etnoantropološki Problemi*, 2(2), 123–135.

Zubieta, L. F. (2016). Animals' Role in Proper Behaviour: Cheŵa Women's Instructions in South-Central Africa. *Conservation and Society, 14*(4), 406–415. JSTOR.

Zuolo, F. (2016). Dignity and Animals. Does it Make Sense to Apply the Concept of Dignity to all Sentient Beings? *Ethical Theory and Moral Practice, 19*(5), 1117–1130. https://doi.org/10.1007/s10677-016-9695-8

The Author, photograph by Raj Kamal Aich, 2019